Vagueness and Thought

Vagueness is the study of concepts that admit borderline cases: the property of being bald is vague because there are people who are neither definitely bald, nor definitely not bald. The epistemology of vagueness concerns the sorts of attitudes we ought to have towards propositions we know to be borderline. Is it possible to discover whether a borderline bald man is bald? Could two people with access to the same facts reasonably disagree about whether he is bald? Does it matter, when making practical decisions, whether he is bald?

By drawing on such considerations, Andrew Bacon develops a novel theory of vagueness in which vagueness is fundamentally a property of propositions, and is explicated in terms of its role in thought. On this theory, language plays little role in explaining the central puzzles of vagueness. Part I of the book outlines some of the central questions regarding the logic and epistemology of vagueness, and criticizes some extant approaches to them. Part II concerns issues in the epistemology of vagueness, touching on the ramifications of vague thoughts on the study of evidence, ignorance, desire, probability theory, and decision theory. By examining the effects of vague information on one's beliefs about the precise, a positive theory of vagueness is proposed. Part III concerns the logic of vagueness, including the interaction between vagueness and modality, vague identity, and the paradoxes of higher-order vagueness. Bacon suggests that some familiar philosophical notions—including the concept of a fundamental proposition, a possible world, and a precisification—need to be revised.

Andrew Bacon is Associate Professor of Philosophy at the University of Southern California. His work typically applies the methods of philosophical logic to issues in metaphysics, epistemology, and the philosophy of language. Recently he has worked on vagueness, the semantic paradoxes, and logic and epistemology of conditional statements.

Vagueness and Thought

Andrew Bacon

OXFORD
UNIVERSITY PRESS

OXFORD
UNIVERSITY PRESS

Great Clarendon Street, Oxford, OX2 6DP,
United Kingdom

Oxford University Press is a department of the University of Oxford.
It furthers the University's objective of excellence in research, scholarship,
and education by publishing worldwide. Oxford is a registered trade mark of
Oxford University Press in the UK and in certain other countries

© Andrew Bacon 2018

The moral rights of the author have been asserted

First published 2018
First published in paperback 2022

Published in the United States of America by Oxford University Press
198 Madison Avenue, New York, NY 10016, United States of America

British Library Cataloguing in Publication Data
Data available

Library of Congress Cataloging in Publication Data
Data available

ISBN 978-0-19-871206-0 (Hbk.)
ISBN 978-0-19-285608-1 (Pbk.)

Links to third party websites are provided by Oxford in good faith and
for information only. Oxford disclaims any responsibility for the materials
contained in any third party website referenced in this work.

For Helen

Preface

According to orthodoxy the study of vagueness belongs to the domain of the philosophy of language. According to that paradigm, solving the paradoxes of vagueness involves investigating the nature of words like 'heap' and 'bald' in English, and parallel words in other languages.

By contrast, the theory I advocate in this book is a theory of *propositional* vagueness. While I certainly recognize a notion of vagueness applicable to sentences and other linguistic expressions, these notions are to be explained in terms of propositional vagueness and not the other way around. Instead of understanding vagueness in terms of language, the view of this book places the study of vagueness squarely in epistemological terms, situating it within a theory of rational propositional attitudes.

Let me begin with a bit of autobiography. This project started off in 2009 as a 30,000 word essay written for the thesis portion of the Oxford BPhil examination. Over the next eight years the project turned into my DPhil thesis, and eventually this book. However, the project changed a great deal over that period. Many of the views criticized in the following pages are views I once held, and the end result no doubt bears the marks of this. Indeed, the original purpose of the thesis was to investigate the epistemology of vagueness in the context of a linguistic theory of vagueness. While my primary focus was on the epistemological aspects of vagueness, I was working with the background assumption that vagueness was fundamentally a linguistic phenomenon. My hope was that it would be possible to use game-theoretic concepts—specifically, certain sorts of mixed strategy equilibria—within a broadly Lewisian account of the conventions of language to explain why we are ignorant about the vague. However, about a year into my DPhil, I realized that this aspect of the project was a dead end. While I could explain why we are ignorant about a certain specialized class of linguistic matters—such as *whether the sentence 'Harry is bald' is a true sentence of English*—I could not explain ignorance about more mundane, but more pervasive, vague matters—such as *whether Harry is bald*. Part of the problem had to do with issues of fineness of grain. According to some linguistic theories, there's no such thing as propositional vagueness: all propositions are precise, and so the proposition that Harry is bald is identical to a precise proposition about hair number. Once you are knowledgeable about all the precise propositions, there are no propositions left to be ignorant about. However, even if we posit vague propositions in addition to vague sentences, explaining ignorance in the former in terms of facts about the latter turned out to be highly non-trivial.[1]

---

[1] It is also worth noting that, at the same time, I was working on a paper on the liar paradox (Bacon [7]). There, for mostly independent reasons, I came to the conclusion that the notion of indeterminacy suitable

Thus, despite the prevailing orthodoxy that vagueness has at least something to do with language, I found the alternative non-linguistic picture had a lot to recommend itself. What follows is particular way of spelling out that picture.

Here is a brief overview of the structure of the book. In chapter 1, I argue against the nihilist responses to the sorites paradox and against responses that reject classical logic, setting up one of the fundamental presuppositions of this book: that sorites sequences have cutoff points. In chapter 2, I move on to the debate over classical theories of vagueness—the two most prominent of which are epistemicism and supervaluationism. I outline some widely discussed questions in the philosophy of vagueness, and look at how epistemicism and supervaluationism are typically taken to engage with them. I argue, however, that these questions on the whole do not really carve the philosophical view-space at its joints and that the classification of views as supervaluationist and epistemicist is often not always that insightful. I suggest, in chapter 3, some questions of my own and give a broad overview of my theory. In chapter 4, I discuss the proper formulation of the distinction between sentential and propositional vagueness and distinguish between approaches to vagueness that take one or the other to be the central theoretical concept. In chapter 5, I present my argument against linguistic accounts of vagueness; I argue that only a propositional theory of vagueness can explain ignorance due to vagueness.

My positive view starts in chapter 6. This chapter tells us what the *role in thought* of a vague proposition is, and introduces the **Principle of Plenitude** that states, roughly, that there's a vague propositions for every possible role in thought. Chapter 7 takes an in-depth look at an alternative, but related, view that is due to Hartry Field. One of the primary differences between the views is that, unlike Field, I accept probabilism—the thesis that rational credences are governed by the axioms of the probability calculus. In this chapter I defend that commitment, and discuss the connection between credences and higher-order vagueness. Chapter 8 introduces the principle **Rational Supervenience**, that captures a sense in which all disagreements about the vague boil down to disagreements about the precise. Chapter 9 introduces Jeffrey's decision theory in a context where vague propositions are among your preferences and are among the things you can make true. Here I discuss the thought that attitudes with vague contents are superfluous for practical reasoning. Chapter 10 introduces the principle **Indifference**, stating that it is irrational to care *intrinsically* about the vague.

In the third part of the book I turn to a number of logical and metaphysical issues relating to the nature of propositions, modality, possible worlds and the semantics for propositional vagueness. Chapter 11 outlines a general theory of proposition according to which they are individuated by their epistemic roles, and introduces a broadest modality for reasoning about notions like propositional entailment. Chapter 12 introduces the distinction between vagueness and borderlineness, and argues

for diagnosing the liar paradox cannot be a linguistic predicate of sentences, but must rather be expressed by an operator expression.

that we should take the former as the primitive notion, rather than the latter. Chapter 13 introduces the central notion of a *symmetry*—roughly an automorphism on the algebra of propositions that preserves their role in thought—and shows how it can provide an illuminating analysis of vagueness. Chapter 14 discusses the relation between fundamental propositions, vagueness, and possible worlds, and argues that certain uses of the ideology of fundamentality and possible worlds lead to paradoxes of higher-order vagueness. Chapter 15 discusses the interaction between the determinacy and modal operators, the supervenience of the vague on the precise, and compares the supervaluationist approach to these questions to the present approach. Chapter 16 examines the notion of a vague object, and shows how it fits naturally into my theory of vagueness. Finally, chapter 17 discusses cases of indeterminacy (not vagueness) that seem like they genuinely have to do with language; I argue that they can be naturally accounted for within a theory of propositional vagueness such as my own.

There are several ways to read this book. There is, of course, the most straightforward route: to read it from beginning to end. Readers wishing to get a relatively self-contained, albeit compressed, introduction to my positive account, however, may take the following route:

1. Chapter 3 for a broad overview of my theory.
2. Chapter 6 for the **Principle of Plenitude**.
3. Chapter 8 for the principle **Rational Supervenience**.
4. Chapter 10 for the principle **Indifference**.
5. Chapter 12 for the distinction between vagueness and borderlineness, and problems for a supervaluational approach to propositional vagueness.
6. Chapter 13 for an explanation of symmetry semantics.

For those not already relatively familiar with decision theory, I would also recommend adding chapter 9—at least section 9.1—to this route: the proper interpretation of decision theory when vague propositions are among the things you can make true is a delicate matter. Missing from this route, however, is most of the motivation for this sort of view and discussion of alternative views.

The third part of the book addresses a number of issues concerning the proper logical and semantic treatment of propositional vagueness. These chapters explore ways in which the existence of propositional vagueness forces a radical departure from conventional thinking about concepts such as worlds, precisifications, fundamentality, and similar concepts. One of the running themes of these chapters is that it's not possible to isolate the objective 'worldly' divisions in logical space from those divisions 'merely having to do with vagueness'. One instance of this idea is a natural supervaluational way of modelling propositions as sets of world–precisification pairs. In this framework logical space can be decomposed straightforwardly into 'worldly' and 'non-worldly' components. For readers wanting to skip straight to these topics, one could read chapter 3 for an overview of the theory, and chapters 11–16.

Any book on vagueness will surely have been influenced heavily by the vast literature that has come before it. In my case, the most easily traced influences have been in the work of Tim Williamson and Hartry Field. While I often oppose Williamson's epistemicism, it was reading his work that first made me see clearly how hard it was to avoid some of his main conclusions—that vague predicates have sharp cutoff points, and that we don't know where they lie. Again, although I criticize a lot of Field's work in this book, there are clear commonalities between his approach and mine. Although the view I develop goes in a very different direction, thinking about his approach to vagueness and the liar has greatly helped me sharpen my own ideas, and led me to raise many of the questions I address in this book. It is also worth emphasizing that the literature has evolved a lot as I was writing this book; I have attempted to discuss new work where it is relevant, but there are omissions nonetheless.

This book has been mostly written from scratch. The material on broad necessity in chapter 11 has been developed further into a paper in Bacon [3], and there is a natural companion paper to this book, Bacon [7], exploring an account of propositional indeterminacy in the context of the liar paradox.

Many people, both in print and in conversation, have shaped my thinking on vagueness. I would like in particular to thank the graduate students and faculty at Oxford during my time there and audiences at Barcelona, Bristol, Brown, Dubrovnik, Leeds, Munich, Oxford, USC, Yale, and a symposium on vagueness and belief at the 2015 Central APA, at which I presented various versions of the ideas developed in this book. I received lots of helpful feedback from these people. I also owe a debt of gratitude to two anonymous referees for OUP, both of whom made extremely helpful suggestions about the structure of the book, which led me to significantly reorganize the content.

In 2013, I took part in an online reading group on this book. Participants included Mike Caie, Andreas Ditter, Cian Dorr, Hartry Field, Peter Fritz, Jeremy Goodman, and Harvey Lederman. The results of these discussions greatly improved the book. I'd also like to thank Shieva Kleinschmidt, Ofra Magidor, Jeff Russell, Bernard Salow, Mark Schroeder, and Robbie Williams from whose conversations I have benefited over the years (Robbie was the external examiner for the aforementioned DPhil thesis from which this book evolved, and Ofra, I later discovered, was the examiner for the BPhil thesis; both their comments lead to some important improvements).

I would like, in particular, to thank Mike Caie, who sent me incredibly detailed comments on the book. Apart from providing many on point criticisms of the central philosophical claims of the book, Mike worked his way through all of the technical material, spotting several mistakes and gave me many helpful suggestions about presentation.

Special thanks are also due to John Hawthorne and Tim Williamson. Apart from his influence in print, Tim, as my secondary supervisor for the DPhil, has provided me with lots of insightful feedback and comments on my work during my time at

Oxford; both his teaching and writing have been a great source of inspiration for me. Many ideas in this book (and, indeed, many ideas not in this book) have been filtered through John in one form or another. In 2015, I co-taught a seminar on vagueness with John which covered a large portion of this book. His comments during this seminar improved the book in a great number of matters of detail, but also helped me see the book in the context of the bigger picture.

My deepest debt of gratitude is to my supervisor and mentor, Cian Dorr. Although he holds completely opposing views on the topic of vagueness, his influence on this project is too pervasive to be properly credited. His mark can be found on most of the central ideas in this book.

I would also like to thank Peter Momtchiloff for his support on this project. I'd especially like to thank Tanya Kostochka and Brian North: Tanya for spotting countless errors and offering several helpful suggestions about making the content more accessible, and Brian North for his meticulous work in copy-editing the book.

Lastly, thanks to the Pasadena branch of Oh My Pan! Taiwanese bakery, and the LA metro Gold and Expo lines, where a non-trivial portion of this book was written.

Gratitude of another kind is owed to my wife, Helen, without whom I would have probably never finished this thing.

Contents

Part III. Logical Matters

List of Figures

List of Tables

PART I

Background

1

Non-Classical and Nihilistic Approaches

Consider the following valid piece of reasoning:[1]

1. Anyone with 100,000,000 cents (a millionaire) is rich.
2. For any n, if a person with $n + 1$ cents is rich, a person with n cents is rich.
3. Therefore anyone with 1 cent is rich.

This argument can be formalized and logically proven from the rule of modus ponens (from A and 'if A then B' infer B) and universal instantiation (from 'everything is F' infer 'a is F')—both of which are principles accepted by pretty much everyone who has participated in this debate. Anyone who accepts the premises of this argument and closes their beliefs under modus ponens and universal instantiation will therefore accept the conclusion. Yet the premises are seemingly acceptable and the conclusion, seemingly, is not.

When I present this paradox to students for the first time, or when I am explaining what I do to friends and family, I commonly receive the following dismissive response:

There is no paradox here, it all just depends on how you define the word 'rich'.

I am sure many reading this are familiar with this response. How exactly does this remark solve the paradox? Does it recommend accepting the conclusion, and if not which premise does it recommend denying? After a little thought some will elaborate as follows:

Once you have said what you mean by 'rich' it will become clear which premise to deny. If you stipulate 'rich' to mean having more than 20,000,000 cents ($200,000) then the second premise fails, but there is nothing puzzling about this—read this way the second premise entails 'if a person with 20,000,001 cents has more than 20,000,000 cents, then a person with 20,000,000 cents has more than 20,000,000 cents' and this is quite clearly false.

Of course, there is nothing particularly special about the second premise. You might also add that if you stipulate 'rich' to mean having more than a trillion trillion cents

---

[1] A number of simplifying assumptions have been made to make the argument below easier to parse. For example, I have assumed that whether someone is rich depends on how many cents they have. Of course this assumption is wildly unrealistic—one can have wealth in other currencies and assets that are not monetary.

then nobody is rich and the first premise fails. If you stipulate that 'rich' applies to everything then the conclusion holds. Whatever one takes 'rich' to mean there is no paradox; on some readings the premises aren't acceptable (they don't even *seem* acceptable) and on others the conclusion isn't unacceptable (and doesn't even *seem* unacceptable).

I hope that everybody agrees that this 'solution' is not satisfactory. Whether someone is rich or not does not depend on what you have decided to stipulate the word 'rich' to mean. There is a quite straightforward empirical test one can perform at home to demonstrate this: log into your bank account, stipulate away, and observe that you do not become one bit richer. Your balance will remain exactly the same—if you were rich before you will remain rich and if you weren't you will remain that way too. This is a get-rich-quick scheme that will not work.

To think that you have solved the sorites paradox by saying something about the word 'rich' is wrong headed—the word 'rich' and the way it is used has nothing intrinsically to do with being rich. There is, of course, another sorites paradox not about rich people but about people to whom the word 'rich' applies when it is being stipulated that 'rich' means such and such. However, deflating this paradox does very little to address the original.

This is the explanation I give to my students, and for what it is worth, it is the explanation I gave before I endorsed the views defended in this book. It is also, I hope you'll agree, good old common sense.

It should be extremely surprising, then, to discover that almost all contemporary accounts of vagueness commit something like the conflation noted above. Invariably the conflation is less blatant, but it is there nonetheless. According to these theorists, vagueness—the phenomenon responsible for the sorites paradox—is just a feature of the way that linguistic communities use words like 'rich'. By one popular account (but by no means the only one satisfying this description) the use of the word 'rich' by English speakers is not specific enough to allow it to latch on to a single property and to consequently draw a single boundary. On each way of drawing that boundary it is sharp—a single cent can take you over the boundary—but the use of the word 'rich' in English doesn't determine which cent that is because it doesn't determine a single sharp meaning with which to draw the boundary.

Even with such a preliminary sketch of the view, it is hard not to think that it is ignoring the moral we drew from our discussion of the naïve response to the paradox—that one cannot solve the sorites paradox just by saying something about the word 'rich'. Even if we could convince the entire English speaking population to use the word 'rich' differently, doing so would not make you any richer or poorer. The account therefore does very little to explain why we find it hard to imagine that one person could be rich while another person possessing one less cent isn't. The most it does is explain why it's hard to imagine that the word 'rich' could apply in English to one person without applying to someone with one less cent. This would be fine if we had asked about the variant sorites involving people to whom 'rich' applies in English, but once again we have done little to address the paradox we started with which was about rich people. (Indeed, some semantic indecision theorists are even quite explicit about the centrality of the variant sorites. For example McGee and McLaughlin, after proving there must be a sharp cutoff concerning what looks red to someone, write 'How do we get from the thesis that, for some n the nth tile looks red to you but the $n + 1$th tile does not to the metatheoretical conclusion that the concept expressed by the phrase "looks red to you" has a sharp boundary? This, it seems to us, is the crux of the sorites problem.' (McGee and McLaughlin, [104]).)

To see why such accounts are explanatorily unsatisfactory note that one doesn't need to be familiar with the English language to appreciate the original paradox. A monolingual Russian speaker can easily see the conflict between the idea that millionaires are rich and the idea that small amounts of money cannot make the difference between being rich and not rich. Like an English speaker, she can also feel the intuitive pull of both claims. However, she will be utterly baffled by an attempt to explain this in terms of the conditions under which the English produce tokens of a certain English word.

These preliminary points, I think, will leave many people unconvinced. After all, if there's one thing that almost everyone agrees on—whether you're an epistemicist or supervaluationist or something else—it's that vagueness has *something* to do with the way we use language. And if it isn't linguistic it must be either metaphysical or purely epistemic, and neither of these options seem particularly promising. The primary aim of this book is to outline an account of vagueness that isn't fundamentally a linguistic phenomenon or, indeed, a metaphysical or purely epistemic phenomenon (indeed, this seems to be a false trilemma).

Rather than treat the study of vagueness as a branch of the philosophy of language, the alternative view places the study of vagueness squarely in epistemological terms— here vagueness is characterized by its role within a theory of rational propositional attitudes; specifically belief and desire (see the chapters in part II). I outline the main features of this view in chapter 3.

1.1 Responding to the Sorites

We began this chapter with a paradox. Any solution worth its salt must at minimum say which premise to reject. I reject the second premise; I deny that for any n, if a person with $n + 1$ cents is rich, so is a person with n cents.

Let us say that number (of cents) is a cutoff for the property of being rich if the corresponding conditional is false. This definition generalizes:

DEFINITION: An element, x, of a sorites sequence for F is an interior point if and only if, if x is F then x' (x's successor in the sequence) is F.

DEFINITION: An element, x, of a sorites sequence for F is a boundary, or cutoff point if and only if it's not an interior point. I.e. it is not the case that if x is F then x' is F.

I also think that properties, like being rich, have cutoff points. If one assumes classical logic this fact can be inferred, via contraposition, from the validity of our initial argument. Thus, from the fact that anyone with 100,000,000 cents is rich, and the fact that someone with 0 cents is not rich it follows that for some n, it's not the case that someone with n cents is rich if someone with $n + 1$ is. That is, the property of being rich has a cutoff point.

This means there is a number, n, such that the person with $n + 1$ cents is rich, whilst the person with n cents is not: a single cent can make the difference between being rich and not being rich.[2] If that number is 159,927,821 cents, for example, it follows that people with 159,927,821 cents are rich while people 159,927,820 cents are not rich.

Although this might sound wild, it is important to stress at this point that this conclusion was derived in classical logic only from the claim that millionaires are rich and the claim that people who have nothing are not rich. This conclusion is therefore common to anyone who accepts classical reasoning and these two premises. The existence of cutoff points is not just the domain of epistemic theories of vagueness, it is quite general.

Throughout the rest of this book I shall assume that the reader is familiar with, and is at least able to get into the mindset of someone who accepts this result; unless they do so they will get very little out of this book. It will be worth our while, therefore, to briefly say a bit about the alternatives. The alternatives can, broadly speaking, be divided into two kinds: (i) those that reject classical logic, and (ii) those that reject the first premise or accept the conclusion—i.e. those who maintain that either nobody or everybody is rich. Since the dialectic will be pretty symmetrical I clump both those options into the same category. I shall now argue that these alternatives are even worse than the result that a single cent can make the difference between being rich and not rich.

1.2 Weakening Classical Logic

A perhaps noteworthy aspect of the result that there is a sharp cutoff between the rich and non-rich people is that it is non-constructive. Classical logic tells you there must be some sharp cutoff, but it doesn't tell you which. How is it that you can know the existential claim that there is a cutoff point, without knowing any of the instances? You know because you know the last guy isn't rich and the first guy is, and classical logic guarantees that there must be a cutoff point even if there is no particular point that it guarantees is the cutoff. The possibility of this kind of situation—situations where you can prove the existence of something, even though you can't prove any

---

[2] Again, I am assuming classical logic here. The non-classical logician who distinguishes between $\neg(Fx \rightarrow Fx')$ and $Fx \wedge \neg Fx'$ simply has two different notions of a cutoff point. A non-classical logician attempting to avoid sharp cutoff points must therefore be careful to ensure that they avoid both kinds of cutoff point.

instance—is a feature distinctive to classical logic, and there are alternative logics that do not have it.

A related feature of classical logic that you might also find puzzling is that it commits you to the law of excluded middle (LEM):

LEM.   Either A or it's not the case that A

In the particular case at hand this entails that everyone is either rich or not rich. This is surprising, for let us suppose that Janet has that borderline amount of money in which it is neither clear that she's rich nor clear that she's not rich. The law of excluded middle tells us that even here, either Janet is rich or she isn't. Of course, logic alone doesn't tell us which, but what's surprising is that logic alone delivers disjunctions whose disjuncts seem to be in principle unknowable.

All of this might suggest that it is the non-constructive nature of classical logic that is responsible for this paradox, particularly the law of excluded middle which seems to play a special role in the derivation of the sorites paradoxes. However, one might argue that the instances of the law of excluded middle responsible for the paradoxes don't seem to be particularly plausible. To prove that there's a cutoff for richness one will inevitably have to appeal to instances of excluded middle that have borderline disjuncts, and these don't seem to have much independent appeal.

Indeed, if the dispute just boiled down to the plausibility of the law of excluded middle I think that non-classical responses to the paradoxes would have the clear upper hand. The loss of this non-obvious theorem of classical logic is, to my mind, of little importance when compared to the puzzling phenomenon of cutoff points.

The idea that the law of excluded middle is the culprit is supported by the fact that its presence suffices for a derivation of the existence of cutoff points, against some relatively natural background logic. Given the role of the law of excluded middle in the non-constructive proof of a cutoff point the most obvious alternative to classical logic to consider would be intuitionistic logic. Not only does this logic fail to have the law of excluded middle as a theorem, it has the more general property that a disjunction is only provable if one of the disjuncts is, and an existential formula is provable only if one of its instances is, making it suitable for constructive reasoning in general.

Unfortunately, although excluded middle is one way to prove the existence of a cutoff point it is not the only way. For example, it's been known for a while that one can in a certain sense prove the existence of a cutoff point in intuitionistic logic.[3] This derivation makes no use of the law of excluded middle since it is not present in intuitionistic logic. Thus the assumptions we must relinquish to avoid the existence

---

[3] Although this fact depends somewhat on how one formalizes the claim that there are cutoff points. One can prove the negated conjunction 'it's not the case that: 1 is an interior point and 2 is an interior point and . . . '. One can also prove that it's not the case that there are no cutoff points. Although the intuitionist might be able to make their peace with these conclusions, other formulations of the existence of sharp cutoff points can arise if we add natural principles about truth and justification. See Williamson [157]; see also Read and Wright [118] for more discussion.

of sharp cutoff points must include more than just the assumption that everyone is either rich or not rich. What, then, must we give up in addition to excluded middle?

There are no doubt many intuitive principles that must be given up, but let me focus on a couple that I think are particularly striking:

CS. $((A \to B) \wedge (B \to C)) \to (A \to C)$

PMP. $(A \wedge (A \to B)) \to B$

These are sometimes called conjunctive syllogism and pseudo modus ponens respectively. In both cases one can prove from these principles, along with some relatively modest background logic, that there is a sharp cutoff point. For the proof, and a more detailed description of the background logic, see appendix 18.1.[4]

The kinds of instances of PMP that we need to appeal to can be interpreted as follows. Imagine you have two people, Alice and Bob, with n and $n+1$ cents respectively. Then to avoid sharp cutoffs the non-classical logician must reject instances of PMP like the following:

(1)   If Alice is rich and, moreover, Bob is rich if Alice is rich, then Bob is rich.

Unlike the claim that either Alice is rich or she isn't, the denial of this principle invites the incredulous stare.[5] If Bob is rich if Alice is rich, and moreover Alice *is* rich, how on earth could Bob fail to be rich? To deny (1) borders on incoherence.

A similar puzzle can be raised against denials of CS. Imagine now a third character, Charlie, with $n - 1$ cents. The proof that there are sharp cutoff points now appeals to the following kinds of instances of the principle:

(2)   If Bob is rich if Alice is, and Charlie is rich if Bob is, then Charlie is rich if Alice is.

Once again, I cannot fathom how (2) could be denied, or how one could take the phenomenon of vagueness to motivate the denial of this claim. If Bob's rich if Alice is, and Charlie is rich if Bob is, then how on earth could Charlie fail to be rich if Alice is rich?

Unsurprisingly, these principles are not derivable in the logics that are standardly appealed to in the context of the sorites paradox (this includes the 3-valued Kleene and Łukasiewicz logics, infinite valued Łukasiewicz logic, and various logics that Field has appealed to solve the paradoxes of vagueness).[6] The above shows that this is not

---

[4] The logical assumptions I make are modest in the sense that they are shared by all the non-classical logics for dealing with vagueness on the market: Łukasiewicz logic, Kleene logic, and various logics that Hartry Field has applied to the paradoxes of vagueness.

[5] Note that in denying instances of these principles these theorists are not necessarily committed to asserting the negation of those instances. Such theorists typically take denying to be a *sui generis* speech act not reducible to assertion.

[6] See, for example, Field [54]. Interestingly the first principle, CS, does not appear to be as straightforwardly problematic when combined with the semantic paradoxes (see Bacon [5]).

really an accident, since these principles feature essentially in derivations establishing the existence of sharp cutoff points.

Let me mention there are certain paraconsistent logics that do accept CS and PMP: the paraconsistent logic **LP** based on the dual of the three valued Kleene logic, for example.[7] Since this logic accepts these principles and the background logic required to derive the existence of sharp cutoffs, these logicians must accept the existence of a last rich person in a sorites for richness. Thus this kind of paraconsistent logician is no better off than the classical logician with respect to the existence of sharp cutoffs, and they are probably worse off, for this logic does not have the rule of modus ponens.[8] Unlike the classical logician, these theorists are *also* committed to the claim that there are *no* sharp cutoffs in a sorites for richness. But it is unclear how this extra belief helps make their original commitment to sharp cutoffs any more palatable; indeed it arguably makes it less palatable due to its contradictory nature.[9]

It should be acknowledged at this juncture that there are non-classical logicians who have made their peace with failures of principles such as PMP, and are quite upfront about it. Field is probably the most prominent example.[10] One interesting thing to note about this is that these theorists are often engaged in a more ambitious project: that of producing an all-purpose non-classical logic that not only deals with the sorites paradoxes but also accommodates the liar paradox. Field, for example, draws a number of parallels between the two paradoxes in Field [54], for example. Given this background project some comfort can be derived from the fact that there are independent reasons to be suspicious of the principle PMP—it is susceptible to a version of the liar paradox known as the 'Curry paradox'.

Be this as it may, it is worth noting that there is no similar independent argument against CS. There are several consistent approaches to the liar paradox that can recover a significant amount of reasoning about truth without relinquishing CS.[11]

My main point, which I think most can take on board, is that the costs of the classical/non-classical debate are frequently mischaracterized as a debate about the status of the law of excluded middle. I am not dogmatic about the law of excluded

---

[7] See Priest [111]. The precise details of this logic need not concern us here.

[8] Modus ponens is the rule: from A and $A \rightarrow B$ infer B. This is distinct from the principle PMP, which is a single sentence and not a rule. One cannot derive the rule of modus ponens from PMP without applying the rule of modus ponens itself. (Conversely, one cannot derive PMP from modus ponens without applying conditional proof—conditional proof is therefore missing in the logics we have been considering, which contain the rule of modus ponens but not PMP.)

[9] One might think that the paraconsistent logician can say something about her commitment to sharp cutoffs to make it sound better: unlike the epistemicist, she will typically be able to say that there are lots of sharp cutoffs, and so in some sense the feeling of arbitrariness is curtailed. But again, this is also a belief she has in addition to one she shares with the epistemicist—for just like the epistemicist, this kind of paraconsistent logician thinks that there is exactly one cutoff point. This seems to me to be the very thing we find so puzzling about epistemicism.

[10] See Field [54], for example.

[11] See Bacon [5], for example. Brady also develops some weakenings of relevant logic that deal with the liar paradox but contain CS in [18].

middle, and I can conceive of situations where it would be reasonable to revise it. However, the real cost of such accounts, in my opinion, are that they typically give up principles like pseudo modus ponens and conjunctive syllogism.[12]

A final issue worth highlighting is that even if a non-classical logic can give us a satisfactory account of the sorites paradox, there are other puzzles relating to vagueness that need to be addressed before we have a fully adequate theory. One of these further puzzles is the 'problem of the many' (see Unger [145])—a problem not evidently directly related to the sorites. There are no doubt a number of different issues that need to be addressed here, but here is something that seems initially puzzling about the non-classical theories we have been considering so far. Firstly, it seems as though the property of being at least 29,000ft is a precise property. Although the classical and non-classical logician disagree about much, one might have thought that they could surely agree about the subject matter of their investigations: which properties are vague and precise. Thus, one might have thought that the non-classical logician can agree with the classical logician about the preciseness of this property. Since precise properties satisfy the law of excluded middle we have:

Everything is either at least 29,000ft tall or not at least 29,000ft tall.

Now, all of the logics we have considered so far allow us to infer from a universal generalization, 'everything is F', a specific instance, 'a is F' (indeed many endorse a stronger axiom form of this inference: that if everything is F that a is F). Thus in all of these logics we can infer from the above sentence:

Either Mt Everest is at least 29,000ft tall or Mt Everest is not at least 29,000ft tall.

Where F is being understood as the disjunctive property of either being at least 29,000ft or not at least 29,000ft, and a as Mt Everest. Now evidently the non-classical logician should not accept this conclusion because it is borderline. It follows that the non-classical logician must either make some revisions to the ordinary conception of precision, or make some revisions to the quantified logic beyond those that have to be made to accommodate the sorites paradox (which tend to be modifications to the propositional logic).

1.3 Nihilism

If we are granting classical logic, then the only other way to avoid the existence of sharp cutoffs is to either deny the first premise—that anyone with 100,000,000 cents is rich—or accept the conclusion—that someone with 1 cent is rich. Call these views nihilism and universalism respectively. For whatever reason, philosophers who make

---

[12] The failures of PMP and CS do not exhaust the problems. Related principles are also problematic: contraction—the inference from $A \to (A \to B)$ to $A \to B$—and conditional proof—the meta-inference from $\Gamma, A \vdash B$ to $\Gamma \vdash A \to B$—must also be dropped to avoid the existence of sharp cutoff points.

these kinds of responses usually make the nihilist response and so I shall focus on that response.[13] The two views are, however, very closely related and for most of the things I say about nihilism there is an analogous thing that can be said about universalism.

Note that the nihilistic response, if it is to serve as a general response to the sorites paradox, must generalize in a few ways. It must firstly extend to people who have more than 100,000,000 cents—thus no-one, whatever their wealth, is rich. Secondly, if the response is to be general it must extend to other properties that can be the subject of a sorites sequence; thus a nihilist thinks that nobody has ever been to the moon, learnt to swim, kissed, flipped a pancake, and so on, for clearly each of these properties is susceptible to a sorites sequence. Note that this is more radical than accepting a mere conspiracy theory, for the nihilist not only thinks that nobody has been to the moon, they also think that nobody has even appeared to have gone to the moon, no-one has seemed to have flipped a pancake, and so on, for these subjective properties are no less susceptible to sorites reasoning than their original counterparts. Thus straightforward observational facts about how things seem to us also go out of the window if nihilism is to be accepted. Symmetrical things must be said about universalism.

How, then, do nihilists explain these wacky sounding claims? Typically they do so by invoking some peculiar feature of the language that is used to express the strange sounding claims—according to nihilism vague language is defective in some way. Perhaps sentences involving vague predicates simply always fail to express a proposition, or perhaps they always express a false proposition—presumably the inconsistent proposition, to avoid arbitrariness. Thus in response to the argument we opened with, the nihilist will say things like the following:

NIHILISM:

  – The predicate 'rich' does not apply to anybody.
  – Sentences of the form 'S is rich' are never true.
  – Belief tokens corresponding to 'S is rich' are never true.

They will of course say similar things about other vague predicates.

At this point it will be useful to distinguish two different ways one could be a nihilist. The first way, which I take it is the most common way, is to accept the semantic claim about the word 'rich' but to nonetheless retain one's common-sense beliefs about who is rich and who isn't. Let's call this the *semantic nihilist*. The other kind of nihilist—the *radical nihilist*—not only comes to think that all of their ordinary beliefs are untrue, they come to reject those beliefs as well.

Let me start with the semantic nihilist. The thinking behind it goes as follows: the belief that Warren Buffett is rich, while strictly speaking semantically defective and thus untrue, is nonetheless an incredibly useful belief to have and thus we should retain the belief that Warren Buffett is rich despite this semantic defect. Without

---

[13] Peter Unger [144] is probably the most famous instance of this response. For more recent defences, see Braun and Sider [20] and Beall [13].

beliefs like these it would be impossible to go about our ordinary business. I don't know exactly how much money I have, but I have a rough idea. Since my beliefs about how much money I have aren't precise, they're untrue—but I shouldn't drop them altogether, otherwise I wouldn't be able to make economic decisions: I wouldn't know what I can and can't afford.

Parallel things can be said about assertion on this view. Although this type of nihilist will maintain that uttering the sentence 'Warren Buffett is rich' involves uttering an untruth, they will still retain the practice of uttering this sentence provided they believe that Warren Buffett is rich. The thought is that in uttering strictly untrue sentences one communicates beliefs that, while also strictly speaking false, are useful.

Note that we have to be careful about what we mean by the word 'true' in these contexts. The nihilist I've described believes both that Warren Buffett is rich and also that their belief that Warren Buffett is rich is untrue. Let us say that a belief is 'Ramsey-true' if it is a belief that P and, in fact, P.[14] Thus any moderately reflective person who believes that Warren Buffett is rich believes that their belief that Warren Buffett is rich is Ramsey-true. They believe (i) that Warren Buffett is rich, and since they are moderately reflective they realize that (ii) this belief is a belief that Warren Buffett is rich and therefore are in a position to conclude (from the definition of Ramsey-truth) that this belief is Ramsey-true. Thus the nihilist I am describing is not talking about Ramsey-truth when they talk about their vague beliefs being untrue. These theorists rather think that there is another property, more substantial than Ramsey-truth, that plays an important role in semantics and they moreover believe that their vague beliefs do not have this property.

For this kind of nihilist it is important to distinguish sharply between the following two questions:

1. Are there any rich people?
2. Does the word 'rich' apply to anyone in English?

Whether or not you are a nihilist, these are quite clearly *not* the same question. If you are not a nihilist they both have the same answer, but they might not have done. To see this imagine a world in which English speakers use the word 'rich' in such a way that it applies only to round squares—let's say, they use the word 'rich' in roughly the same way we actually use the phrase 'round square'. Let's also suppose that, much like the actual world, this world suffers from widespread wealth inequality. In this world the answer to whether there are rich people is clearly 'yes' while the answer to the second question is clearly 'no'—the word 'rich' applies to nothing.

This example is actually not that far-fetched: the semantic nihilist thinks the actual world is like the world described above in the relevant respects. Due to the way we actually use the word 'rich', and other vague predicates, it ends up expressing an

---

[14]   This notion is due to Ramsey: see [116].

inconsistent property (or perhaps no property at all). However, the semantic nihilist maintains that despite the fact that the word 'rich' does not apply to anyone in English, this is not a sufficient reason to abandon the common-sense belief that there are rich people. Thus according to the beliefs of the semantic nihilist the actual world is a bit like the world described above: there are rich people, but the word 'rich' doesn't apply to anyone.

This highlights the most important feature of this kind of nihilist: they have a completely standard conception of what it is to be rich—they retain the common-sense belief that billionaires, like Warren Buffett, are rich and that people with only a few cents to their name are not. It is their understanding of 'truth in English' and 'applies in English' that is non-standard: they will agree with everyone about who is rich and who isn't, and only disagree about who the word 'rich' applies to in English. In particular, they think that Warren Buffett being rich does not suffice for the truth of the sentence 'Warren Buffett is rich', since they believe that Warren Buffett is rich, but that the sentence 'Warren Buffett is rich' is not true in English.[15]

These nihilists are guilty of the very mistake I raised at the beginning of this chapter. Consider the following argument, which is somewhat analogous to the argument we opened with:

1. The word 'rich' applies to anyone with 100,000,000 cents (a millionaire).
2. For any n if 'rich' applies to a person with $n + 1$ cents, it applies to a person with n cents.
3. Therefore 'rich' applies to anyone with 1 cent.

In response to this sorites it is completely clear which premise this nihilist denies: since the word 'rich' does not apply to anything, premise 1 is denied.

However, we took care to distinguish the above paradox from the paradox we opened with, which did not assume that the word 'rich' applies to millionaires; it simply assumes that millionaires are rich. And we have noted already that the nihilist I have been describing has a completely standard conception of what makes someone rich. Thus the semantic nihilist grants the first premise of our original argument, and we are back to where we started. If the nihilist retains the common-sense beliefs about richness, then they accept the first premise and reject the conclusion. So given classical logic, they must accept that there is a last rich person in a sorites sequence for richness. Of course, there is no last person to which the word 'rich' applies, but this is of little comfort if you already have to accept the existence of a sharp cutoff point for richness.

---

[15] In fact, their account of truth in English, and their overall metasemantics, will be much more radical than this. They maintain that the practice of making assertive utterances of the sentence 'Warren Buffett is rich' is perfectly legitimate amongst people who believe that Warren Buffett is rich. They will therefore agree that this sentence is used to communicate the belief that Warren Buffett is rich. Many people simply take such patterns of usage to indicate that 'Warren Buffett is rich' means that Warren Buffett is rich. The nihilist will not—they will not take these patterns of usage to be sufficient for this sentence to mean that Warren Buffett is rich in English, and consequently, will not take the richness of Warren Buffett (which is something they accept) to suffice for the truth of this sentence in English.

This type of nihilism is one instance of the general tendency in the philosophy of vagueness of changing the subject from puzzles about richness, baldness, and so on, to questions about language. In doing so we often leave the original puzzles unanswered.

These problems stemmed from the fact that the semantic nihilist kept her common-sense false beliefs. There is a much more radical way of being a nihilist—not the kind that is typically endorsed—that seems be in a better position to resist the original sorites paradox we opened this chapter with. This is the kind of nihilist who not only declares common-sense beliefs to be untrue, but also abandons those beliefs. The radical nihilist therefore not only thinks that the word 'rich' does not apply to anybody, she also thinks that nobody is rich and is therefore well placed to answer our original paradox by denying the first premise. (Strictly speaking, the nihilist could abandon the belief that millionaires are rich by being agnostic, but then she would have to be agnostic about the existence of cutoff points; I shall therefore focus on the view that there are no rich people, etc.)

While it is consistent for a radical nihilist to continue the practice of assertively uttering vague sentences, whatever the reason she gives for that practice, it is presumably not about passing along useful beliefs. Presumably the only beliefs the radical nihilist has are the kinds of beliefs you can state in logico-mathematical language, and perhaps the language of fundamental physics, since vagueness is pervasive in other realms of discourse. Thus in uttering 'Warren Buffett is rich' the only contingent belief I could be passing on is a belief about fundamental physics, and it seems relatively clear that I do not learn anything about fundamental physics upon hearing utterances like these.

However, a little further reflection reveals that the radical nihilist doesn't really *need* to explain why English speakers go about uttering vague sentences because, according to the radical nihilist, they don't. In fact, radical nihilists believe that there aren't any English speakers and there is therefore no need to explain their practices.

Unfortunately, this type of reasoning highlights why it's almost impossible to be a radical nihilist—for if you are a radical nihilist it is unclear why you need to do *anything*. A radical nihilist should feel that she has no need to do the groceries because she believes that there are no grocery stores, and moreover no need to eat since not only is eating impossible, but there is no food to eat and no one ever dies of starvation anyway.[16] If a radical nihilist does decide to go grocery shopping, surely it is not her logico-mathematical beliefs that explain this action, nor her beliefs about fundamental physics—and if it is none of these, it is unlikely to be a precise belief, for very little is precise outside of those realms.

One response you could make, at this juncture, is that there is another attitude, not belief, which our hypothetical radical nihilist holds towards the vague. Maybe this attitude is something like 'pretending that *P* for such-and-such purposes', or 'believing

---

[16] Here I am appealing to the facts that eating, being food, and dying of starvation are all soritesable.

that according to the fiction of grocery stores, P' or maybe it is just a *sui generis* attitude that plays a particular causal functional role. Whatever this attitude is, perhaps it is this attitude and not belief that explains action: when the agent goes to the grocery store this is because she has this attitude towards the proposition that there is food in the grocery store whilst desiring to have food. And perhaps it is the passing on of this attitude that communication achieves.

It is clear that this attitude, whatever it is, pretty much plays the role that belief plays according to a common-sense view. Once we have introduced an attitude that plays the belief role, then I think we have just introduced belief under another name; radical nihilism collapses into the more moderate nihilism we rejected earlier. In the important respects this kind of nihilist agrees with me, for the attitude that rationalizes action is one which is held towards the proposition that millionaires are rich, and the proposition that people with nothing are not rich. Moreover, assuming that this attitude is belief-like enough to be subject to the norms of classical reasoning, the attitude must also be held to the proposition that there is a last rich person in a sorites for richness. In the sense in which this nihilist can be said to be committed to anything at all, this nihilist is committed to sharp cutoff points.

2

Classical Approaches
An Overview of the Current Debate

Having discussed and ruled out the nihilistic and non-classical approaches to the sorites paradox we must follow logic where it leads: sorites sequences have cutoff points. There was a first nanosecond at which a person becomes old, a single hair that makes the difference between being bald and not bald, and so on and so forth. While this may sound admittedly wild at first, we saw that the alternatives fared little better.

Of course, even though a single nanosecond can make the difference between being old and not old, it's a vague matter which that nanosecond is: people at the cutoff age are *borderline* old. According to the naïve conception of this notion of borderlineness, the properties of being old, being not old, and being borderline old divide people into three mutually exclusive classes. However, a corollary of accepting classical logic is that this naïve conception must be abandoned. For given excluded middle, even borderline old people are either old or not old; thus there are either borderline old people who are old, or borderline old people who are not old (presumably both). Some classical logicians play down this consequence, insisting that even though there are old people who are borderline old (or non-old people who are borderline old), there is still something incoherent about the *assumption* that a person is old and borderline old.[1] Nonetheless, the fact that borderline oldness does not preclude oldness is a surprising consequence of classical logic that I think one ought to be upfront about.

However, while classical logic alone settles many of the questions of interest— the existence of cutoff points being a prime example—there are many questions it leaves open. Providing answers to these further questions fleshes out one's theory of vagueness. The purpose of this chapter and the next is to sort through some of these questions and to delineate the important points of dispute in the philosophy of vagueness for a more detailed treatment in later chapters. In chapter 3, I'll give a swift overview of the main theses of this book, and try to situate them within the context of more traditional debates in the philosophy of vagueness.

---

[1] According to some, it is incoherent in the sense that it is an assumption from which anything whatsoever can be inferred. See section 2.5.

That said, a few words of warning are in order. It is my view that many of the questions that have traditionally occupied philosophers of vagueness do not really carve up the space of possible views at its philosophical joints. Thus, if we are to get a feel for what the distinctive positions in the philosophy of vagueness are, and indeed to understand where the position favoured in this book falls, it will be important to spend some time clarifying these issues. This might require unlearning many of the things we think we know about the space of views consistent with classical logic. For instance, the two most well-known classical theories of vagueness, at least as they are usually represented, will not carve out distinctive positions, if my characterization of the philosophical landscape is correct.[2]

2.1 Epistemicism and Supervaluationism

Perhaps the most famous classical accounts of vagueness are epistemicism and super-valuationism.[3] As we shall see later, both of these terms are fairly slippery: different theorists emphasize different things, and the clearest indication as to when one of these labels can be applied is when the theorist herself identifies one way or the other. Rather than attempting to give precise, and inevitably contentious, definitions of these positions let us start by giving an introductory presentation of the views of the sort you might find in a textbook.[4]

The epistemicist, it is said, takes the argument for sharp cutoffs from classical logic *at face value*. However initially compelling the thought that a nanosecond cannot make the difference between being young and not young, the epistemicist reasons, it is false: logic sometimes teaches us surprising things. The epistemicist therefore embraces sharp cutoff points. The other distinctive feature of epistemicism is that it tries to explain vagueness in terms of our *ignorance* about the location of these cutoff points. Since there is a last nanosecond at which I ceased to be young, and it is unknown which nanosecond that was, and since similar ignorance about the locations of cutoff points is common to all sorites sequences, some epistemicists have suggested that vagueness just *is* a kind of ignorance. No doubt this ignorance is unusual: while it is easy to find out at what age one is eligible to vote, it is hard or impossible to find out at what age one stops being young. However, the epistemicist maintains, ignorance about vague cutoff points is fundamentally the same sort of thing as ignorance about other ordinary matters such as eligibility to vote.

Just as the sorites paradox teaches us something surprising about youngness, similar conclusions about language are also often emphasized by the epistemicist.

---

[2] The idea that the traditional ways of carving out the main debates in the philosophy of vagueness are in need of revision has been around for a while: see, for example, Dorr [34], Field [53], Williams [153], and Williamson [158]. Not all these authors draw quite the same morals as I do, however.

[3] See Sorensen [138] , Williamson [156] and Horwich [71] for defences of the former and Fine [56], McGee and McLaughlin [104] and Keefe [78] for the latter.

[4] See, for example, Keefe and Smith [80] chapter 1 and Sorensen [137].

There is a last nanosecond at which the word 'young' applies to me in English, so that our use of English must determine a specific enough interpretation of 'young' to determine that cutoff point given facts about my age in nanoseconds: according to the epistemicist, we can retain classical semantics even in the presence of vagueness. There is a unique classical interpretation of English, determined by the linguistic practices of English speakers, which fixes the cutoff point for 'young' given the age-in-nanoseconds facts. Moreover, since a sentence is either true or false in this interpretation, the epistemicist endorses the principle of *bivalence*: for any meaningful sentence, either '*P*' is true or '*P*' is false.[5]

Supervaluationists, by contrast, are not directly motivated by the sorites paradox and, to the extent that they acknowledge the existence of sharp cutoff points, do not usually engage in the project of explaining them. Rather, they are frequently motivated by general considerations concerning language use. In contrast to epistemicism, supervaluationism maintains that there are lots of ways of assigning precise (classical, bivalent) interpretations to predicates like 'young' that accord with the linguistic practices of English speakers. Our practices are determinate enough to exclude some interpretations—for example, interpretations in which 'small' applies to the sun, and other similarly uncontroversial cases are ruled out—but not determinate enough to narrow the possible interpretations down to a unique one. Rather, a vague language gets associated with a collection of interpretations: what are called the *admissible precisifications*, or *admissible interpretations* of that language. An admissible precisification is roughly a classical interpretation that fits with the practices of the language users in the right sort of way.

Unlike the epistemicist, the supervaluationist appears to reject classical semantics. An urgent question then arises: How should one recover theoretically important semantic concepts, such as truth and falsity, on this conception of a vague language? For according to supervaluationism, different admissible interpretations may deliver differing verdicts of truth and falsity for some sentences. According to orthodox supervaluationism a sentence is *supertrue* if it is true according to all admissible classical interpretations of the language. A sentence is similarly *superfalse* if it is false relative to all admissible interpretations. Since the admissible interpretations may differ regarding the cutoff for 'bald'—some may draw the line at 1,023 hairs, others at 1,024—a sentence like 'Harry is bald' will be true on some admissible interpretations and false on others if Harry has 1,023 hairs. In which case, this sentence is neither supertrue nor superfalse: it is a (super)truth value gap.

---

[5] I am suppressing some important caveats here: for example, given the existence of context sensitivity, it is, presumably, utterances or sentence-context pairs that are the real bearers of truth and falsity, not sentences. One might also countenance failures of bivalence for reasons unrelated to vagueness, whilst still counting as an epistemicist about vagueness.

Unlike the epistemicist, then, the orthodox supervaluationist rejects the law of bivalence. For given the identification of truth with supertruth, the law of bivalence becomes:

BIVALENCE: For every meaningful sentence P, either P is true in every admissible interpretation or P is false in every admissible interpretation,

and we have just seen that this has counterexamples. Note, however, that we must draw a sharp distinction between bivalence and the truth of the law of excluded middle. The claim that the law of excluded middle is true, given this account of truth, becomes:

TRUTH OF LEM: The sentence 'P or not P' is true according to every admissible interpretation.

Since admissible interpretations are just special kinds of classical interpretation, it follows that the law of excluded middle is true according to all admissible interpretations. More generally, every classical tautology will be true in all admissible interpretations for exactly the same reason. To illustrate how bivalence and the law of excluded middle come apart, recall that 'Harry is bald' is neither true nor false, but since every admissible interpretation makes either 'Harry is bald' true or makes it false, the disjunction 'either Harry is bald or he isn't' is true on all admissible interpretations. Indeed, this is a true disjunction composed of untrue disjuncts, illustrating the failure of truth functionality on this conception of truth.

What I have described above is the orthodox version of supervaluationism.[6] Not all supervaluationists accept the identification of truth with the notion of supertruth defined above—there are other reasonable candidates out there.

Note that it is always possible to introduce a predicate of sentences that behaves disquotationally and which, assuming classical logic, is bivalent. Let us introduce the artificial term 'dtrue', and stipulate that it applies to the sentence 'Harry is bald' if Harry is bald, and doesn't apply if he isn't. Since we've acknowledged that Harry is either one or the other this definition is perfectly well-defined on this sentence. By making lots of similar stipulations we can ensure that dtruth behaves disquotationally for every sentence of our original language. (Of course, in making these stipulations I can't use the word I'm trying to introduce within those stipulations: thus I can't make stipulations that use the word 'dtrue', but this will matter little for our purposes since we are not primarily interested in disquotational sentences involving the word 'dtrue'.)

To see that dtruth is a bivalent concept, let us adopt the usual convention of identifying falsity with having a true negation: say that a sentence is *dfalse* if its negation is dtrue. Without loss of generality, we can show that the sentence 'Harry is bald' is either dtrue or dfalse. By excluded middle either Harry is bald or he isn't, by the disquotational nature of dtruth it follows that either 'Harry is bald' is dtrue or 'it's not the case that Harry is bald' is dtrue, and so 'Harry is bald' is either dtrue or dfalse.

---

[6] See, for example, Fine [56] and Keefe [78].

Note, however, that the claim that 'Harry is bald' is dtrue is just as borderline as the claim that Harry is bald (after, our stipulations were designed so as to make these two claims materially equivalent). Thus, even though the sentence 'Harry is bald' is either dtrue or dfalse, it's borderline which of the two possible dtruth values it assumes.

Some proponents of supervaluationism have suggested that dtruth is a better candidate than supertruth for doing the philosophical work of truth (see McGee and McLaughlin [104]). For this brand of supervaluationism borderline sentences do not lack truth values, but they lack what you might call a 'determinate truth value': while borderline sentences can be true, we shall say that a sentence is *determinately* true only if it is true without being borderline.

If we were to reduce both views to a one line slogan, they might go something like the following:

EPISTEMICIST SLOGAN: Vagueness is ignorance.

SUPERVALUATIONIST SLOGAN: Vagueness is lack of truth value (or, determinate truth value).

Note that epistemicists do not identify vagueness with just any kind of ignorance, but with the special kind of ignorance associated with borderline cases—the details of which will depend on the specific version of epistemicism in question.

There are several theories of vagueness on the market that accept the existence of cutoff points, but epistemicism and supervaluationism in particular have enjoyed pride of place in the literature. Indeed, if one is in the business of cataloguing the important questions in the philosophy of vagueness one might expect one or more of these disputes to be at stake between the epistemicist and the supervaluationist since, in the minds of many, supervaluationism and epistemicism are the two main competing classical theories. In this section I shall use the supervaluationist/epistemicist debate as a springboard for my discussion of some of the traditional questions in the philosophy of vagueness.

Although the dispute between supervaluationism and epistemicism has dominated much of the debate on classical theories of vagueness, a few philosophers have begun to question how substantive the dispute actually is.[7] Indeed, despite the seemingly straightforward differences between the views outlined above, I'm inclined to think the accepted classification of classical views into 'epistemicism' and 'supervaluationism' to be thoroughly unhelpful: the issues that are sometimes thought to characterize the epistemicist and supervaluationist positions either turn out to be issues on which both sides agree, or issues that have few repercussions for the central problems in the philosophy vagueness. In my view the most important disputes in the philosophy of vagueness lie elsewhere. What follows is a discussion of several of these non-disputes; I'll return to the questions I take to be more central in chapter 3.

---

[7] See Dorr [34], Field [53], Williamson [158].

2.2 Does Vagueness Involve Ignorance?

Let us identify our first dispute:

IGNORANCE: Does borderlineness exclude knowledge?

A positive answer to this question has for a long time simply been taken for granted: that if Harry is a borderline case of baldness then one can't know whether Harry is bald. However, this consensus has recently been challenged by Cian Dorr and David Barnett, who have advocated for a distinctive approach to vagueness that denies that vagueness excludes ignorance. (We shall treat these types of views in chapter 5.)

On what side of this debate do the supervaluationist and epistemicist fall? Let us begin with the slogans. Epistemicism is notorious for making a pair of counterintuitive claims. According to epistemicism there's a nanosecond during which I stopped being a child, and nobody knows (or could come to know) when that nanosecond occurred. It is often suggested that this consequence of epistemicism is particularly outrageous, and that an alternative, such as supervaluationism, must be sought in its place. But a little inspection reveals that supervaluationism does no better in this regard. We have shown earlier that classical logic alone commits us to the existence of this critical nanosecond.[8] But once this is accepted, the only point at which we can disagree with the epistemicist is over whether it is known (or possibly known) when that last nanosecond occurred. To disagree here is to maintain that not only is there a last nanosecond during which I was a child, but it is furthermore known (or at least possible to know) when this nanosecond occurred. Questions quickly multiply: if it is known, then by whom is it known? How was it discovered? If it is merely possibly known, how does one go about discovering it? Of course supervaluationists are no better positioned to answer these questions than anyone else, and most supervaluationists therefore agree with the epistemicist that we don't and can't know where the cutoff points are in sorites sequences.

One might think that while epistemicists and supervaluationists agree about the relation between borderlineness and ignorance, only the epistemicist *identifies* vagueness with ignorance. This way of distinguishing the views puts a lot of weight on the difference between identity and necessary equivalence. This is a fairly fine point of disagreement, but more importantly, it is not even clear if actual epistemicists do identify vagueness with (vagueness related) ignorance. Timothy Williamson, for example, develops a theory of vagueness in which it corresponds to a certain kind semantic plasticity—vague expressions have different meanings in nearby worlds—and the connection between vagueness and ignorance looks like it is at best a derived fact.

---

[8] Sometimes people say that epistemicism is committed to '*sharp* cutoff points' whereas supervaluationism is not. If the nanosecond at which I stop being a child counts as a sharp cutoff, the supervaluationist is committed to sharp cutoffs in exactly the same way. Of course, neither view accepts the existence of a nanosecond that is determinately the last nanosecond during which I was a child (reading 'determinately *p*' as '*p* but it's not borderline whether *p*'). Thus neither view accepts the existence of sharp cutoffs in the second sense, maintaining parity between the views.

2.3 Does Vagueness Involve Truth Value Gaps?

Let us turn to another question that has occupied the attention of many philosophers of vagueness:

TRUTH GAPS: Do borderline sentences have truth values?

This is, of course, the central point of contention between the ordinary supervaluationist on the one hand and the epistemicist and disquotational supervaluationist on the other. But is it a substantive question? Hartry Field thinks that it is not:

[. . .] let us set aside a verbal issue. Proponents of [determinately operators] differ as to their preferred use of 'it is true that p': some take it as equivalent to 'p' (call this *the weak reading*), others take it as equivalent to 'it is a determinate fact that p' (*the strong reading*). Obviously nothing can hang on whether we use the term 'true' in its weak or its strong sense. Perhaps the safest policy is to introduce two distinct words, 'true$_w$' and 'true$_s$', for these notions.

(Field [53])

What is clear from Field's presentation is that the point of contention is not over the coherence of either of these two notions of truth. As noted above, one can always introduce a disquotational notion of truth stipulatively. Similarly, I can always introduce a stronger notion of determinate truth by defining it as a (weak) truth that's not borderline. Since everyone theorizing about vagueness ought to have the notion of borderlineness at their disposal, the notion of determinate truth, so defined, will always be available.

One question we might ask is: Which of these two notions best fits our use of the word 'true' in English (parallel questions could be asked about the word 'vrai' in French, 'wahr' in German, and so on). One way to settle this question would be to look at a body of data concerning the way that 'true' is used in English and see which fits best. I strongly suspect that the answer to this question is neither: most uses of 'true' in English apply not to sentences but to propositions (for example 'everything Alice said is true'), and is thus not the kind of property of sentences that a supervaluationist is attempting to capture. Of course we can talk about true sentences in English, even though it is less common, but it is well known that context sensitive sentences are neither true nor false on their own but only relative to a context and the typical speaker is not particularly careful about these subtleties. Thus the question becomes about which of the two notions of truth best fits our use of the somewhat technical notion of 'truth in English relative to a context c'. This is no longer a question about ordinary language but about philosophers' technical vocabulary.

A better way to approach these questions is to outline a wider theoretical role that our philosophical notion of truth is supposed to play, and to see whether disquotational truth or determinate truth plays it. This wider role presumably relates the concept of truth to assertion, reasoning, the possibility of knowledge, necessity, belief, evidence, disquotational principles, compositionality, and so on. It is unlikely, however, that either concept satisfies all of these connections: disquotational truth is compositional, disquotational, and is presumably more useful for semantic theorizing, whereas determinate truth presumably is more closely connected to the possibility of knowledge, being a prerequisite for proper assertion, and so on. The question of which role to associate with the word 'true' is no more substantive than the question we began with.

Perhaps one can disagree about whether determinate truth or disquotational truth plays a specific part of this role (for example, one might maintain that a good argument only needs to preserve determinate truth, or one might insist that this is not enough and that good arguments must also preserve disquotational truth). Once this move is made the debate is no longer isolated to sentences involving the word 'true'. Depending on the wider role, we may find ourselves disagreeing about what kinds of inferences are permissible, about the kinds of doxastic attitudes we should have about the borderline, what rational credences in vague propositions should be like, and so on.

The thing to note here, however, is that these further questions do not need to be formulated in terms of truth at all, and can be formulated directly as questions about the relation between borderlineness and these other concepts such as good reasoning, and rational belief. Indeed, I will argue shortly that some of these debates are in fact substantive, non-verbal disputes: disputes that aren't purely about which notion to call 'truth'.

Relating our discussion back to supervaluationism and epistemicism, note also that while the question about truth has the appearance of being central to that dispute, these further questions about rational credences and reasoning are not obviously part of that debate. No consensus exists among supervaluationists about how these further questions should be answered: there is, for example, no consensus among supervaluationists about what kind of doxastic and credential attitudes we ought to have towards the borderline (contrast, for example, Field [53], Williams [152]).

I am thus inclined to agree with Field: the dispute about whether borderline sentences have 'truth values' is not substantive if it is a debate isolated purely to questions involving the word 'true'. If one attempts to widen the dispute to questions about belief, reasoning, and so on, it might be better to just ask questions directly about the relation between borderlineness and belief, reasoning, and so on, instead of making the detour through the debate about 'truth'. After all, presumably there are roles for both disquotational truth and determinate truth, so the question of which role to give the honorific title 'truth' does not seem like an important issue.

Before we move on, it is worth noting that a completely parallel dispute can be raised about other semantic vocabulary, such as 'refers' and 'expresses'. The analogue of the view that a borderline sentence doesn't have a truth value is that a borderline sentence doesn't express exactly one proposition, and that a vague name doesn't refer to exactly one thing. A traditional supervaluationist will usually chalk this up to non-uniqueness rather than non-existence: borderline sentences express many propositions, and borderline names refer to many objects. But as before, we can also introduce notions that behave disquotationally on a limited fragment of the language: we can stipulate that 'Harry is bald' expresses the proposition that Harry is bald and nothing else, that 'Harry' refers to Harry and nothing else, and so on. Thus unless this dispute about the words 'expresses' and 'refers' is hooked up to further questions it begins to look no more substantive than the question about truth.

Since the choice is for the most part terminological, I shall use the words 'true', 'expresses', and 'refers' to denote the disquotational notions: I shall adopt this convention throughout this book unless I indicate otherwise. When I discuss orthodox supervaluationism I shall reserve the technical term 'supertruth' for the non-disquotational notion of truth.

2.4 Many Interpretations or One?

The epistemicist and supervaluationist pictures also appear to disagree about the extent to which the use of a language determines the meaning of that language. According to the epistemicist there is exactly one interpretation of the language compatible with the way a vague language is used, yet according to the supervaluationist there are many. The epistemicist's position here is often thought to be highly counterintuitive, and the dispute is often taken—incorrectly, I shall argue—to be a straightforwardly substantive one.

> ADMISSIBLE INTERPRETATIONS: Are there many admissible interpretations of a vague language, or is there exactly one?

At a rough gloss, an interpretation of the language is admissible if it doesn't get anything determinately wrong. An admissible interpretation won't include anything that's determinately not red in the extension of 'red': thus, for example, no admissible interpretation of 'red' includes the ocean. Conversely, an admissible interpretation won't *exclude* anything from the interpretation of 'red' that's determinately red either: thus, for example, no admissible interpretation of 'red' excludes the Golden Gate Bridge.

Let us investigate this idea a little further. Recall that given classical logic it follows that there's a first rich person in a given sorites sequence. Moreover, there is a last person in the sequence who is determinately not rich, and a first determinately rich person as well: in terms of wealth, the first rich person lies somewhere between the two. Since an admissible interpretation is an interpretation that doesn't get anything determinately wrong (i.e. doesn't classify anyone as rich who is determinately not rich, and doesn't classify anyone as not rich who is determinately rich) it follows that there are lots of *admissible* interpretations of the word 'rich'. Any interpretation which assigns the cutoff point somewhere between the last determinately not rich person and the first determinately rich person will be an admissible interpretation (Figure 2.1).

Now assume temporarily that interpretations are just possible extensions for the word 'rich'. Notice, then, that there is exactly one interpretation that locates the cutoff

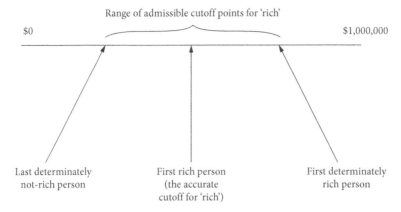

Figure 2.1. The admissible cutoff points for richness.

for 'rich' at the first rich person. In summary: there is exactly one interpretation that gets the interpretation of 'rich' exactly right (to be exactly the rich people) and several interpretations that don't get the extension determinately wrong.

If we wanted to be completely pedantic about it, we can make our argument that there's exactly one set (candidate extension for the word 'rich') containing exactly the rich people more explicit as follows. Suppose that the set X contains all and only rich people, and similarly for Y. It follows that an individual belongs to X if and only if that individual is rich, and that that individual is rich if and only if it belongs to Y. So X and Y are coextensive—an individual belongs to X if and only if it belongs to Y—and so (by the axiom of extensionality) X and Y are identical.[9]

Our argument above relied on the idea that interpretations are just extensions for the word 'rich'. Note, however, that the conclusion that there's exactly one 'correct' interpretation doesn't really rely on this assumption. When properly formalized it's a theorem of classical logic and the axiom of extensionality that there's exactly one set containing all and only the rich people. Thus it's necessary that there's exactly one set (candidate extension for the word 'rich') that contains exactly the rich people.[10] Thus there's exactly one function from possible worlds to extensions that maps each world to the set of people who are rich at that world. So even if we relaxed the assumption that interpretations of 'rich' are extensional, and treated them as functions from possible worlds to extensions, we'd still get the conclusion that there's only one interpretation of 'rich' that necessarily locates the cutoff for 'rich' at the first rich person. Call an interpretation like this an *accurate* interpretation. We have just shown

---

[9] That there even *is* a set of rich people is guaranteed by the axiom schema of separation for sets, provided the schema is understood to include instances involving the predicate 'rich'. The existence of this set also can be proven from instances of separation that only involve precise vocabulary: consider the precise property of owning at least n cents—where n is the cutoff for being rich.

[10] Note that, although it's necessary that *there is* a unique set of rich people, it's not necessary which that set is: different people are rich at different worlds.

that if there is more than one accurate interpretation the differences between them must be purely hyperintensional. Indeed, I shall ignore this caveat in what follows and just assume there is exactly one accurate interpretation.

Given all this, our notion of an admissible interpretation can now be defined more simply as an interpretation that isn't determinately inaccurate. It is easy to see that if there is something that is borderline red, there are at least two interpretations of 'red' that aren't determinately inaccurate: one of which will include the borderline red object, and another that excludes it. It thus follows from the existence of vagueness that there are multiple admissible interpretations.

These facts were derived without appealing to any distinctive ideology: the notion of an accurate interpretation is given by the matching of the predicate 'rich' to rich people, and so on, and the notion of an admissible interpretation relies only on one having the locution 'it's borderline that A' (from which one can define 'it's determinate that p') and the above notion of accuracy at one's disposal.[11] Since we can extend these notions beyond the interpretation of the predicate 'red' to the whole language, ADMISSIBLE INTERPRETATIONS does not seem to carve out an interesting debate, since every classical logician must accept the existence of a unique accurate interpretation and multiple admissible interpretations of the language.

Note, in particular, that supervaluationists must accept the existence of a unique accurate interpretation, and epistemicists must accept the existence of many admissible interpretations.

2.5 Is Validity Local or Global?

Once one has the non-disquotational notion of determinate truth on the table (being true and not borderline true) in addition to disquotational truth, one might go on to ask what arguments count as valid. If one maintains that a valid argument need only preserve determinate truth one gets a distinctive logic: one that agrees with classical logic about all of its theorems, but counts further inferences as valid which classical logic does not (and consequently must relinquish some *meta*-inferences to compensate—principles stating that if one sort of inference is valid, so is another). An inference that preserves determinate truth is called a *globally valid* inference, whereas an inference that furthermore preserves disquotational truth is called a *locally valid* inference. More explicitly, we say that an argument from premises Γ to conclusion A preserves determinate truth if the following material conditional holds of logical necessity:

If each of the premises in Γ is determinately true, then A is determinately true.

---

[11] It should be noted that some epistemicists do not employ the *word* 'determinately', since it is often associated with supervaluationism. However, they have the concept it expresses, since it is something that is defined in terms of borderlineness—a concept they are happy to theorize in terms of.

Similarly, we say that an argument preserves disquotational truth if the corresponding conditional involving disquotational truth holds of logical necessity.

The global account of consequence is typically associated with traditional super-valuationism, where the notion of 'supertruth' is intended to replace the usual role of truth in reasoning and in other domains: on this picture the identification of a valid argument with an argument that preserves supertruth is extremely natural. However, there is an ambiguity in the literature: some supervaluationists define a globally valid inference as one that not merely preserves determinate truth (i.e. supertruth), but additionally preserves determinately determinate truth, determinately determinately determinate truth, and so on. Thus on this alternative conception a globally valid argument Γ entails A when the following conditional holds of logical necessity:

If each of the premises in Γ is determinate at all orders, then A is determinate at all orders.

The different ways of understanding global validity can lead to conflicting verdicts. The inference from A to 'it's determinate that A' is taken to be a globally acceptable inference in this alternate sense even though it does not preserve determinate truth. It is immediate that if A is determinate at all orders then so is the claim that A is determinate, so the inference preserves determinacy at all orders. The inference does not preserve determinate truth, however, because A can be determinate without it being determinate that A is determinate. (To deny this would be to endorse the S4 principle for determinacy, an extremely contentious logic for determinacy. We will return to this in chapter 7.)

Here and throughout we will discuss inferences that involve the notion of de-terminacy itself. We represent this formally by introducing a sentential operator, Δ, into our language, where ΔA informally means 'it's determinate that A'. When Γ is a set of sentences, and A a sentence of this language, we will write $\Gamma \models A$ to mean that the inference from Γ to A is valid.

The inference from A to 'it's determinate that A' is globally valid on one under-standing of 'globally valid' but not the other. However, there are inferences which are uncontroversially globally valid—globally valid on both disambiguations—that are not locally valid. Thus I can, for the most part, draw the conclusions I need by focusing on these inferences. Here is one example of an inference that is globally valid on either understanding:[12]

$$(\Delta ECQ)\ A, \neg\Delta A \models B.$$

This is effectively a strengthening of the rule of explosion. If both A and $\neg\Delta A$ were determinate, then by assumption A would be determinate, and by the factiv-ity of determinacy, A would not be determinate. Thus it is simply impossible for the premises of this argument to be determinate, and *ipso facto*, impossible for the

---

[12] Note, though, that the related inference $A, \neg\Delta\Delta A \models B$ does not preserve determinate truth, even though it preserves determinacy at all orders.

premises of this argument to be determinate at all orders. Thus in either sense of global validity these premises validly entail everything.

Accounts of consequence that accept the above inference cannot be closed under certain meta-inference rules commonly associated with classical logic. For example, the inference rule of reductio would allow one to infer from the validity of the above inference and conjunction elimination that $\neg(A \wedge \neg\Delta A)$. This is equivalent to the absurd principle $(A \rightarrow \Delta A)$, which entails, via excluded middle, that everything is either determinately true or determinately false. This same conclusion could also be inferred using conditional proof with some other uncontroversial logic. Thus conditional proof must be relinquished as well.[13] For this reason people have called global accounts of consequence 'semi-classical'.

Just as we argued that ΔECQ preserves determinate truth, we can also argue that it doesn't preserve disquotational truth. For if ΔECQ preserved disquotational truth that would mean that A and $\neg\Delta A$ could never be both disquotationally true together, which means in particular that all instances of the problematic schema we just encountered, $A \rightarrow \Delta A$, would be always be disquotationally true.

Thus we have another question that one must make one's mind up about:

VALIDITY: Is the consequence relation global or local?

One might be tempted to think that a disagreement about logic is automatically to be taken seriously.

But this is too fast, for it is far from clear that VALIDITY actually corresponds to a disagreement about what the correct logic is. One could imagine two people who, when restricted to sentences not involving the word 'valid', agree about everything and are willing to reason from the same sets of sentences in exactly the way. But one could also imagine that one of the disputants counts no inference as 'valid' for spurious philosophical reasons (choose your favourite). This looks like it is purely a disagreement about which inferences to call 'valid', and not about logic at all. (Since we may even suppose that both disputants are classical logicians: that they both assent to sentences like 'either Harry is bald or he isn't', and they draw exactly the same conclusions from these sentences. It's just that while one considers this sentence to be a logical validity, the other merely counts it as a truth that is not valid.)

Note that there are genuine questions about how best to use the word 'valid'. Should we apply it to inferences that preserve truth in all models that keep the meanings of the truth functional connectives and the quantifiers fixed, or should we apply it to inferences that preserve truth in models that keep the interpretations of other expressions fixed as well (such as the identity symbol, the modal operators, generalized quantifiers, and so on)? Should we count the theorems of higher-order logic as logical validities? Are contingent but *a priori* sentences like 'I'm here now' valid?

---

[13] See Williamson [156], chapter 5, for a more comprehensive discussion of the meta-inferences that fail.

Presumably the answers to these questions won't be resolved by simply reflecting on our pre-philosophical notion of validity, but on the slightly more technical notion that is of interest to philosophers.

Such questions are often not as productive as they might first appear. It could just be that there are lots of different properties that arguments can have, none of which is more important for theorizing about reasoning than any other (see, for example, Beall and Restall [14]). In which case it may just be a matter of preference which to accord the title of 'valid argument'. But whether this is merely a matter of preference or not, the important point here is that these disputes are importantly different from the dispute between the classical and non-classical approaches to vagueness, where the truth of certain principles of logic, such as the instance of excluded middle mentioned above, are actually in contention. Those who maintain that the schema 'if it's necessary that p then p' is not a logical truth because they do not consider modal operators to be logical constants, are nonetheless in full agreement with everyone else about the fact that what's necessary is true—they are not endorsing a non-standard logic of necessity, only a non-standard view about which truths are logical. One must take care, then, to distinguish a dispute about logic from a dispute about 'logic': the classical and non-classical logician are having the former kind of dispute, whereas someone who denies that the truths of modal logic are logical validities is having the latter.

Notice, then, that in the dispute about local and global validity both participants of the dispute agree about the following: ΔECQ preserves determinate truth but not disquotational truth (recall that both of these conclusions were derived from assumptions that both sides accept: that the schema $\Delta A \rightarrow A$ is generally true, and that the converse schema $A \rightarrow \Delta A$ is not). Thus the dispute looks initially as though it is merely a dispute about which of these inferences to call 'valid', and not a dispute about the inferences themselves. In particular, someone endorsing a global account of validity need not be endorsing a revisionary logic.

3

An Outline of a Theory of Propositional Vagueness

The disputes we have discussed in chapter 2 do not draw the lines you'd initially expect. We saw that IGNORANCE, for example, does not straightforwardly separate the epistemicist from the supervaluationist. Opposing views about truth, the multiplicity of admissible interpretations, and validity similarly don't (on their own) seem to be as striking or profound as they might first appear. Indeed, without connecting these debates up to other questions, they begin to look like side issues whose bearing on the fundamental puzzles of vagueness remains obscure.

However, there are a cluster of questions just beneath the surface that do seem to map out important terrain in the philosophical landscape. To set my own theory in context I'll raise a number of questions of my own, and briefly explain how my theory supplies a certain combination of answers to them.[1]

There are many open questions concerning the *epistemology of vagueness*. Closely related to disputes about truth and validity, there are important questions about how to correctly reason with vague propositions, and about the proper doxastic attitude to bear to propositions you know to be vague—should one be agnostic, or something else? Once these are settled there are further questions about how to align one's bouletic attitudes—how should one *care* about the vague?—and about how to make decisions when one's information is vague. There are also many open questions about the *logic of vagueness*: How does the determinacy operator iterate? What is the combined logic of vagueness and modality? How does vagueness apply to other semantic types?

Theories of vagueness ought to be individuated at least partly by the verdicts they deliver to these questions. I do not think, however, that there is such a thing as the official 'supervaluationist' or 'epistemicist' line on any of these questions, and so these

---

[1] A couple of points of caution are in order. Firstly the questions that I take to be central to the philosophy of vagueness, and which I subsequently focus on, may not turn out to be the questions supervaluationists and epistemicists typically find themselves disagreeing about. Secondly, certain combinations of answers to these questions may not fit neatly into the supervaluationist/epistemicist dichotomy, even if some answers are more naturally aligned with the supervaluationist or epistemicist slogans than others. It is for these sorts of reasons I'm inclined to think that the terms 'supervaluationism' and 'epistemicism' and the dispute as to which is correct is not really carving out a well-defined dispute on its own.

theories remain silent on many of the issues I take to be most central. The theory of vagueness I propose in this book arose out of an attempt to answer these questions. In what follows I outline the basic features of the view. I aim here to give no more than a rough sketch of the basic ideas, so as to give an impression of where the book is headed: readers should consult the relevant chapters for more rigorous treatments and explanations of the key concepts and ideas.

3.1 Is Vagueness Linguistic?

The central thesis of this book is:

> **Propositional Vagueness.** Vagueness is fundamentally a property of propositions, objects, and other non-linguistic entities.

Propositional Vagueness should be contrasted with the usual assumption that vagueness is fundamentally a property of sentences, names, and other linguistic items. This thesis will prove indispensable to framing many of the disputes that this book is concerned with.

Almost everyone writing on vagueness has assumed that vagueness has something to do with language. Indeed, when this assumption is not held without reservation, it is often because one wants to make room for the possibility of 'metaphysical vagueness' in addition to linguistic vagueness; few have argued that ordinary vagueness has little directly to do with language.[2] I spell out the main differences between linguistic and non-linguistic theories of vagueness further in chapter 4.

One might think that since the epistemicist slogan is that vagueness is ignorance, and since the objects of knowledge and ignorance are propositions, epistemicism ought to maintain that vagueness is propositional and thus not linguistic. However, matters are not so clear cut, since two of the most prominent epistemicists in fact fall on the linguistic side. Timothy Williamson [156] argues that vagueness arises when a linguistic expression is semantically plastic: there are nearby worlds where slight changes in the use of the term have caused it to have a slightly different meaning. Paul Horwich [71] argues that vagueness arises for sentences in a mental language when the constitutive rules of use for that sentence preclude both its application and the application of its negation.

Most supervaluationists fall on the linguistic side of the fence, maintaining that vagueness arises when there are multiples interpretations of the language compatible with linguistic conventions (see, for example, Keefe [78]). Note, however, that not

---

[2] There are a few notable exceptions. In response to certain puzzles about attitude reports Schiffer [126] has argued that vagueness is not linguistic; see also Barnett [11], Field [53], and even possibly Fine [56]. Defenders of 'vagueness in the world' typically do so to account for special kinds of indeterminacy, such as indeterminacy about the future or indeterminacy due to quantum mechanics: such accounts do not obviously extend to vagueness about richness.

every supervaluationist talks this way. Kit Fine, for example, employs an operator locution 'it's determinate that' rather than a predicate applying to linguistic items, which might naturally be cashed out in metaphysical language (e.g. 'being grounded by reality'). Of course it may turn out that the correct interpretation of a language is not determined by the world in this inflationary sense, and that a *sentence* is vague to the extent that the world fails to determine what its meaning is. But it is clear that this sentential notion of vagueness is being spelled out in terms of the non-linguistic notion of worldly indeterminacy, and not the other way around.

As may be expected, in addition to accepting the ideology of propositional vagueness I also endorse the existential claim:

Vague Propositions. There are vague propositions.

(It should be noted, by contrast, that although I maintain that the ideology of a vague object is more fundamental than that of a vague name, I'll remain for the most part neutral about whether there are any vague objects in this book. The status of vague objects is discussed further in chapter 16.)

Philosophers who take vagueness to be primarily linguistic might naturally be associated with the view that there are no vague propositions. And indeed many do have this view, maintaining that a vague sentence expresses (in some sense) an array of different but closely related precise propositions. But this association is not forced: it could be that there are vague propositions, but that sentential vagueness is the basic notion and that propositional vagueness is to be somehow explained in terms of it.

3.2 Booleanism

One of the central theoretical commitments of this book is a thesis about the structure of propositions:

Booleanism. The set of propositions form a Boolean algebra.

We assume, moreover, that propositions form a *complete, atomic* Boolean algebra:

Completeness. This Boolean algebra is complete.

Atomicity. This Boolean algebra is atomic.

What follows in this section is something of a digression, but since concepts from the theory of Boolean algebras appear frequently throughout this book, this is a good opportunity to introduce them. Readers already familiar with Boolean algebras should simply skim the definitions, and skip ahead.

To understand what these principles say, we shall look at each part separately. A Boolean algebra is a set of objects B—in this case, our set of propositions—with two elements $\top, \bot \in B$, two binary operations $\vee, \wedge$ taking two elements of B to an element of B, and a unary operation $\neg$ mapping B to B subject to certain constraints. Intuitively, these constraints say that two propositions are identical whenever the corresponding

Table 3.1. Axioms for Boolean algebras

IDENTITY	$p \vee \bot = p$	$p \wedge \top = p$
DISTRIBUTIVITY	$p \vee (q \wedge r) = (p \vee q) \wedge (p \vee r)$	$p \wedge (q \vee r) = (p \wedge q) \vee (p \wedge r)$
COMPLEMENTARITY	$p \vee \neg p = \top$	$p \wedge \neg p = \bot$
COMMUTATIVITY	$p \vee q = q \vee p$	$p \wedge q = p \wedge q$
ASSOCIATIVITY	$(p \vee q) \vee r = p \vee (q \vee r)$	$(p \wedge q) \wedge r = p \wedge (q \wedge r)$

biconditional is provable in the classical propositional calculus. Thus, for example, $p = \neg\neg p$ and $p \wedge q = q \wedge p$, since the corresponding biconditionals are provable in the propositional calculus. A more compact way of axiomatizing these identities is given in Table 3.1.

In addition to these axioms, it is normally stipulated that $\top \neq \bot$. It should be emphasized that Booleanism is not the same as the assumption of classical logic: while Booleanism entails classical logic, it does not follow from the assumption of classical logic.[3] According to theories in which propositions are *structured* in a way that parallels the sentences that express them, identities like $p \wedge q = q \wedge p$ fail, for the left-hand-side and right-hand-side are structurally different. Yet it is open to those who endorse structured theories to accept classical logic. Booleanism is thus a further, substantive, commitment that takes us beyond a commitment to classical logic.

According to some theories, the things that play the proposition role are just linguistic constructions of some sort—perhaps sentences in some idealized language of thought. However, Booleanism is not consistent with this interpretation of propositions either, since there are distinct, but tautologically equivalent sentences. The contrast between propositional and linguistic accounts of vagueness, on this conception of propositions, is thin. The assumption of Booleanism thus registers one important sense in which propositions, as I understand them, are not derivative on language: my theory of propositional vagueness is a genuinely *non-linguistic* theory.[4]

The simplest example of a Boolean algebra is the algebra that has only two values—the True and the False—where $\wedge$, $\vee$ and $\neg$ are interpreted by the evident truth functions. So one version of Booleanism is the view that there are only two propositions. Frege arguably held this view, but few philosophers since have defended it. Probably the most well-known Booleanist view is the one that propositions are sets of possible worlds, where $\wedge$, $\vee$, and $\neg$ are interpreted as set intersection, union, and complementation respectively. According to this view metaphysically necessarily equivalent propositions are identical. Booleanism alone does not force us to individuate propositions this coarsely either. There are many necessary equivalences

---

[3] According to Booleanism, for every classical tautology ϕ, we have $\phi = \top$. Since $\top$ is true, ϕ is true by Leibniz's law. We can make this argument slightly more carefully by employing a propositional identity connective in the object language; such a device is introduced in chapter 11.

[4] Field and Horwich, for example, both hold views in the vicinity of the view endorsed in this book. However, they both endorse some version of the view that propositions are linguistic constructions.

that are not provable in the propositional calculus, such as the equivalence between the proposition that water is wet and the proposition that H_2O is wet: even though these propositions are necessarily equivalent, this is not something one can prove in the propositional calculus alone. Booleanism, therefore, does not force us to make these identifications. We shall see in chapter 11, that if we are to make good on the idea that the vague supervenes on the precise, we must also distinguish necessarily equivalent vague propositions.

Every Boolean algebra comes with a natural ordering of entailment over its elements: say that p *entails* q iff $p \wedge q = p$. In the possible worlds theory, p entails q just in case p is a subset of q, since $p \cap q = p$ iff $p \subseteq q$. A Boolean algebra is *complete* iff every set of elements, X, has a least upper bound under the entailment ordering: an element a such that $b \leq a$ for every $b \in X$, and moreover $a \leq a'$ whenever $b \leq a'$ for every $b \in X$. It is possible to show that in a complete Boolean algebra, every set X also has a greatest lower bound, namely, the least upper bound of the set of propositions that entail all members of X. In the possible worlds framework, for example, the least upper bound of a set of sets of worlds is just their union; least upper bounds are thus like disjunctions of (possibly infinite) sets of propositions. (Similarly, greatest lower bounds correspond to infinite conjunctions.) It follows that propositions in the possible worlds theory form a complete Boolean algebra.

We now come to an important definition:

MAXIMALLY STRONG CONSISTENT PROPOSITION: A proposition, i, is a *maximally strong consistent* proposition iff:

(i) It is consistent: it is not the element $\perp$.
(ii) For any other consistent proposition p, if p entails i then $p = i$

A maximally strong consistent proposition is sometimes called an *atom*.

A Boolean algebra is called *atomic* iff every consistent proposition is entailed by a maximally strong consistent proposition. According to the possible worlds account of propositions, a singleton set, $\{w\}$, containing exactly one world will be a maximally strong consistent proposition. It follows that according to the possible worlds theory, the propositions also form an atomic Boolean algebra.

Indeed, any theory in which propositions are sets of things—whether they be sets of worlds, world–precisification pairs, world–time pairs, etc.—is one that accepts Booleanism, Completeness, and Atomicity. Conversely, it can be shown that every complete atomic Boolean algebra is isomorphic to a Boolean algebra of this sort: in particular, p corresponds with the set of maximally strong consistent propositions that entail p. It follows, given our assumptions, that it is always possible to think of propositions as sets of *indices*, where an index is just (yet) another name for a maximally strong consistent proposition.

A *Boolean* subalgebra of a Boolean algebra B is a non-empty subset $A \subseteq B$ that is closed under $\wedge, \vee$ and $\neg$: if $p, q \in A$ then $p \wedge q, p \vee q, \neg p \in A$. It is easy to check that A is itself a Boolean algebra if B is. The notion of a Boolean subalgebra is useful in the context of discussions of vagueness, since it is natural to think that the precise propositions form a Boolean subalgebra of the propositions: if p is precise, so is $\neg p$, and if p and q are precise, so is $p \wedge q$ and $p \vee q$. Indeed, throughout this book I shall assume the stronger thesis:

Boolean Precision. The precise propositions from a complete atomic Boolean algebra.

Presumably infinite disjunctions of precise propositions are precise, ensuring completeness. The atomicity of the precise propositions is less obvious, but can be motivated as follows. If I conjoin all the precise truths, the resulting proposition is precise by completeness. Moreover, the conjunction of precise truths is surely also consistent, for it ought to be true given that its conjuncts are. Thus, the conjunction of precise truths is an atom of the algebra of precise propositions.[5] This sort of reasoning ought to be necessarily true, if true at all. So, on the assumption that every precise proposition is metaphysically possible, it follows that every precise proposition is entailed by an atomic precise proposition.[6]

The preceding discussion introduced another crucial concept that will appear regularly throughout the book: the notion of an atom of the algebra of precise propositions.

MAXIMALLY STRONG CONSISTENT PRECISE PROPOSITION: A proposition p is a maximally strong consistent precise proposition iff:

(i) p is precise.

(ii) p is consistent (i.e. it's not $\bot$).

(iii) For any other consistent precise proposition, q, if q entails p then $q = p$.

A maximally strong consistent precise proposition settles all precise questions, in the sense that for any precise proposition q, it either entails q or it entails $\neg q$. However, if there are vague propositions, a maximally strong consistent precise proposition will not settle all questions: they may leave open questions about vague matters.

We can get an intuitive picture of the precise propositions by appealing to the representation of a proposition as a set of indices. Roughly, the precise propositions will partition logical space—represented by the set of all indices—into a bunch of *cells*: a collection of sets of indices that do not overlap each other, and are such that every index is contained in some cell. Each cell corresponds to a maximally strong

---

[5] Suppose p is the conjunction of all precise truths. If another consistent precise proposition q entailed p, q would have to be true. For if q were false, $\neg q$ would be a precise truth and would thus be entailed by p. Since q entails p it would follow that q entailed $\neg q$, contradicting the assumption that q was consistent. Since we have established that q is a precise truth, it follows that p entails q (p is the conjunction of precise truths), and so p and q are identical (as q also entails p). This establishes that p is an atom of the precise propositions.

[6] Thanks to Jeff Russell for discussion here. Without the assumption that every precise proposition is metaphysically possible, an alternative argument is available using a broader notion of necessity; we will introduce this notion in chapter 11.

consistent precise proposition. An arbitrary precise proposition consists of a union of cells. We will regularly invoke this sort of picture-thinking in later chapters.

3.3 The Epistemology of Vagueness

Once one has acknowledged that there are vague contents, questions about their role in thought and practical reasoning quickly become urgent. Indeed, answering such questions will pave the way for a positive account of propositional vagueness.

For example, one might wonder whether one could directly come to learn a vague proposition—whether by perception or some other means—or if we only ever know vague contents by inferring them from precise things we've learned directly.[7] In chapter 6, I argue that the totality of an agent's evidence at any given time is almost always vague. Our investigation builds on the similarities between the effect of learning a vague proposition and the effects of information obtained by imperfect perceptual faculties. Indeed, by studying this relationship we are able to provide the beginnings of a positive account of propositional vagueness.

Central to this positive account is a Principle of Plenitude governing the existence of vague propositions. The principle says, roughly, that for any potential effect an inexact experience might have on one's credences about the precise, there is a vague proposition such that learning it has that effect. The 'effects' of learning a vague proposition can be likened to what's known as a *Jeffrey conditioning* on a certain partition of precise propositions, the coefficients of which determine what we call an *evidential role*. In these terms, the Principle of Plenitude can be informally put as saying that:

> **Principle of Plenitude.** For every evidential role, there is some vague proposition occupying that role.

(The precise formulation of this principle will, of course, have to wait until chapter 6, as will the proper explanation of key formal concepts such as Jeffrey conditioning, and the notion of an evidential role.)

Once one has acknowledged that our information is often vague, it is natural to wonder whether this information could ever be useful. Could vague information, for example, play an essential role in your decision making—could it effect an outcome of your practical reasoning that couldn't have been arrived at by someone with the same attitudes towards the precise who lacked that information? In chapter 9, I argue that, in deciding what to do, you must sometimes take into account your beliefs and desires (and attitudes more generally) about vague matters. In that chapter we consider, for

---

[7] It is important to distinguish vagueness from borderlineness here: the proposition that I am not bald is a vague proposition even though it is determinately true. It is very controversial whether one can learn (and thus know) a borderline proposition—the above question concerns whether one can come to directly learn a vague proposition. I discuss this further in section 6.2.

example, the possibility of agents who agree about the precise, act rationally, but behave differently due to their opinions about the vague.

It is worth re-emphasizing at this juncture the importance of the ideology of propositional vagueness to the formulation of these questions. If, for example, all propositions were precise there would be no vague propositions to be uncertain about, to be learnt, to desire, to decide to make true, and so on.

The above aspects of the view I am going to advocate for suggest that attitudes with vague contents play a fairly rich role in thought. It is natural to wonder, then, how autonomous our opinions about the vague are. Are they independent of our beliefs in the precise, similar to the way that opinions about the weather and the economy are independent (there are no *a priori* constraints connecting our opinions about these subject matters)—or is there a tighter rational connection between them? Indeed, one might wonder whether one of these mental states reduces to the other: is having a credence in a vague proposition just a matter of having certain credences in precise propositions?

This issue relates indirectly to the issue of whether vague propositions carve out real factual distinctions. In a parallel debate, philosophers sceptical of genuine conditional facts—*expressivists* about conditionals—have argued that to have a credence in a conditional is just to have a certain distribution of credences in non-conditional propositions; similarly philosophers sceptical of moral facts have suggested that having a doxastic attitude towards a moral proposition is just a matter of having certain sorts of bouletic attitudes towards non-moral propositions.

Assuming that evidence can be vague, it is natural to think that opinions about the vague are autonomous to a certain degree: one could imagine two people with exactly the same precise evidence (and the same priors) but different vague evidence. Such people could in principle exhibit a rational difference of opinion about the vague without a difference of opinion about the precise. In chapter 8, I argue that there is nonetheless a sense in which vague beliefs are not autonomous, vindicating one aspect of the expressivist picture: for rational *ur-priors*—probability functions representing the *a priori* opinions of a completely uninformed person who is not subject to any conceptual confusions—the probabilities of the vague are completely fixed by the probabilities they assign to the precise. That is, I argue that:

Rational Supervenience.   Any two ur-priors that agree about the precise propositions agree about every proposition.

A similar set of questions arises concerning the autonomy of our desires about vague matters. If there are vague propositions, it will presumably be possible to have desires with those propositions as contents. It is extremely natural to wonder whether caring about some vague matter amounts to distributing your cares about the precise in a certain way, or whether fixing which precise things you care about determines the vague things you care about. That is: is it possible to rationally care *intrinsically* about the vague? In chapter 10, I argue that there's a sense in which it is not possible to be rational and care intrinsically about the vague. Putting it roughly, I argue:

> **Indifference.**   Every rational set of preferences is indifferent between any two vague propositions that settle all precise matters in the same way.

Here a vague proposition *settles* a precise one if it entails it or its negation, and two propositions settle a third in the same way if they both entail that proposition or both entail its negation.

Putting these three theses about the epistemology of vagueness together puts us in a position to give an account of what a precise proposition is squarely in terms of their role in thought. More specifically, the notion of a precise proposition can be defined in terms of the set of rational prior probability and utility functions. For example, **Indifference** says that a maximally strong consistent precise proposition will always have the feature that any two propositions entailing it receive the same value, according to any of the permissible utility functions. Assuming that the collection of rational utilities is sufficiently rich it can be shown that the maximally strong consistent precise propositions are exactly the propositions with this feature. Once you have identified the maximally strong consistent precise propositions, you have identified the precise propositions (their disjunctions). A similar such characterization can be given in terms of priors, appealing instead to **Rational Supervenience** (see chapter 13).

Of course, this characterization of precision in terms of rational utilities and priors is a bit ham-fisted. In chapter 13, I show that the notion of a precise proposition can be defined succinctly in terms of the notion of a *symmetry* of a set of probability functions and utility functions. A symmetry is a Boolean automorphism of the space of propositions—a one-to-one function that preserves the Boolean operations—that also preserves the values and probabilities of every proposition. A precise proposition turns out to be exactly a proposition that is mapped to itself by every automorphism that preserves the values and probabilities of every proposition:

> **Symmetry.**   A proposition is precise if and only if it is fixed by every symmetry.

A similar characterization of determinacy is possible as well: a proposition p is determinate if every proposition p is mapped to by a symmetry is true. (Again we must wait until chapter 13 for a proper explanation of these definitions.)

3.4 Probabilism

Questions in the epistemology of vagueness can also be related to some of the disputes we discussed earlier about truth and validity.

Recall that to have a disagreement about logic (as opposed to a disagreement about which truths to count as logical) involves having a difference of opinion about a principle of logic—perhaps a disagreement about the claim that either Harry is bald or he isn't. It is a first-order disagreement about Harry and the status of his head, and not merely a metalinguistic disagreement about whether to call this sentence 'valid'. One might hope to put the debate between global and local validity in a similar standing by identifying a first-order disagreement like this.

In the global/local debate about validity both sides agree about which *sentences* are valid—the theorems of classical logic. It is only about inferences with non-empty premise sets that an apparent disagreement arises. What would a genuine, first-order disagreement about an *inference* amount to? Presumably it wouldn't amount to a straightforward disagreement about whether to accept some sentence or proposition, such as in the dispute between the classical and paracomplete logicians, but it might involve a difference of opinion about when one can correctly infer a conclusion from one's beliefs.

One can formulate this idea more precisely in a quantitative framework that assigns degrees of belief to propositions. A good inference on that theory is one in which a drop in credence from the premises to the conclusion is not rationally permitted.

In this setting the debate about the goodness of ΔECQ concerns whether one's confidence in $A \wedge \neg\Delta A$ cannot exceed that of an arbitrary proposition. In other words, whether my credence in $A \wedge \neg\Delta A$ must be 0. But assuming probabilism, this would mean that one's credence in the negation, which is logically equivalent to the conditional $A \to \Delta A$, must be 1. This is absurd, for it entails that one cannot believe in vagueness: since, for any A, one's credence in $A \vee \neg A$ is 1, certainty in the conditionals $A \to \Delta A$ and $\neg A \to \Delta \neg A$ would engender a credence of 1 in $\Delta A \vee \Delta \neg A$ for any A.

Indeed, consider the following principle that is weaker than probabilism and plausible independently of it: if you are certain in A your credence in $A \wedge B$ should just be your credence in B. Note that since the claim that it's borderline whether A entails that it's not determinate that A, it follows that your credence in $A \wedge \nabla A$ must also be 0, writing ∇A to mean 'it's borderline whether A'. So it follows that if you are certain that ∇A then your credence in A must be identical to your credence in $A \wedge \nabla A$ (which is 0) and, by parallel reasoning, your credence in $\neg A$ must be identical to your credence in $\neg A \wedge \nabla \neg A$ (also 0). Thus: if you are certain that A is borderline your credence in A and in $\neg A$ must be 0, in clear violation of the probability calculus.[8]

A similar conclusion can be reached by reflecting on truth theoretic considerations. One of the roles that truth is supposed to play connects it closely to assertion, belief, and their graded counterparts. According to this idea, one is in a position to assert that A and believe A only if one is also in a position to assert and believe that A is true. In a quantitative framework this idea can be generalized to the claim that one's credence in A should always be the same as your credence that A is true. For those who take supertruth to play this part of the truth-role we arrive at a similar conclusion: when your credence that A is true is 0 and your credence that A is false (i.e. that $\neg A$ is true)

---

[8] Williams [152] presents a similar argument for this conclusion by appealing to the inference from A to ΔA. However, as we noted in section 2.5, this inference is not globally valid according to one disambiguation of 'globally valid'.

is also 0—as typically happens when you are certain that A is neither true nor false, according to this view—your credence in A and in $\neg A$ must also be 0.

This consequence strikes me as a deep and important one, although it is now fairly far removed from our original questions about validity and truth. We might as well leave that question behind, and focus on the following one:

> **Probabilism.**   Do rational degrees of belief in vague propositions obey the probability calculus?

Hartry Field, for example, has argued for exactly the conclusion we outlined above: when you are certain that it's borderline whether A your credence in A and its negation must be 0. We shall abbreviate this by saying that you must be *anti-certain* in A and in $\neg A$. By contrast, probabilism is an important component of the theory of propositional vagueness I prefer—it is indispensable to the formulation of the **Principle of Plenitude** and **Rational Supervenience**, for example. I elaborate and defend the assumption of probabilism more generally in chapter 7.

Note that by taking a stand on probabilism, I am not thereby committed to the position that validity is local rather than global. Preservation of disquotational truth and preservation of determinate truth are two perfectly fine notions by my lights, and nothing I say will turn on which we choose to call 'validity'. Probabilism does commit us to the result that there are globally valid arguments (such as $\triangle$ECQ) for which a drop in credence is permitted from premise to conclusion. But that does not mean that global validity is not a useful or important property of arguments. Globally valid arguments have a different epistemic property: if you are *certain* in the premises of a globally valid argument you should be certain in the conclusion. For example, no one can ever be rationally certain in the premises of $\triangle$ECQ, for it can be known *a priori* that their conjunction is at best borderline, and at worst false. Since it is impossible to be rationally certain in the premises of $\triangle$ECQ, this inference preserves rational certainty. (In chapter 7, we will develop the tools to generalize this conclusion.)

It is worth relating the question of probabilism back to the supervaluationist/epistemicist debate. It's quite natural to associate with epistemicism the view that vagueness-related uncertainty is no different from other kinds of uncertainty. In which case probabilism about the vague is as on as firm a footing as it is more generally. On the other hand, the denial of probabilism is not usually associated with supervaluationism. Neither of the most prominent anti-probabilists—Field and Schiffer—identify as supervaluationists. By contrast many versions of supervaluationism have been developed that accept probabilism (see Kamp [76], Lewis [89], Williams [152]), and there are classical views that identify neither with supervaluationism or epistemicism that accept probabilism (Edgington [39]). However, note that some theorists have noticed that the denial of probabilism fits fairly naturally with traditional supervaluationism with a global conception of validity, and with the identification of truth with supertruth (see Williams [152]).

3.5 Logical Features

What are the main logical differences between this approach to propositional vagueness and its rivals? As may be expected, the central notions of this sort of theory are most aptly captured using operator locutions (as in Fine [56], for example) as opposed to metalinguistic predicates of sentences (as in, e.g. McGee and McLaughlin [104]). In chapter 4, we discuss some ways in which this can affect the logic of vagueness.

Other differences, however, require more emphasis. Throughout this book we draw a sharp distinction between being *borderline* and being *vague*. The distinction is best dramatized by examples. Suppose that we have a sorites sequence consisting of individuals of gradually increasing heights. Every proposition stating that one of these individuals is tall is vague, for they all ascribe the vague property of being tall. On the other hand, not all of these propositions are borderline—only the propositions ascribing tallness to individuals towards the middle of the sequence are borderline, the rest are determinately true or determinately false. On the intended understanding of vagueness, a sufficient condition for a proposition to be vague (but not an analysis) is for it to be possibly borderline, and each individual is possibly borderline tall even if they're in fact not borderline tall.

The majority of work on the logic of vagueness has focused on the notion of being borderline, and the cognate notion of being determinate (where p is determinate if and only if p and it's not borderline whether p). Indeed, the main formalisms for dealing with vagueness—such as the theory of precisifications associated with supervaluationism (see chapter 12)—are all geared towards providing analyses for determinacy and borderlineness operators. It is far from clear how this work bears on the notion of propositional *vagueness*, and the cognate notion of propositional *precision* (where a proposition is precise if and only if it is not vague).

The usual strategy is to try and reduce vagueness to borderlineness by means of some kind of modal notion—perhaps a proposition is vague if it *could* have been borderline. Although I think there are general problems with this reduction, we will see in chapter 12 that the theory of propositional vagueness I outline above requires a primitive notion of precision and vagueness which cannot be defined in terms of propositional borderlineness and metaphysical modality. On this sort of view vagueness must instead be taken as primitive, and borderlineness must be defined in terms of it.

This raises a wider question about the suitability of the usual formalism—the theory of precisifications—for thinking about vagueness in connection with this view. Without the modal reduction of vagueness to borderlineness this formalism has nothing to say about the vague/precise distinction. To model these notions a new formalism is needed, and in chapter 13, I argue that the theory of symmetries advertised in section 3.3 fits the bill. The resulting theory does away with the traditional formalism for dealing with vagueness involving possible worlds and precisifications, and uses an alternative formalism formulated in terms of a set of propositions and a certain group of automorphisms, the *symmetries*.

As noted, on this theory the central notions are formalized using operators, by analogy with the operators used to formalize metaphysical modalities. Given the formal similarity, many natural questions about the interaction of vagueness-theoretic operators and modal operators arise. One such question is the status of a certain supervenience principle:

Supervenience.   The vague truths supervene on the precise truths.

Here the notion of supervenience is to be explicated in terms of metaphysical modality. This principle is formulated precisely and adopted in chapter 15.

One consequence of Supervenience for our theory of propositional vagueness is that the theory of propositions must be a *hyperintensional* one: there must be necessarily equivalent but distinct propositions (indeed, given Boolean Precision, Supervenience guarantees that every vague proposition is necessarily equivalent to a precise one). A broadly supervaluationist take on vague propositions might identify them with certain sets of ordered pairs consisting of worlds and precisifications. It will be seen that this sort of theory of propositions secures the supervenience of the vague on the precise in a particularly simple way. It is also clearly a hyperintensional theory, since propositions are more fine-grained than sets of worlds.

However, there are a number of substantive assumptions hard-wired into this broadly supervaluationist account of vague propositions, some of which are highly contentious and ought to be brought into the open. One of these is the commitment to certain principles concerning the interaction of the modal and determinacy operators. For example:

It's determinately necessary that A if and only if it's necessarily determinate that A.

It's possibly determinate that A only if it's determinately possible that A.

In chapter 15, I show how these principles fall out of a supervaluationist semantics for determinacy and modality, and present some difficulties for these consequences. Another issue is that, given a certain resolute attitude towards the use of possible worlds in supervaluationist semantics, the theory is subject to paradoxes of higher-order vagueness (see chapter 7). These sorts of considerations indicate the need for an alternative hyperintensional theory of propositions, which I provide in chapter 11.

Although my main objective in this book is to spell out a theory of propositional vagueness, the more general perspective of this project is that vagueness is a property of non-linguistic entities; propositions being merely the non-linguistic analogue of a sentence. But there are also non-linguistic analogues of other linguistic items: objects correspond to names, properties to predicates, and so on. Given that there can be vagueness in names, predicates, and so on, it is worth wondering how to generalize a theory of propositional vagueness to other types. I explore this topic more systematically, in the context of type theory, in chapter 16.

3.6 Vagueness-Related Uncertainty as a Special Sort of Psychological Attitude

The study of vagueness is traditionally taken to be a branch of the philosophy of language. This is not the approach of this book: the treatment of vagueness outlined here falls squarely within the purview of epistemology. Propositional vagueness is characterized completely in terms of the *role it plays in thought*—in this case, the role it plays in a theory of rational belief and desire. The important questions about propositional vagueness are epistemological ones, and the substance of the theory is given by settling those questions.

The further question of whether the view outlined should be considered as an elaboration of epistemicism or supervaluationism strikes me as an unproductive one. Like the epistemicist, I accept the conclusion that there is a cutoff point for the property of being rich, and it is unknown where that cutoff is. But, as we noted, this does not distinguish the view from the supervaluationist either. If we follow through the definitions in section 2.4, it will similarly turn out that there is exactly one accurate interpretation and many admissible ones. But this is not surprising since, as we noted there, these were consequences of minimal assumptions (modulo definitions) that are shared by most classical approaches to vagueness, including the epistemicist and supervaluationist.

However, the view does make some distinctive predictions that are in tension with the orthodox versions of epistemicism and supervaluationism. We have briefly outlined some of the difficulties with reconciling the present view with a supervaluationist formalism in section 3.5 above. Unlike the epistemicist, this view maintains that vagueness is not *merely* a matter of being uncertain about the locations of cutoff points. There is a bouletic element as well: you shouldn't care intrinsically about where the cutoff points lie. Moreover, uncertainty about the vague is in some sense derivative on uncertainty about the precise, as encoded by **Rational Supervenience**. On this sort of view, then, rational disagreements about the vague (in credence and in value) tend to boil down to disagreements about the precise. If vague matters were truly factual, as some epistemicists maintain, it's hard to see why vague beliefs and desires should depend in this way on precise ones.

In some respects the nearest theory of vagueness is neither supervaluationism nor epistemicism, but an approach to vagueness that has recently been endorsed, with differing details, by Hartry Field, Stephen Schiffer, and Crispin Wright. According to these theorists there is a distinctive psychological attitude that is characteristic of taking a proposition to be borderline, and it is in the study of this attitude that the proper nature of vagueness is to be uncovered. For example, as we saw above in Field's theory, taking A to be borderline consists in being anticertain in both A and in $\neg A$. Although Field does not offer an explicit definition of vagueness in terms of this attitude, the thesis that you should be anticertain in propositions you are certain are borderline serves as a constraint on the conceptual role of the borderlineness operator, which helps distinguish the operator from one that merely expresses ignorance (as in an epistemicist theory). Different characterizations of this state are offered by Schiffer and Wright, but they agree that the distinctive features of vagueness are to be explained in terms of it.

In contrast to the present account, these theories are formulated in a non-classical setting: Field [54] and Schiffer [126] in a certain kind of contractionless logic (see section 1.2), and Wright [167] in intuitionistic logic (although an earlier version of Field's theory [53], discussed in chapter 7, is thoroughly classical). But more importantly, the view just outlined assumes a completely classical Bayesian epistemology. According to this theory, the uncertainty that vagueness generates is not a distinctive attitude. It is exactly the same sort of attitude you have when you are uncertain about where you left your keys, about the goings-on in other galaxies, and so on, and can be characterized quantitatively in the usual way, in terms of having a middling credence in a proposition.

PART II
Epistemological Matters

4

Vagueness and Language

As is often the case in philosophy, competing philosophical theories, purportedly about the same subject matter, can be couched in very different vocabulary. In such cases it is often quite hard to state, in a neutral way, exactly what the disagreement is— choice of ideology can sometimes contain assumptions of its own. Sometimes it is not even clear whether two theories are in competition or whether they are just talking about two different things.

As a case in point, consider the two distinctions that will be the focus of this chapter:

SENTENTIAL BORDERLINENESS: Some sentences exhibit features that are correlated, in some way or other, with the presence of vagueness: call this feature 'sentential borderlineness'. Examples of English sentences that are sententially borderline include 'if a glass is two thirds full it is pretty full', 'France is hexagonal', and 'Janet is rich', where Janet is someone of moderate wealth. Examples that are not sententially borderline include the sentences 'there are electrons', 'there are at most three bald people on the planet Earth', and 'the smallest tall person is tall'.

PROPOSITIONAL BORDERLINENESS: Some propositions exhibit features that are correlated, in some way or other, with the presence of vagueness: call this feature 'propositional borderlineness'. Examples of propositions that are propositionally borderline include the proposition that if a glass is two thirds full it is pretty full, the proposition that France is hexagonal, and the proposition that Janet is rich. Examples that are not propositionally borderline include the proposition that there are at most three bald people on the planet Earth and the proposition that the smallest tall person is tall.

Related to these distinctions are parallel distinctions between vague propositions and sentences. The sentence 'there are at most three bald people on Earth' is sententially vague; but it is not sententially borderline because it is determinately true. Similarly, if there are any vague propositions at all, presumably the proposition that there are at most three people on Earth is propositionally vague, even though it is similarly not propositionally borderline. Throughout this chapter when I talk about giving 'an account of vagueness', whether sentential or propositional, I shall generally mean the project of giving an account of *both* vagueness and borderlineness. The difference between giving an account of vagueness and giving an account of borderlineness may actually be more subtle than it might first appear, but we will defer that discussion until chapter 12 and treat them as if they stand or fall together for the moment.

I take it to be uncontentious that the first of these distinctions draws a non-trivial line: there are sentences which are sententially borderline and sentences which are not sententially borderline. There is, for instance, a distinctive reason not to assert the sentence 'Janet is rich' when you know that Janet falls within the borderline region, and it clearly has something to do with vagueness. And there are other sentences, like 'Janet has $200,000' for which there is no distinctive obstacle to appropriate assertion.

The non-triviality of the second distinction, however, is a bit more contentious. On the face of it, it might seem obvious that there's a principled distinction to be drawn between, say, the proposition that Janet is rich, and the proposition that Janet has at least $200,000. For example the truth of the former proposition seems not to be amenable to investigation, whereas the truth of the latter is. However, some philosophers maintain that this apparent difference is an illusion and that these seemingly distinct propositions might even be identical, for all we know. By analogy, one might note that there is no fundamental difference between the largest small number, whichever it might be, and the number 5—they are both precise entities. There is no vague number—*the largest small number*—in addition to the numbers 1, 2, 3, and so on; there is just vagueness concerning the relation between these precise entities and the vague description 'the largest small number'. Perhaps the entities denoted by 'the proposition that Janet is rich' and by 'the proposition that Janet has at least $200,000 dollars' are similarly just as precise as one another, although the relation between the former *expression* and the proposition it picks out is more complex than it seems. There is one important disanalogy between numbers and propositions, however: for each n the proposition that Janet has at least n is something we appear to be in a position to know, whereas the proposition that Janet is rich doesn't seem to be like this, blocking a straightforward identification of the latter proposition with one of the former precise propositions.

In chapter 5, I shall argue that the best way to make sense of our ignorance is to accept both sorts of distinction: the distinction between sentences and the distinction between propositions, or the objects of thought whatever they may be. For someone who accepts the existence of both distinctions, however, it is natural to wonder what the dispute between the linguistic and propositional approaches to vagueness and borderlineness amount to. Are both sides not just giving an account of different things?

The answer, I take it, is that those espousing one or the other of these approaches typically take themselves to be doing more than just exploring a side effect of vagueness, whatever it may be. They take their project to be that of giving an explication of *what vagueness is*, and providing the means to address the fundamental problems of vagueness. Thus I take it that the issue at stake concerns the following two hypotheses:

1. Hypothesis one: it is sentential borderlineness (or, perhaps, a related property of some other linguistic category) which is at the root of, and will feature in the explanations of all the fundamental puzzles associated with vagueness.

2. Hypothesis two: it is propositional borderlineness that is at the root of, and will feature in the explanations of all the fundamental puzzles associated with vagueness.

The fundamental puzzles of vagueness, I take it, include at least the sorites paradox, and the problem of explaining the puzzling kind of ignorance that comes along with vagueness. Given the first hypothesis, our theoretical investigations ought to be focused around a metalinguistic relation to languages: 'sentence S is vague in language L', or something similar. Propositional vagueness should be reduced to and explained in terms of sentential vagueness. According to the latter hypothesis it is rather a notion of propositional vagueness that plays a more basic role in explaining the relevant puzzles and we should instead theorize about this monadic property of propositions, which we might express using a predicate taking a that clause, or more commonly, an operator taking a sentence. Sentential vagueness and the phenomena associated with it should be explained in terms of propositional vagueness.

The linguistic approach to vagueness, characterized by its preoccupation with the aforementioned relation between sentences, utterances, or more generally linguistic items, is pervasive within the philosophy of vagueness. The following prominent theories, for example, all appear to subscribe to something like the first hypothesis:[1]

- Semantic indecision (Dorr [34], Keefe [78], Lewis [92], McGee and McLaughlin [104], Rayo [117])
- Semantic plasticity (Williamson [156])
- The inconsistency theory (Eklund [40])
- Contextualism (Graff Fara [45], Raffman [115], Shapiro [131])
- Ambiguity (Braun and Sider [20])
- Use theories of meaning (Horwich [71])
- Nihilism about vague language (Unger [144]).

Although not complete, I take it that this list is fairly representative of the state of the literature on vagueness at present. Proponents of propositional vagueness are by contrast less abundant but include, among others, Salmon [125] and Schiffer [127].[2]

---

[1] Horwich perhaps has some affinities with the second hypothesis. However, since propositions for Horwich are, for most intents and purposes, just sentences, it's clear that he belongs on the linguistic side of the fence.

[2] There are some theorists who are naturally interpreted as being primarily concerned with propositional vagueness, but who are not explicit about it. Barnett [11] and Field [53], for example, do not appear to ascribe to a linguistic theory. More controversially, Fine [56], might best be understood as a non-linguistic theorist, although he certainly relates his theory to language. There are also theorists who are not primarily concerned with ordinary vagueness, but provide non-linguistic accounts of metaphysical indeterminacy such as those that arise in, for example, quantum mechanics and the open future; these theories are not my primary concern here.

An analogy with a related debate might be illuminating here. Historically, linguistic accounts of modality were quite pervasive. According to that view the fundamental issues surrounding contingency and necessity were to be best understood by looking at the way we use language, and it was common to theorize about modality by drawing a distinction between certain sentences.[3] Although that view is held by hardly anybody today, it is also surely true that nobody denies the difference between the sentence 'Hesperus is Phosphorus' and 'Hesperus is bright'. The modern view is not that it was a mistake to draw this distinction between sentences, but that it is a mistake to think that an account of this distinction exhausts all there is to be said about modality. By contrast, once we have an adequate account of propositional necessity, either a theory of a propositional necessity predicate, or (as is more common) of a necessity operator, the distinction between sentences can be explained in terms of the distinction between propositions: in the above example the former sentence expresses a necessary proposition whereas the latter expresses a contingent proposition.

Critical to the failure of the linguistic approach was the fact that the converse reduction does not seem to be possible: one cannot explain propositional necessity in terms of sentential necessity. Contingency concerning the brightness of Venus itself is a contingency Venus would have had even if there hadn't been any language users, and it looks on the face of it as though no linguistic account of necessity could account for this.

The situation is similar in the case of vagueness. I of course acknowledge the existence of sentential borderlineness, but think that it is not a theoretically central notion: sententially borderline sentences merely express borderline propositions, and it is in elucidating the latter notion that the action lies. Accounts that theorize purely in terms of sentential borderlineness, I will argue, do not have the resources to explain propositional borderlineness and are therefore, at best, incomplete.

Indeed, the analogy between the two debates is much closer than I have suggested above. Much like the linguistic account of modality, linguistic accounts of borderlineness are subject to a number of technical difficulties that need to be worked out before we can compare them. The purpose of this chapter is to describe the best version of each kind of formalism, so that we will have a more precise target in the later chapters.

4.1 Grammar

In chapter 1, we drew a distinction between people based on how much money they had. Not the distinction between being rich and not rich, but a slightly more demanding one—that of being clearly rich as opposed to not clearly rich. The most

---

[3] See, for example, Ayer [2], Carnap [24], and Malcolm [100]. Of course, matters were slightly muddled by the fact that these authors often conflated analyticity, necessity, and knowability *a priori*. Presumably they were right about analyticity.

natural way to express this distinction in English is to prefix, as I just have done, an adverb such as 'clearly' or 'definitely' to a verb or verb-phrase, as in 'clearly rich', 'clearly richer than', and so on. These expressions fall under the same grammatical class as many other words of interest to logicians such as 'not' and 'possibly'; compare 'Harry is clearly bald' to 'Harry is not bald', and 'Harry is possibly bald'. These adverbial expressions play an important role in the way that many philosophers informally talk about vagueness.

It is common among logicians to regiment adverbial modification in formal contexts using operators—things which result in a sentence when prefixed to a sentence— so that 'Harry is not bald' becomes the slightly more awkward 'it's not the case that Harry is bald' and 'Harry is possibly bald' becomes 'it's possible that Harry is bald'. For our purposes this regimentation is innocuous and I shall frequently adopt it in what follows. The operator locutions can further be divided according to whether they take the complementizer 'that' ('it's possible that') or 'whether' ('it's contingent whether'). The most common ways of drawing the distinction I have just alluded to all use expressions that fall into one of these classes.

ADVERBS: Harry is definitely/clearly/determinately/as a matter of fact/borderline/indeterminately tall.

'THAT' OPERATORS: It's clear/determinate/definite/a fact that Harry is tall.

'WH' OPERATORS: It's borderline/it's indeterminate/it's vague/there's no fact of the matter/it's unclear [whether/when/how Harry became tall]/[who/where/what the size of/which the oldest tall person is].

A common way of representing these operators in logical notation is to use Δp for the operators expressed in the second class (e.g. 'it's determinate that p') and ∇p for the operators in the third class (e.g. 'it's indeterminate whether p'). It is typically assumed that the latter can be defined in terms of the former by writing $\neg \Delta p \wedge \neg \Delta \neg p$: it is indeterminate whether p iff neither p nor its negation is determinate. Given a similar assumption, the former can be defined from the latter by $p \wedge \neg \nabla p$: it is determinate that p iff p is true and isn't indeterminate. I shall adopt this notation and the two assumptions throughout the book.

One notable thing about expressions in these classes that distinguish them from, say, verbs is that adverbs and operators can iterate. One can combine different adverbs with other adverbs, such as with 'clearly not bald' and 'not clearly bald', as well as iterating them, such as in 'not not bald' and 'clearly clearly bald'. Analogous remarks apply to operators.

This way of speaking does not obviously relate to the metalinguistic attributions of vagueness that the linguistic theorist is prone to make. If I say that it is not the case that Harry is bald, for example, it is not at all obvious I have said anything about the English sentence 'Harry is bald'. More generally, attaching an operator to a sentence does not involve saying something about the sentence that it precedes; it would be analogous to suggesting that appending 'runs' to 'Alice' results in a sentence that says something about the name 'Alice'—the resulting sentence is not about names at all, it is about the person Alice.[4] This suggests, in particular, that when I say 'it is clear that Harry is bald' I have not said anything about the English sentence 'Harry is bald'. I have said something about Harry and the status of his hairline; this way of speaking seems, therefore, to be fundamentally unsuited to a linguistic theory of vagueness.

If we want to talk about the sentence rather than whatever that sentence is about it is useful to have some canonical way of referring to that sentence; the use of quotation marks serves this purpose well. A linguistic theorist must reject the thought that the adverbial and operator expressions listed above play a central role in the theory of vagueness; she might therefore prefer to theorize with a *predicate*—something which attaches to a name to form a sentence such as '*S* is definitely true' (see McGee's [102]; many other linguistic theorists use this locution as well). She may then combine this with the relevant quotation names of sentences to draw distinctions that at least seem to correspond to the distinction we introduced earlier, thus saying, for example, 'the sentence "Patrick Stewart is bald" is definitely true' instead of 'Patrick Stewart is definitely bald'.

One thing to note about these particular locutions is that they do not completely do away with the adverbial notion of clarity. In the sentence '*S* is clearly true', 'clearly' appears as an adverb and the verb it is modifying is a linguistic truth predicate. We can regiment this in terms of operators with the sentence 'it's clear that *S* is true'; as we mentioned already, this does not appear to be about the sentence '*S* is true' but what that sentence itself is about. In this case the sentence happens to be yet another sentence, *S*, but in general uses of the adverb 'clearly' need not be so. Thus, while the distinction between clear and unclear truth is perfectly acceptable to the adverbialist, it is not obvious whether linguistic theorists ought to be using the locutions 'definite truth' and 'clear truth' to state their theories; if anything is linguistic in this example it is the truth predicate and not the adverb 'clearly' which is modifying it. Better, then, that the linguistic theorist adopt some primitive simple predicate, 'is vague' or 'is definite', to do the work that she wants; it will be this linguistic notion of vagueness that this theorist will be attempting to analyse.

Another structural difference between the preferred terminology of the linguistic theorist and that of the adverbialist comes out with the point about iteration. When one applies an operator to a sentence one gets a sentence back, so that an operator can be grammatically applied again and again. By contrast, predicates take something of one grammatical type as input—a singular term—and output something of a different type—a sentence. Predicates thus cannot be iterated—one cannot, for example, say 'Harry is tall tall'—so we must find some other way to paraphrase apparently coherent talk involving iterations of the above adverbs and operators. While one can say, for

---

[4] As Prior put it in the context of tense operators, 'when a sentence is formed out of another sentence or other sentences by means of an adverb or conjunction, it is not *about* that sentence, but about whatever those sentences are themselves about' [113].

example, that the sentence S is vaguely vague, the word 'vaguely' here is an adverb which modifies the predicate 'vague' but does not predicate anything of a sentence.[5] The linguistic theorist might capture something like the original thought by instead saying S' is vague, where S' is a name for the sentence 'S is vague' (for example, S' could be the quotation name "S is vague"). Something similar to adverbial iteration is achieved by ascribing vagueness to a sentence, 'S is vague', which in turn ascribes vagueness to another S.

4.2 Parameters

However the linguistic theorist cashes out this linguistic notion, it is not obvious that an adequate theory can be stated using a simple monadic predicate such as 'S is borderline' or 'S is vague'. For one thing there is no sense in which the sentence 'Harry is tall', on its own, is borderline without further qualification. There are a range of heights such that if Harry is within that range then, in ordinary contexts, it is not permissible to assert the sentence 'Harry is tall' or its negation; relative to these contexts, the sentence is intuitively borderline. In a context where we are talking about basketball players, however, it may be perfectly acceptable to outright assert the sentence 'Harry is not tall'. Presumably the sentence 'Harry is not tall' is not borderline in this context, even though it is in the former context: whether a sentence is borderline can depend on the context of utterance. Thus a linguistic theorist cannot take as her basic theoretical term a simple monadic predicate; she must take a *relation* which holds between a sentence and some parameters that includes, at the very least, the context of utterance.

What must these parameters include other than the context of utterance? It is natural to think that we must also relativize to disambiguations of a given sentence. The sentence 'some banks treat their customers well' might be borderline if 'bank' means a financial institution but definitely false if it means the bank of a stream or river.[6]

Presumably we must also specify a language; if 'Harry is tall' means that 1=1 in some language L at context c then this sentence is borderline in English at c but not in L at c. Similarly, sentences within a single language change their meaning over time— some terms start off vague and become more precise, and vice versa.[7] For similar reasons one ought to relativize to dialects. Last, but not least, we must relativize to

---

[5] The word 'borderline' seems to be able to function as both a verb and an adverb; thus one *can* say 'S is borderline borderline'. However, just as with 'S is vaguely vague', we have an uneliminated adverbial use of the word 'borderline'.

[6] Unfortunately, this kind of relativization would be disastrous for a linguistic theorist who thought that vague sentences typically have lots of disambiguations, and that each of these disambiguations is completely precise: for if all disambiguations are precise, no sentence could be borderline relative to a disambiguation. But without relativizing to a disambiguation we can't capture the two senses in which 'some banks treat their customers well' is both borderline and determinately false.

[7] Of course, the context provides the time of utterance but this does not tell you what version of English you are speaking. One might still be able to speak ancient versions of English by uttering sentences today.

a world—the sentence 'Harry is bald' could have been used in exactly the same way as the sentence '5 is prime' is actually used. Even though there is an intuitive sense in which the sentence 'Harry is bald' would not be borderline at such a world, it is clear that we need a way of ascribing vagueness to a sentence that does not depend on how that sentence is used in the world of evaluation. In this sense we will say that at the world x, 'Harry is bald' is borderline as used at w and parameters . . . even if this sentence is used in a precise way at x (what this means is that the sentence 'Harry is bald' counts as borderline at x according to the way that sentence is used at w).

4.3 Can One Explain Propositional Borderlineness in Terms of Sentential Borderlineness?

I have so far been characterizing the contrast between linguistic and non-linguistic theories of vagueness as being concerned with which terms will play a more basic role in the explanations of the vagueness-related phenomena: an adverbial or operator locution such as 'it's vague whether' or the metalinguistic predicate 'S is borderline in L relative to parameters . . .'. According to the non-linguistic view, when I say that something is definitely the case I am no more talking about sentences or linguistic items than I would be if I were talking about what will be the case, what could be the case, what is not the case, and so on.

As with many philosophical disputes in which different sides adopt different basic ideologies, it is often desirable to be able to explain the vocabulary of one theory in terms of the other. I have suggested that the notion of a sentence being borderline in language L relative to parameters p can be explained in terms of propositional borderlineness: it is just for the sentence in question to express a borderline proposition in L relative to parameters p. As I mentioned earlier, some linguistic theorists accept both the distinction between sentences and the distinction between propositions. It is crucial, then, that they be able to offer some kind of explanation of the latter notion in their preferred vocabulary.

But even if a linguistic theorist does not accept a non-trivial notion of propositional borderlineness, it is prudent for her to be interested in such paraphrases. Indeed, given all the technicalities to follow, it is tempting for the linguistic theorist to adopt the adverbialist way of talking for the sake of convenience, but to insist that when it gets down to it, this way of speaking is really an innocuous shorthand for something else. By analogy, it is quite common to use the operator locution 'it's ambiguous whether p' to say something about the ambiguity of a particular sentence: strictly speaking this way of speaking makes no sense, but it is presumably harmless provided it can be systematically paraphrased in terms of something linguistic. My suspicion is that a similar attitude towards borderlineness is actually quite common, since the formalism of determinacy operators is ubiquitous in the philosophy of vagueness, yet most of those who employ the formalism officially endorse a linguistic theory of some sort.

The practice of using the operator formalism as shorthand for something linguistic, however, is not innocuous. Indeed, this practice can lead one to think that options open to the non-linguistic theory are also open to the linguistic theory when, in fact, they are not. We will encounter two specific examples of this, relating to the problem of quantifying in (in section 4.4) and Montague's paradox in (section 4.6). However, it is instructive to first see how some natural concrete paraphrasing strategies deliver the wrong results.

Consider a typical example of a statement we might make using an operator:

(H)   It's borderline whether Harry is bald.

As a first stab: it is natural to paraphrase (H), by picking a language, L (along with other parameters), and sentence, S, such that the claim that S is borderline in L (at parameters p) is a reasonable paraphrase of the claim expressed by (H); one of the things this paraphrase must do is play the same role that (H) does within a theory of vagueness. An example of the paraphrase strategy, therefore, would be to choose English as our language, and the sentence 'Harry is bald' as our sentence. Our paraphrase of (H) is thus that the sentence 'Harry is bald' is used by English speakers in whatever way is required to make it a vague sentence of English.

This is effectively the strategy Quine adopts in [114] for paraphrasing modal sentences in which only closed formulae appear in the scope of intensional operators. In the context of vagueness the relevant paraphrase is rarely made explicit. When it is made explicit, such as in Dorr [35] for example, this is the kind of strategy adopted, so it seems like a natural place to start.

An obvious objection is that it is not equivalent to (H): English speakers could have used 'Harry is bald' in a precise way, to mean that 1+1=2 say, yet it would still have been borderline whether Harry is bald provided he has the same number of hairs he in fact has. The equally obvious reply, which we have already pre-empted, is that we should instead paraphrase (H) with the claim that 'Harry is bald' is a vague sentence of English *as it is used in the actual world*.

But even this seems inadequate: a non-English speaker may not know anything about how English is actually used, but still know that the number of hairs Harry has falls within the borderline region for baldness. It seems, in this case, that one can know whether Harry is borderline bald without knowing whether the sentence 'Harry is bald' is vague as it is actually used in English. Conversely, someone who doesn't speak English may have it on good authority that the sentence 'Harry is bald' is vague in English as it is actually used, but have no idea whether Harry has 0 hairs, 1,000,000, or a borderline number, since she might not know what 'Harry is bald' actually means in English. Although these are only hyperintensional differences they are important: it is natural to think that the notion of propositional borderlineness is the notion that regulates our epistemic and doxastic attitudes for both English and non-English speakers. It is irrational, for example, to believe that Harry is bald but not determinately so; it is not irrational, if you were rationally mistaken about how

English is used, to believe that Harry is bald but that 'Harry is bald' is not definite as it is actually used by English speakers.

A more pedestrian worry is the dependence of the paraphrase on the choice of sentence and language: the claim that 'Harry is bald' is borderline in English as it is actually used, and the claim that 'Harry ist kahl' is borderline in German as it is actually used are two completely different paraphrases of (H). One could, for all we know, be true while the other false—they state completely different facts about different sentences in different languages. Without smuggling in any suspect imperialist assumptions, neither paraphrase is superior to the other—if two incompatible paraphrases are equally good, neither are perfect paraphrases. On the other hand, what is asserted by (H) can be said in a number of languages without mentioning English or German. The operator formulation does not seem to be about sentences, the English language, or the German language; it seems to be about Harry and the status of his head.

Perhaps we could do better with language independent paraphrases:

E.   There is some sentence, S, in some language, L, whose linguistic community actually uses S (i) in such a way that it says that Harry is bald and (ii) in whatever way it takes to make S vague in L.

U.   Every sentence, S, in any language, L, whose linguistic community actually (i) uses S in such a way that it says that Harry is bald (ii) uses S in whatever way it takes to make S vague in L.

These paraphrases still mention languages and sentences, unlike (H), but perhaps they do better with regard to our first problem of language dependence.

Unfortunately, according to some linguistic theories (i) and (ii) are incompatible: the vagueness of a sentence relative to its use in a linguistic community precludes that use determining any proposition as being uniquely expressed by S in L.

However, I think the most problematic feature of this strategy is that it's either inadequate or it assumes the distinction between propositions we are trying to dispense with. Consider the following two possibilities.

The proposition that electrons are positively charged is expressed in some language by a vague sentence.

The proposition that Harry is bald is expressed in some language by a precise sentence.

If the first claim was metaphysically possible then E. would be inadequate and if the second statement was metaphysically possible U. would be inadequate.

It should be noted straight off the bat that coarse-grained theories of propositions, such as the view that propositions are sets of worlds, are already committed to both these possibilities. For example, for some N, the proposition that Harry is bald is necessarily equivalent, and thus identical to the proposition that Harry has N

hairs.[8] It follows that the precise sentence 'Harry has less than N hairs' expresses the proposition that Harry is bald, demonstrating the second possibility. An analogous argument can be made for the first possibility. The paraphrases E. and U. are therefore simply not adequate on a coarse-grained theory; such a theorist would have to resort to a language dependent paraphrase.

Fine-grained theories (theories that distinguish the proposition that Harry is bald from the proposition that Harry has n hairs, for each n) do not have this problem. It follows that if the paraphrases are adequate neither statement is possible: if E, is adequate the proposition that electrons are positively charged couldn't possibly be expressed by a vague sentence in some language, and if U, is adequate the proposition that Harry is bald couldn't possibly be expressed by a precise sentence.

But facts like this call out for explanation! What is so special about the proposition that Harry is bald that means it can't even *possibly* be expressed by a precise sentence? Yet it seems hard to see how the difference between propositions which could be expressed by a precise sentence and those which couldn't, could be explained without appealing to the propositional notion of vagueness that we are attempting to eliminate.

To put it another way, while we know that the proposition that Harry is bald is expressed in English by the vague sentence 'Harry is bald' and in German by the vague sentence 'Harry ist kahl', what is it about *the proposition* that Harry is bald that prevents it being expressed by a precise sentence? Couldn't there be a language with a completely precise predicate, such that there is no unclarity about when it applies in that language and when it doesn't apply, which is such that it applies in L only when the object is bald, and doesn't apply otherwise? How are we supposed to distinguish the proposition that Harry is bald from a proposition that can be expressed by a precise sentence, like the proposition that Harry has less than 2,000 hairs, if not by employing the distinction between vague and precise propositions we are trying to eliminate? It thus seems that if we are to explain why the two possibilities raised above cannot arise, we will need something similar to the operator way of talking to distinguish vague from precise propositions.

4.4 Quantifying In

Another technical issue that must be surmounted by a linguistic theorist is related to what Quine calls the third grade of modal involvement: how a linguistic theorist should say what we would normally say by quantifying into the scope of an operator or adverb. Formally speaking, quantifying in is straightforward for the adverbialist, and is semantically well understood from modal logic. The naïve way of translating

---

[8] That is, assuming that whether Harry is bald supervenes on how many hairs he has.

a formula like $\exists x \Delta F x$ into vocabulary that only attributes determinacy to sentences would result in the following piece of nonsense:

$\exists x Def\,'Fx'$

Supposing here that 'Def' represents our linguistic predicate ascribing definiteness to sentences, then this paraphrase suffers from two problems. Firstly, 'Fx' is not a sentence; it is an open formula and our primitive was introduced so as to apply truthfully only to sentences. Secondly, even if a sensible notion of definiteness could be introduced for open formulae, the above quantification is vacuous, since x does not appear free in '$Def\,'Fx'$ '; it only appears free in the sentence that is mentioned by that sentence.

Before one dismisses this as a technical side issue, note that one of the crucial concepts in the study of vagueness is the notion of a *predicate* being vague or having borderline cases. Adverbialists often introduce a parallel notion for properties: a property, F, has borderline cases if there is something which is borderline F. Similarly, a property is said to be vague if it's possible that there is something which is borderline F. This involves exactly the kind of quantification into the scope of an operator that is problematic for the linguistic theorist to emulate.[9]

A natural way to make sense of the above formula is to invoke a substitutional understanding of the quantifiers. For instance, 'something is definitely F', becomes: 'for some name a, "Fa" is a definite sentence'. Similar proposals have been discussed in the context of linguistic accounts of modality (Quine [114]). For that proposal to even get off the ground, one has to restrict the possible substitutions to *rigid* names. Otherwise it would follow that 'there's some number which is necessarily the number of planets' is true, by using the non-rigid designator 'the number of planets' as our substitution for eight, and noting that the sentence 'the number of planets is the number of planets' is necessary.

Provided we restrict ourselves to proper, and hence rigid, names the substitutional analysis is not subject to straightforward counterexamples involving modal contexts. If we were to extend that proposal to quantification into contexts involving definiteness, however, the results would prove much less satisfactory. Recall that a property, F, has borderline cases iff for some x it's borderline whether x is F. The linguistic version of this analysis, understood in terms of substitutional quantification, would count the predicate 'is 29,000ft' as having borderline cases. Since the sentence 'Mt Everest is 29,000ft' is a borderline sentence, it follows that there's a proper (rigid) name, a, for which the sentence 'a is 29,000ft' is a borderline sentence (namely $a =$ 'Mt Everest'). But this seems wrong: 'is 29,000ft' is as good a candidate as any for being a precise predicate.

---

[9] I will take issue with the above characterization of property vagueness in chapters 12 and 16, but the alternative I offer similarly requires quantification into the scope of operators. Roughly, a property F is vague iff for some precise object, o, it's vague that o is F.

Indeed, sceptics of modal logic, such as Kneale and Quine, took similar positions to suggest that modal properties only belong to individuals relative to some way of describing them.[10] The analogous move here would be to say that *things* cannot be borderline or determinately tall in themselves. Rather, a given person, Alice, is borderline tall relative to some names, and determinately tall relative to others. Relative to a name like 'Alice' she is borderline tall. But she is potentially determinately tall relative to other descriptions. Perhaps Alice, despite being borderline tall relative to 'Alice', is the shortest tall person. In which case she would be determinately tall relative to the description 'the shortest tall person'. Or, that Mt Everest is a borderline case of being over 29,000ft relative to the 'Mt Everest' way of selecting it for attention, but not relative to a precise name for the fusion of rock and soil that constitutes Everest.

A much less revisionary move would be to introduce further ideology. Rather than theorizing with a sentential notion of vagueness, one could start instead with a notion that relates an object to a predicate when it is a borderline case of that predicate relative to the relevant parameters: 'x is a borderline case of predicate F at parameters p'. Note that this move does not allow us to simply dispense with the old sentential notion of borderlineness. We must have that too, or we would not be able to draw the sentential distinction any more. We cannot straightforwardly define sentential borderlineness in terms of predicate borderlineness: 'Mt Everest is 29,000ft' is borderline, but we cannot state this in terms of the borderline cases of the predicate 'is 29,000ft', since there are none. If we are to go this route, then, we will need separate accounts of both predicate and sentential borderlineness.[11]

Not only that, the above primitive is only good for monadic predicates. We will need a four-place relation to make sense of cases where it's borderline whether two objects stand in a relation: 'x and y, in that order, are borderline cases of the relation R relative to parameters p'. And we will need a five-place relation to ascribe borderline cases to ternary relations, and so on. (Of course, rather than multiply primitives, one could talk about monadic predicates of tuples.)

Most of the issues raised above arise for plural predicates and plural quantification. An adverbialist can say things like the following: it's borderline whether Tom, Dick, and Harry have enough hair between them to make a hairball, so there are some people such that it's borderline whether they have enough hair between them to make a hairball. A similar theory would have to be constructed for the linguistic theorist to make sense of this sentence.

---

[10] Kneale, for example, says that properties cannot 'be said to belong to individuals necessarily or contingently, as the case may be, without regard to the ways in which the individuals are selected for attention' [81].

[11] Or perhaps we could get by with accounts of predicate borderlineness and referential vagueness for names.

4.5 Vague Objects

While it would not be at all surprising that a theory of quantification and determinacy could be carried out rigorously in a linguistic theory, it is worth being upfront about the complexity of this sort of theory. Note, by contrast, that from a purely formal perspective quantification into the scope of a determinacy operator is no more complicated than quantification into the scope of the negation operator. For the operator approach it is rather the metaphysical commitments that are problematic.

One puzzle that has occupied philosophers concerns issues to do with vague parthood and vague objects. Presumably there are rocks and stones such that it is borderline whether they are part of Mt Everest. Let *a* be one of these rocks. On the other hand, there is also the total lump of rocks and stones—call it Lump—that make up Mt Everest: it is presumably not borderline whether *a* is a part of Lump, since there is nothing more to a lump of rock and stone than the rocks and stones that make it up. It follows by Leibniz's law that Mt Everest and the lump of rock and stone are different.

This conclusion should strike us as extremely perplexing: Mt Everest and Lump coincide in space and time, they have all the same mereological parts, they appear to stand in the same causal relations to things, and so on. Moreover, this puzzle rested on the adverbialist account of borderlineness. The puzzle does not arise on a linguistic conception of borderlineness: the fact that '*a* is a part of Mt Everest' is a borderline sentence, and the fact that '*a* is a part of Lump' is not a borderline sentence, do not together entail that any mountain is distinct from any lump of rock and stone. They at best entail that the names 'Lump' and 'Mt Everest' are different—one is a precise name for a particular lump of rock and stones, and the other is a vague name whose meaning is indeterminate between a range of different candidate lumps of rock and stone. It is not the things made up of rocks and stone themselves that are vague or precise.

How this puzzle relates to the issues of quantifying into determinacy contexts, however, is extremely vexed. According to one sort of approach, inspired by McGee [103], we can say things like *it's borderline whether a is a part of Mt Everest*, but we cannot infer from that things like *there's an x and a y such that it's borderline whether x is a part of y* using existential generalization into a borderlineness context.[12] This is, in some sense, thought to curtail commitment to vague objects. It is unclear, however, to what extent our original puzzle really relied on this instance of existential generalization: the distinctness of Mt Everest from Lump was derived from Leibniz's law and the claim that *a* is a borderline part of Mt Everest but not Lump. The view also has some fairly surprising consequences: one might have naïvely thought that there is a certain mountain which is widely known to be the tallest mountain in the world, that was first climbed in 1953, and so on. Of course it is widely known that Mt Everest has these properties, but without existential generalization you cannot infer that there is something that is widely known to have these properties. Moreover,

---

[12] McGee is explicitly a linguistic theorist and so doesn't apply the theory to operator expressions as I have done here.

since nothing determinately has these properties according to the view in question,[13] we have positive reason to doubt that there is something which is widely known to have those properties, given the widely held view that one cannot know something unless it is determinate.

It is worth drawing further parallels with the case of modality here. Quine, for example, was suspicious of the adverbial view of modality because of the metaphysical implications. As we know, in this debate the adverbialist approach to modality now dominates, metaphysical puzzles and all.

A strikingly similar puzzle for the adverbialist account of modality is the problem of the statue and the lump of clay. The lump of clay out of which the statue is constituted could have been deformed without being destroyed, whereas the statue couldn't; thus the statue, it is concluded, is not identical to the clay for they have different modal properties. If this conclusion is correct then it is not at all surprising that Mt Everest is not identical (even indeterminately) to a precise lump of rock, soil, and snow—this is a conclusion that is motivated by an already established response to the corresponding modal puzzle. Indeed there are many solutions to this puzzle in the modal case that can be applied fairly straightforwardly in the present case.[14]

Yet another strategy, one that is much less attractive in the modal case, but perhaps defensible in the case of vagueness, is to reject Leibniz's law. It would then be possible to say that Mt Everest is identical to exactly one fusion of particles, although it is indeterminate which.[15]

Although this is not the place to defend a particular solution to this conundrum, the metaphysical puzzles that adverbialists find themselves faced with are puzzles that I think we are already committed to on independent grounds. I see no particularly pressing objection here to an adverbialist conception of vagueness that is not equally an objection to an adverbialist conception of modality.

4.6 Montague's Paradox

Despite the pervasiveness of the linguistic approach to vagueness, it is somewhat surprising to see that the logical theory surrounding the study of vagueness focuses almost exclusively on the operator formalism. In that setting two principles are almost universally taken for granted. The determinacy operator is factive, which means:

---

[13] The standard version of this view is usually paired with the view that nothing is a determinate mountain. For if there were something that was a determinate mountain in the vicinity of Mt Everest, it would be distinct from Mt Everest. According to this view, nothing has borderline parts but Mt Everest does (remember that we are relinquishing existential generalization), and so, by Leibniz's law any determinate mountain would be distinct from Mt Everest. But then there would be at least two mountains in the vicinity of Mt Everest (Mt Everest and the thing that's the determinate mountain), which seems like a puzzling result.

[14] One strategy I find particularly promising is the idea that there are many coincident fusions of the particles that fuse Mt Everest with different modal profiles (see, for example, Hawthorne [69]. See also Cotnoir and Bacon [28] for a way of spelling out the mereological picture). A similar view could be adopted to the case of vagueness (see the discussion of vague parthood in Korman [82]).

[15] Similar moves have been made in the modal case—see, for example, Gallois [61].

If it's determinate that p then p

and a rule of proof call necessitation, which guarantees the following:

If 'p' is provable in classical logic with factivity, then 'it's determinate that p' is a theorem.

The rule of necessitation is actually strictly stronger because it can be applied repeatedly; however, the above consequence suffices for my discussion. These two principles, and many more, are naturally modelled by the same formal tools used to study modal logic: the kind of framework involving indices and accessibility relations.

This was a framework whose invention coincided roughly with the rise of non-linguistic theories of modality, and which was surely an important component in their success (see Kripke [84]). As was noted by Richard Montague [107] at that time, however, results concerning the above logic and the corresponding model theory cannot be straightforwardly transferred to linguistic theories of modality. Similar things must also be said about the linguistic approach to vagueness. For example, one might naïvely think that a linguistic theorist could develop a theory completely parallel to the operator theory by adopting the following two analogous principles:

If 'p' is definite then p.
If 'p' is provable in classical logic with factivity, then "'p' is definite' is a theorem.

However, perhaps surprisingly, the above two principles are inconsistent, unlike their operator variants. The reason is exactly parallel to the problems Montague raises against linguistic accounts of necessity. Montague proves it formally within a background theory of syntax represented in arithmetic (in which there is no explicit self-reference), but we can give the informal gist of the argument using self-referential sentences. Let D be the sentence 'D is not definite', then by Leibniz's law we can infer that if D is definite, then 'D is not definite' is definite, and by the factivity principle, that 'if D is not definite' is definite, then D is not definite. Putting these together we get that if D is definite, it isn't definite. Thus D isn't definite, and we have proved this in classical logic with the analogue of factivity. So, by the second principle we may conclude that 'D isn't definite' is definite. But this is just the conclusion that D is definite, which contradicts our earlier conclusion that D isn't definite.

No parallel argument can be levelled at the operator formalism, or even the variant formalism utilizing a predicate of propositions, without making substantial assumptions about propositions. For example, if one represented propositions as simply sets of indices of some kind, it is natural to simply deny the existence of a proposition, p, identical to the proposition that p is not true, just as we are forced to deny the existence of a proposition, p, identical to the proposition that it's not the case that p in this setting, for no set is identical to its set theoretic complement.[16]

---

[16] A common, albeit flawed argument, that there must be propositional paradoxes is that it seems as though it should be possible for Martha, say, to have the following unique favourite proposition: the proposition that Martha's favourite proposition isn't true. From this we can derive a paradox assuming a

It is thus not generally safe to assume that a linguistic account of vagueness can simply piggy-back off the success of formalisms formulated using operators. Few linguistic theorists are careful about this, and simply theorize using an operator assuming that it can safely be reinterpreted within their preferred account vagueness. The most notable exception to this exclusive focus on operators among linguistic theorists is McGee [102], who has done more than anyone to spell out the consequences of using a linguistic definiteness predicate. McGee's theory relaxes the factivity requirement of definiteness and keeps necessitation.

However, the costs of McGee's approach are more than just the failure of factivity. One might think that it's never the case that a sentence and its negation are both definite at the same time. However, even this principle must be relaxed in McGee's theory. Indeed, one of McGee's own limitative results suggests that some concession beyond factivity will have to be made. In the operator formalism it is standard to assume that a determinate conditional with a determinate antecedent has a determinate consequent. Also important, in a first-order theory, is the Barcan principle which says that if it's determinate that everything is F then everything is determinately F. This principle, among other things, helps rule out the possibility of indeterminate existence. However, if we were to adopt the linguistic analogues of these principles,[17] along with the rule of necessitation and the principle that no sentence and its negation are both definite (and the background theory of syntax), the theory would be 'ω-inconsistent': while no contradiction could be derived from it in a finite number of steps, contradictions could be derived if one could perform infinite inferences.

While I by no means think that these kinds of costs are decisive, it does highlight the fact that linguistic theories are usually logically highly complex and cannot be modelled by the simple model theory that has proved so fruitful in the case of the operator approach.

4.7 More Vague Propositions than Sentences

There is another reason not to try to reduce propositional vagueness to sentential vagueness: there are more statements of propositional vagueness to be analysed than statements of sentential vagueness to provide that analysis.

In the parallel modal debate, David Lewis once objected to linguistic theories of modality that identified possible worlds with certain kinds of sets of sentences on

---

propositional variant of the T-schema. But this argument seems in many ways parallel to the following flawed argument. Surely it is possible for Martha to uniquely have the following favourite number: the successor of Martha's favourite number. Again one gets a contradiction, but this time from purely numerical facts and no propositional T-schema. I think in this case it is clear, even if Martha goes about declaring 'my favourite number is the successor of Martha's favourite number', that she hasn't succeeded in making this her favourite number.

[17] One would have to be careful about how one formulates the Barcan principle, given the points made in section 4.5. See McGee [102], for one way of making this precise if you are in the language of arithmetic.

the grounds that there are simply more possible worlds than sets of sentences.[18] A related point applies here. It certainly seems like there are more vague propositions than vague sentences, and there certainly appear to be vague propositions which are not expressed by any sentence of English.

To see this imagine that we encounter an alien race who have evolved on a planetary system orbiting a white dwarf. Consequently the range of light that stimulates their visual system is in the ultraviolet spectrum. Like us they have different names for different segments of that region depending on how things appear to them. For example, just as we're unsure whether to apply the term 'red' to the region of the visual spectrum that is borderline red, there are corresponding regions of the ultraviolet spectrum where the aliens are unsure whether to apply their terms. It seems very natural to say that there is something the aliens can say which we can't. These things aren't synonymous with the precise claims we can make about the ultraviolet spectrum in our language, and therefore encode pieces of information not expressible in English. If the vague propositions exist independently of us and the way we happen to speak, just as precise propositions do, the vague propositions we described would exist even if there weren't any aliens speaking this way. It seems particularly hard to imagine how the unknowability of these vague propositions could be explained by any sentence in English being borderline.

We can massage this point into a much more general worry about the prospects of giving a reductive account of propositional borderlineness in terms of sentential borderlineness. The problem is that there are simply not enough sentences to account for all of the vague propositions. There are, for example, at least uncountably many different vague propositions. Take the colour spectrum and divide it into three adjacent connected regions such that (i) the two outermost regions are each roughly the same width as the range of colours that we are unable to know are red, and (ii) the inner region is roughly the same width as the range of colours we are able to identify as red. There are uncountably many such divisions, but it is natural to think that each of these could be the range of unknowability and knowability of some other vague property much like the property of being red except shifted. There are simply not enough sentences of any spoken language to account for all such properties.[19]

4.8 Vagueness and the Objects of Thought

This chapter has provided a quick tour of the primary differences between the linguistic and non-linguistic approaches to vagueness, and a look at some of the

---

[18] See Lewis [93] §3.2.

[19] More precisely, it follows that if there are at most countably many spoken languages each with at most countably many sentences, then there are at most countably many sentences of any spoken language. (Note that for each real number, γ, we have a name for, there is the property of being a colour γ hertz above red. But since there are only countably many reals nameable in English, this does not provide us with a straightforward way for generating uncountably many sentences.)

outstanding technical issues. I want to close the chapter by looking a bit more closely at the latter sort of theory, and clearing up some potential issues concerning the involvement of propositions in it.

The view I am defending in the present book belongs to the non-linguistic camp: propositional vagueness will take centre stage in our theory. While sentential vagueness is a perfectly good notion—it is just a sentence that expresses a vague proposition—it will play little theoretical role in the theory. It is very natural to associate the view I am going to defend, according to which language-independent entities—*propositions*—are the primary bearers of vagueness and precision, with the view that the adverbial and operator ways of speaking about vagueness are basic.

I do in fact endorse both these ideas, but it is important to be clear about how they are different. For example, it ought to be obvious that the operator and adverbial ways of speaking about negation are theoretically more basic than the linguistic analogue of negation—the falsity predicate.[20] We should be non-linguistic theorists about negation. Yet it's clear that a nominalist about propositions can perfectly well accept this view, and freely employ the negation operator, whilst resisting the further claim that propositions can be negated, on the grounds that there are no propositions.[21] There is a non-trivial step between employing certain operator expressions and accepting the existence of analogous properties belonging to a special class of entities: propositions.

Nominalists could in principle avoid commitment to propositions, but paraphrase many of the things I say in this book by using the language of higher-order logic, in which one can directly quantify into the position that a sentence occupies (as, for example, Prior does in [112]). This way of talking does not commit you to propositions, or any kind of singular entity, in the same way that quantification into the position of a plural term does not commit you to special kinds of set-like singular entities. Although I am no nominalist, I am sympathetic to the framework of higher-order logic as a framework for systematically investigating many philosophical questions about propositions and properties, and in many cases I think the questions I am concerned with could be less ambiguously stated using quantification into sentence position. However, since I do not want to make this viewpoint a prerequisite for engaging with the ideas in this book, I shall stick to talking about propositions for the most part.

I shall therefore generally help myself to these entities and leave it to the nominalists to do whatever they need to in order to make sense of what I say. Once we have helped ourselves to these entities we must get clear on the kinds of things they are supposed to do. In my view there are a number of different roles that philosophers

---

[20] To see this, try to explain what it would mean for a sentence to be false in a language at a context without using the negation operator.
[21] It is not quite as awkward to speak of propositions as bearers of vagueness as it is awkward to say that propositions are the bearers of negation.

want propositions to play. For some philosophers, these entities are, like 'facts' or 'states of affairs', supposed to do heavy-duty metaphysics. For others, propositions are the semantic values of sentences and provide their truth conditions. Yet another important characterization, the one I will adopt, identifies propositions with the denotations of that-clauses and the objects of propositional attitudes. It may turn out that no single kind of entity can play all these roles, in which case a dispute can arise about which entities are the *real* propositions. These disputes are verbal—if entities filling each of the roles in question exist, then so long as we are careful about which things occupy which roles, no issue of substance will turn on which of these entities we choose to call 'propositions'. For my purposes, a proposition will be just whatever the denotation of a that-clause is. Since propositional attitudes are grammatical relations taking names and that-clauses as arguments, this view might naturally be expressed by the slogan that propositions are *the objects of thought*—the objects of propositional attitudes such as believing, knowing, desiring, and so on. Given the operator approach, it's natural to think that propositions, qua denotations of that-clauses, are also the bearers of vagueness, and indeed negation, necessity, and so on. However, as this book continues it will be clear that it is the role that propositions play as the objects of thought that define them, so I shall keep to the slogan that propositions are the objects of thought.

This way of talking about propositions suggests that they are abundant. Whenever there is a sentence, for example 'Harry is bald', there is also a singular term 'that Harry is bald', or more perspicuously, 'the proposition that Harry is bald', which denotes a singular entity, a proposition, that is true if and only if Harry is bald.[22]

It might seem that the theory that merely says that propositions are the denotations of that-clauses tells us very little about their nature. On the contrary, we can deduce quite a lot about them. Indeed, one can assume that they satisfy what I'll call 'the proposition role' described below. Each of these principles simply falls out of the stipulation that propositions are the things expressed by that-clauses.

THE PROPOSITION ROLE:

> One believes (knows, desires, asserted, said, etc.) that *P* if and only if one believes (knows, desires, asserted said, etc.) the proposition that *P*.
>
> It's necessary that *P* if and only if the proposition that *P* is necessary.
>
> It's true that *P* if and only if the proposition that *P* is true.
>
> A sentence means that *P* if and only if it means the proposition that *P*.
>
> It's determinate that *P* if and only if the proposition that *P* is determinate.

---

[22] The thesis that the proposition that ϕ is true if and only if ϕ is importantly different from disquotational schemata which are well known to be problematic ('the sentence "ϕ" is true in English at c if and only if ϕ' is a disquotational schema; our thesis does not mention sentences or involve quotation marks at all). Unlike the disquotational variants, the propositional T-schema I have just presented is known to be consistent.

From these facts we can deduce many things, including that propositions can be said and asserted, are the objects of our attitudes, are necessary and contingent, are the meanings of our sentences, and the objects of truth and falsity, and perhaps also determinate truth and falsity.[23]

In choosing to use the word 'proposition' in this way I have made at least one substantive commitment—that the proposition role is consistent, and that there are entities that occupy the role. However, with the assumption that there are entities playing this role, the dispute over whether they are propositions can be no more than a verbal dispute about how to use the word 'proposition'.

Of course, I don't mean to suggest that this is a complete theory: more concrete theories of propositions tell us whether propositions are structured entities, whether they are set theoretic constructions, whether they have a Boolean structure (see section 3.2), and so on. For all I've said the things playing the proposition role are just linguistic entities—sentences of a particular language, or equivalence classes of sentences from different languages. However, these theories must all agree that these entities satisfy the proposition role if they are engaging in the project of describing the denotations of that-clauses, and it is only facts like those described in the proposition role that I shall need in the following chapters; it is unnecessary to be more specific than this.

Note that a consequence of the proposition role, if you accept the claim that it's borderline whether Harry is bald (and thus the adverbialist way of drawing these distinctions), is that there are vague propositions. For if it's not determinate that Harry is bald then, by THE PROPOSITION ROLE, *that Harry is bald* (a proposition by our lights) is not determinate. If it is furthermore not determinate that Harry is not bald, then we may also infer that *that Harry is bald* is a borderline proposition.

One issue I want to address is whether a non-linguistic view—one that accepts the operator and adverbial ways of speaking about vagueness as basic—has to be a view in which vagueness is, in some sense, 'in the world' (or, alternatively, whether I am committed to 'metaphysical vagueness'). The answer to this question will depend on how the question is posed. On a very deflationary use of the word 'fact' there will be vague facts. For if a fact is just a true proposition, and furthermore, if to say that the proposition that p is true is equivalent to simply saying that p, then we can show that there are vague facts as follows. Suppose that the proposition that Harry is bald is borderline (and thus, it follows, that its negation is also borderline). If Harry is bald, then the proposition that Harry is bald is true, and is thus by definition a fact. So in this case we have a vague fact. If Harry is not bald then the proposition that Harry is

---

[23] It should be stressed that there is more to the story here than I am letting on. For example, it seems as though one can fear that p without fearing the proposition that p, and one can hope that p, but it is not even grammatical to 'hope the proposition that p'. These complications can be avoided mostly by substituting 'the proposition that p' for 'that p' in the following discussion, although for the sake of readability I shall not do this. It is also worth noting that most of the theses in this book can be reformulated by quantifying directly into sentence position, eliminating talk of propositions altogether.

not bald is true and borderline, so again we have a vague fact. Either way there is a vague fact.

This is just one thing that 'metaphysical vagueness' might mean; it might mean other things, and these too might require clarification. At any rate, I suspect that once it is clear what I mean by 'fact' in the above argument many philosophers will find the sense in which I am committed to there being 'vagueness in the world' an uninteresting one.

The real issue at stake, in my view, is whether the entities which serve as the objects of thought are vague in a way that cannot be reduced to the relations they stand in to public (or private) language sentences. The role that vague propositions play in thought will be examined in chapters 6, 8, 9, and 10, where we connect vagueness to evidence, uncertainty, decision, and desire. We begin in chapter 5, however, by looking at the relation between vagueness and knowledge. This is of course one of the most puzzling issues in the philosophy of vagueness; I shall argue that the role that propositional vagueness plays in explaining vagueness-related ignorance is so fundamental that this ignorance cannot be equivalently explained by a relation to a public language.

5

Vagueness and Ignorance

According to a widely held intuition there is an important, systematic connection between borderlineness and knowledge. One instance of this general thought can be described as follows: there are certain heights such that when a person is that height it is impossible to know whether they're tall (even if we know their exact height). Moreover, the fact that we are ignorant in cases such as these has *something* to do with vagueness, and a good theory of vagueness ought to explain this ignorance. For the linguistic theorist, however, this connection seems at least initially mysterious. What is special about the people with these particular heights, according to that theorist, is that certain words in certain languages bear a special relation to them. But it is extremely natural to wonder how this special relation to words could explain why we can't know whether these people are tall. In this chapter I shall develop this sort of thought into an argument that vagueness-related ignorance cannot be explained by a linguistic theory of vagueness.

5.1 In Favour of Vague Propositions

Note that this connection between vagueness and ignorance sheds light on the view, discussed in chapter 4, that the notions of propositional vagueness and precision are trivial, or even meaningless. Although one could in principle reject the distinction because one thinks that all propositions are vague, the most popular version of this line is that the distinction is trivial because all propositions are precise. The proposition that Harry is bald is a precise proposition picked out, in this instance, by a vague piece of language, namely the description 'the proposition that Harry is bald', just as a precise number can be picked out by the vague description 'the largest small number'. Indeed, according to this view, the proposition that Harry is bald just *is* the proposition that Harry has no more than N hairs (for the critical cutoff, N), and so the very same proposition can also be picked out with precise language, e.g. with the description 'the proposition that Harry has less than 1,000 hairs'.

However, if you think that there are some propositions (such as the proposition that Harry is bald) which in certain circumstances enjoy a kind of incurable ignorance, and other propositions (such as the proposition that Harry has less than 1,000 hairs) which do not, and you can reliably distinguish between the cases where the incurable ignorance is due to vagueness, then you have done everything except verbally accept the distinction between propositions that is being claimed to exist. You might think that the distinction is a product of a more basic phenomenon associated with vague language, but you must at least accept that the distinction between propositions exists and is non-trivial. To go beyond a verbal rejection of the distinction you must rather maintain that the propositions being grouped together as 'vague', such as the proposition that Harry is bald, do not really exhibit these epistemic features, or argue that the propositions being classified as 'precise', such as the proposition that Harry has less than 1,000 hairs, do.

I should mention straight away that while the connection between vagueness and ignorance is certainly widely accepted, some philosophers have recently attempted to resist it. I consider these philosophers in section 5.3. However, I will begin by treating the majority of philosophers who do accept the basic intuition behind the ignorance idea gestured at above. Given this assumption, there are certain heights such that we are unable to know whether people with those heights are tall. This fact calls out for an explanation, and moreover since this ignorance manifestly has something to do with vagueness, the explanation ought to be couched in one's preferred theory of vagueness.

In our preliminary characterization of the phenomenon, we said that there are certain heights such that it is impossible to know that a person is tall when they are that height, a certain number of hairs such that it is impossible to know that a person is bald when they have that number of hairs, and so on. One issue that needs to be addressed is how to characterize these heights, hair numbers, and so on. A partisan way of doing this would be to talk about the heights of people who are *borderline* tall, or the number of hairs that people who are *borderline* bald have, and so on. This way of characterizing the cases, of course, uses the adverbial way of talking about vagueness and carves out a distinction between people. Some linguistic theorists would want only to draw a distinction between sentences, leaving the distinction between people derivative at best, and so will say something different in its place. As we discussed in section 4.4 on quantifying in, there are several different ways to do this and perhaps other ways I haven't considered. In what follows I shall leave it to the reader to fill in those details, in whichever way they see fit.

My argument against linguistic theories will revolve around the following example (adapted from Dorr [34]):

Before us is a glass of water that is filled so that it is exactly 70% full. There is a large international team of people ready to inspect the glass, armed with many different measuring devices for calculating every possible dimension of the glass and the water in it. Some of these people speak multiple languages, some of them only speak one, and perhaps some of them (let us suppose) do not speak any languages at all. Each person has been given the task of determining every truth they can about the glass. After they have performed whatever measurements they need, they have been instructed to signal to me whether the glass is pretty full.

It should be obvious to everyone that an unqualified positive or negative answer to this question would be inappropriate, even among those who happened to have measured all the relevant precise facts about the exact volume and shape of the glass, the exact volume of water in it, and so on. Modulo a small number of dissenters, mentioned earlier, most philosophers think that the explanation for this fact is that they simply do not know that the glass is pretty full, despite the fact that they know numerous precise facts, such as the proportion of the glass that is filled with water.

Once we have conceded that, for example, the English speakers do not know whether the glass is pretty full, it becomes pretty hard to imagine that speakers of other languages are in a better position to know than the English. Indeed, it becomes hard to imagine that anybody is better placed to discover whether the glass is pretty full—even the people who do not speak any languages at all. Thus I think we have good reason to accept the following:

IGNORANCE: Nobody knows whether the glass is pretty full.

We can also make it explicit that this is not because they are ignorant about how full the glass is:

KNOWLEDGE: For each $0 \leq n \leq 100$, somebody knows whether the glass is at least $n\%$ full.

This should also be obvious assuming, as I have been, that among the people measuring the glass are people who have measured the exact percentage of the glass that is filled. We could extend KNOWLEDGE to include knowledge of other precise facts without changing the case.

IGNORANCE, I take it, is somehow or other a result of vagueness, and a very puzzling result at that. Each person in our team of measuring experts measures the glass but without fail, each comes away ignorant about whether the glass is pretty full. This isn't just an accident, it is a general fact that calls out for an explanation, and a theory of vagueness ought to surely provide one, or ought at least be capable of providing one within the resources it invokes.

We must be careful to distinguish IGNORANCE from LINGUISTIC IGNORANCE, which is surely also true:

LINGUISTIC IGNORANCE: Nobody knows whether the sentence 'the glass is 70% full and pretty full' is true in English in 2014, at context c, world w (and . . .)

Without a doubt, not one of the measurers knows this fact either. However, it is important to bear in mind the difference. Among our international team, we may suppose, are monolingual Chinese speakers who are ignorant of the second fact for reasons that have nothing to do with vagueness. In general, if you do not know what the sentence 'the glass is 70% full' means in English (at context c and . . .) then you may not know whether it is true. Conversely, you might know that 'the glass is 70% full' is true in English (at c and . . .) without knowing whether the glass in question is 70% full. A competent English speaker knowledgeable about the glass might have told the monolingual Chinese speaker that the sentence in question is true, without telling her that the glass is 70% full.

There are therefore two distinct things we are ignorant about. In fact, there are more than two things: there are countless other languages with sentences like the English one mentioned above whose truth statuses we do not know. But at any rate, the point is that they are all different things to be ignorant about: we are ignorant of the semantic status of a number of different sentences in different languages, and then we are ignorant about whether a particular glass is pretty full.

It is this latter fact, the ignorance about whether the glass is pretty full, that I think is hard for the linguistic theorist to explain and it is this fact that I shall concentrate on in what follows. If the linguistic theorist is right about the nature of vagueness it would be fairly easy to come up with explanations for LINGUISTIC IGNORANCE. However, doing so does not exempt her from the burden of addressing one of the central issues in the philosophy of vagueness: explaining IGNORANCE.

It's worth mentioning that other facts about propositional attitudes need explaining, and would have served equally well as the basis of my criticism in the following. For example, I will argue later in the book that both of the following are true:

BOULETIC: If you know exactly how much water there is in the glass (and any other precise things that you care about), you should not further care whether it is pretty full or not.

DOXASTIC: If you are rationally certain that it's borderline whether the glass is pretty full you cannot be rationally certain that it's pretty full.

Similar questions arise for the linguistic theorist concerning how to explain these truths. For now I will focus on the more familiar principle IGNORANCE. Like IGNORANCE, there are linguistic versions of these principles that we should take care to distinguish.

What follows from the conjunction of IGNORANCE and KNOWLEDGE? An important consequence of these two assumptions is that, given a natural supervenience thesis, propositions are more fine-grained than sets of worlds. Moreover, this increase in the fineness of grain is a result of vagueness. One way to gloss this result would be to say that there are *vague propositions* in addition to precise propositions. All we mean by this is that there are some propositions, like the proposition that the glass is 70% full, whose truth the team of measurers have no problem discovering, and other propositions, such as the proposition that the glass is pretty full, whose truth the measurers have difficulty discovering, and moreover the propositions from the former class are not identical to propositions in the latter class.

To see why IGNORANCE and KNOWLEDGE require this level of fine-grainedness, take any set of worlds that is putatively identical to the proposition that the glass is pretty full—let's say, the set of worlds where the glass is at least n% full. (This choice is natural once we've make the simplifying assumption that whether the glass is pretty full supervenes on the percentage of the glass that is full.) According to KNOWLEDGE somebody knows whether the glass is at least n% full, yet by IGNORANCE nobody knows whether the glass is pretty full. Thus, applying the proposition role specified in chapter 4 and Leibniz's law, the proposition that the glass is at least n% full (i.e. the set of worlds at which the glass is at least n% full) is not identical to the proposition that the glass is pretty full. Here is the argument explicitly (assume, without loss of generality, that the glass is in fact at least n% full):

1. Somebody knows that the glass is at least n% full. (By KNOWLEDGE.)
2. Nobody knows that the glass is pretty full. (By IGNORANCE.)
3. Thus nobody knows the proposition that the glass is pretty full, and somebody knows the proposition that the glass in at least n% full. (Applying the proposition role.)
4. Therefore the proposition that the glass is pretty full is not the same as the proposition that the glass is at least n% full. (Leibniz's law.)

Let me clear up two possible misunderstandings about this argument. The first concerns philosophers who reserve the word 'proposition' for sets of worlds, states of affairs, or for some other coarse-grained entity. Such philosophers can verbally reject any conclusions one might draw from this argument involving the word 'proposition' by divorcing proposition talk from that-clause talk (for example, they might reject the locution 'that snow is white' as a term for denoting the proposition the sentence 'snow is white' expresses—they will end up talking in convoluted ways, but there is nothing that in principle stops them from making this move).

Such philosophers are therefore denying that propositions play what I called the 'proposition role'. But of course, *something* has to play the proposition role—at least the way we use that-clauses in English suggests that something does—and so this is just a disagreement about which entities we should grant the honorific title 'proposition' to. No conclusion that matters to us here can only be stated using the word 'proposition'—the important upshot of the above argument concerns the objects of thought, the denotations of that-clauses—i.e. the things *I* have been calling 'propositions'.

The second misunderstanding concerns a certain approach to the semantics of attitude reports. A common response to Frege puzzles involving Leibniz's law, such as the one above, is to maintain that knowledge and belief, despite their surface form, are fundamentally three place relations between a person, a proposition, and a 'mode of presentation' which represents the *way* in which you come to believe that proposition (see Crimmins and Perry [30] and Richard [120]). According to this view no one ever simply stands in this relation to a proposition—they stand in this relation to a proposition relative to a way of entertaining that proposition—a mode of presentation. One might object that in making this argument I have begged the question against these theorists by not being explicit about the mode of presentation.

I think that whatever that view says about the fundamental psychological structure of propositional attitudes, it still has to account for ordinary language belief reports which have a binary structure and make no explicit mention of modes of presentation. The most natural way to do this is to suppose that a natural language belief report which on the surface appears to be a binary relation in fact states a ternary connection between a person and a proposition (supplied by the referents of the subject and the that-clause respectively) and a contextually supplied mode of presentation which does not appear grammatically as a third term on the surface.[1]

The point to stress here is that my argument was not stated using the fundamental ternary relation; it was stated using the ordinary language binary relation that is expressed in the present context by the verb 'knows'; which binary relation this verb expresses is context sensitive on this view, but that is not to say that I didn't express a particular binary relation when I stated IGNORANCE and KNOWLEDGE. The correct response to this argument for a contextualist is not to reject the validity of the argument—it was literally an application of Leibniz's law, along with a stipulation about what I meant by 'proposition'—but to reject one of the premises. Suppose that the proposition that the glass is pretty full is, in fact, identical to the proposition that the glass is at least $n\%$ full—call this proposition p. If the contextually salient mode of presentation is the vague one then, presumably, nobody stands in the three-place knowing relation to p relative to that mode of presentation. In this context KNOWLEDGE expresses a falsehood. On the other hand, if the contextually salient mode of presentation is the precise one then presumably the people who measured the percentage of the glass that is full do stand in the three-place knowing relation to p relative to the precise mode of presentation. In this context IGNORANCE expresses a falsehood. Either way the argument is valid, and it is one of the premises that fails.[2]

Thus, I contend, anyone who accepts both IGNORANCE and KNOWLEDGE in the same breath, without changing the context, must acknowledge the thesis that there are vague propositions.[3]

---

[1] There are those who accept the ternary analysis of belief but are not contextualists (see Salmon [124]). For Salmon an ordinary belief report merely states that one stands in the ternary relation to a proposition relative to *some* mode of presentation. This type of view simply rejects the principle IGNORANCE altogether; the validity of the above argument is therefore not in question on Salmon's view—it is the truth of the premises. I shall therefore set Salmon aside along with others who reject IGNORANCE—I will come back to these theorists in section 5.3.

[2] A different objection one might have is that according to some views Leibniz's law has to be rejected, at least when stated as a schema involving proper names like 'Hesperus' and 'Phosphorus'. According to these views, however, one cannot existentially generalize on names appearing within the scope of attitude reports. That said, I think that a good case can be made that Leibniz's law is valid when restricted to proposition terms ('that p' or 'the proposition that p') since we need to be able to existentially generalize on these terms in order to do the kind of theoretical work we need to put them to (e.g. as prescribed by the proposition role). If there are entities fine-grained enough to be distinct whenever the psychological facts require it, and if by 'proposition' we just mean whatever satisfies the proposition role, then we may always quantify out on proposition terms and apply Leibniz's law to them. This strategy is not available for concrete entities like Venus.

[3] Note that for all I've said vague propositions are just ordered pairs of sets of worlds and vague modes of presentation. Although this view appears only to deviate minimally from the contextualist mode of presentation view, it is different in the respects that matter, namely, that IGNORANCE and KNOWLEDGE are

I will treat those who deny the conjunction of IGNORANCE and KNOWLEDGE in section 5.3. The simplest way to deny the conjunction is to deny IGNORANCE flat out. However, the contextualist variant of the no-ignorance type view, briefly described above, allows certain attitude reports to depend on a contextually salient mode of presentation. There will be some contexts where they behave like the straightforward kind of no-ignorance theorist by denying IGNORANCE (and asserting KNOWLEDGE). However, there will be many other contexts in which they can assert IGNORANCE (albeit, in these contexts they must deny KNOWLEDGE) thus allowing themselves more flexibility than the simplest no-ignorance view. Of course, at no context will both IGNORANCE and KNOWLEDGE express a truth. In section 5.2, however, I shall simply assume that IGNORANCE and KNOWLEDGE are being granted, and that propositions are therefore somewhat fine-grained.

Before we move on, let me stress that the question we are dividing our discussion around concerns how coarse- or fine-grained we take the objects of attitudes to be. However, it is evident that settling this question still leaves a number of different possibilities regarding the relation between sentences and propositions open. The two most salient options to choose between are the views that vague sentences (i) express exactly one proposition and (ii) that they express several.[4]

Both views about the relation between sentences and propositions can be combined with both the coarse- and fine-grained view about propositions. If we were to combine a coarse-grained account of propositions with the view that 'Harry is bald' expresses a unique proposition, then it presumably has to be a view in which we cannot know which proposition it expresses (presumably because it is vague). For it would be hard to maintain that we can know that 'Harry is bald' expresses, say, the coarse-grained proposition that Harry has at most 1,023 hairs, whilst also maintaining that we can't know whether 'Harry is bald' is true or not.[5] On the other hand, on a more fine-grained view in which there is the proposition that Harry is bald in addition to propositions like the proposition that Harry has at most 1,023 hairs, there is a particularly natural candidate for the proposition that 'Harry is bald' expresses: *the proposition that Harry is bald*. In which case we can know what proposition 'Harry is bald' expresses. We can also combine the coarse- and fine-grained views in various

---

jointly true in the same context and so propositions are fine-grained enough to be the objects of vagueness and precision.

[4] A variant of (ii) would be that sentences don't express propositions *simpliciter*, but only relative to a precisification, much like a context-sensitive sentence only expresses a proposition relative to a context. I haven't considered the more radical view that borderline sentences don't express propositions at all, since this view is subject to straightforward problems. For example, the sentence 'Either Harry is bald and electrons have charge or Harry is not bald and electrons have charge' is a precise sentence equivalent to 'electrons have charge', and should therefore express a proposition. However, it is hard to see how to compute that proposition compositionally according to this account, given that neither disjunct expresses a proposition.

[5] Note that although some self-described linguistic theorists adopt the latter idea, it is stated in the adverbialist vocabulary so it is unclear whether the latter theory is open to a linguistic theorist.

ways with option (ii). The most natural view is a coarse-grained view in which a vague sentence expresses a bunch of precise propositions, but one could also maintain that vague sentences express a collection containing both precise and vague propositions, or even exclusively vague propositions. My purpose, in bringing this up, is to stress that the subsequent discussion relies only on the stance we have taken towards the possibility of ignorance in the vague, and on how coarsely or finely we individuate the objects of ignorance; the status of the relation between sentences and propositions will, for the most part, be absent from this discussion, and nothing I say turns on how we ultimately settle that question.

5.2 Explaining Ignorance about the Vague

IGNORANCE is an instance of one of a tightly knit cluster of phenomena that we associate closely with borderlineness. There are systematic connections between facts like IGNORANCE and borderline cases—facts like this just call out for some kind of general explanation. In our toy scenario, for example, not one of our international team of glass measurers was able to determine whether the glass was pretty full or not; this kind of thing is puzzling and seems to be just the kind of phenomenon a theory of vagueness is supposed to explain.

According to the theory I ultimately defend, that explanation takes the form of a general principle, which in turn falls out of a general theory of vague propositions:

EPISTEMIC: Necessarily, if it's borderline whether p, then it's not known whether p.

A linguistic theorist who has no paraphrase for EPISTEMIC in her own ideology cannot accept it as an explanation. If she is to offer a general explanation of IGNORANCE which still has something to do with vagueness she must instead seek to explain IGNORANCE in terms of her favoured vocabulary: S is borderline in L relative to context c and other parameters. But if the fact that some sentence, S, is borderline as used in language L, by community C, at context c, at time t, and world w explains why no one in the international team of glass measurers knows whether the glass is pretty full (i.e. explains IGNORANCE), we must ask:

(a) Which language L and linguistic community?
(b) Which sentence of this language is such that its borderlineness explains IGNORANCE?
(c) At what time must the sentence be borderline?
(d) What context must the sentence be borderline in?
(e) At what world must the sentence be borderline?

Before we move on let me stress that the demand is not just to explain why the English or Spanish or Mandarin speakers among our team don't know whether the glass is pretty full. The demand is to explain why *nobody* knows this, no matter how much they have inspected the glass, no matter what language they speak, or whether they even speak a language, no matter what the architecture of their brains and so on.

So let us begin with the first question, (a). Suppose, without loss of generality, that the sentence is simply the English sentence 'this glass is pretty full', uttered in a context where 'this' refers to the 70% full glass in question. Could the fact that this particular sentence is borderline in English (relative to the relevant context and other parameters) explain why *nobody* knows that the glass in question is pretty full? Could it, for example, explain why none of the monolingual Spanish speakers who inspected the glass know that the glass is pretty full? The answer to this question seems to be obviously 'no'—the linguistic practices of people in English-speaking countries can do nothing to prevent monolingual Spanish speakers from knowing whether the glass is pretty full.

The best we can do is explain *my* ignorance by appealing to the borderlineness of an English sentence, a monolingual Spanish speaker's ignorance by appealing to the borderlineness of a particular Spanish sentence, and so on. Although it is unclear to me whether even these explanations are possible, it is natural to object that even if they were, they would be incomplete—they say nothing of intelligent creatures that do not speak a public language. Would it be easier for me to find out whether the glass is pretty full if I didn't speak a language? And even if I do speak a language, why should my co-speakers' linguistic habits bear on whether I can find out whether the glass is pretty full?

More importantly, it is not clear that we have an explanation of the fact that *nobody* who tried to determine whether the glass was pretty full succeeded. What we have here is just a bunch of distinct and very local explanations of one-off facts. Mary doesn't know whether the glass is pretty full because the English use the sentence 'this glass is pretty full' in a certain way, whereas Lucía doesn't know whether the glass is pretty full because the Spanish use the sentence 'este vaso está bastante lleno' in a certain way. But, one might ask, what is the general reason that neither Mary nor Lucía know, or could come to know, whether the glass is pretty full? Also, if one didn't speak Spanish (or English) one might be puzzled by the explanation of Lucía's (or Mary's) ignorance—why is it the way the Spanish use 'este vaso está bastante lleno' that prevents Lucía from knowing, and not some other sentence? For the explanation to be explanatory it must also include a description of what these sentences mean; but what holds the sentences used in these explanations together cannot be that they all express a borderline proposition, for that is to concede these explanations to the adverbialist.

Perhaps the general fact is that both Mary and Lucía speak a language containing a sentence which both expresses the proposition that the glass is pretty full, and is moreover used in whatever way suffices to make a sentence borderline in that language. This explanation is adequate only on the assumption that every language under consideration has a sentence which expresses the proposition that the glass is pretty full, and which is used in that special way that makes the sentence bor-derline. But some of the languages in question might fail to have a sentence which expresses that proposition; maybe there is no perfect translation of 'this glass is pretty full' into German, for example, even if there might be something pretty close. But this should not make it any easier for a German to find out what we cannot in this situation.

These remarks cast some doubt on the possibility of answering our second question, (b). Even if we were just concentrating on explaining Mary's ignorance we'd still need to supply a borderline sentence. If I'm not around to point at the glass I would have to describe the glass in some way or other, and so there will be lots of different sentences to choose from. However, we can't explain Mary's ignorance in terms of the borderlineness of the sentence 'John is tall'—presumably, it must be a sentence which expresses the proposition that the glass is pretty full in *L*. As mentioned already, there might be languages in which no sentence expresses the proposition that the glass is pretty full. This would not make it easier for Mary to find out whether the glass was pretty full. Furthermore, what if, as some theories claim, vagueness *prevents* a sentence's expressing a unique proposition? Must the proposition that the glass is pretty full be merely *among* the propositions *S* expresses?

Finally, it seems we must specify a time, world, and context at which the sentence is borderline. Words often become more precise as language evolves. If the selected sentence *S* ('this glass is pretty full' say) had once been determinate it would presumably still be impossible to know whether the glass is pretty full. Conversely, if 'electrons have negative charge' had once been borderline that shouldn't prevent me from finding out that electrons have negative charge. Maybe *S* has to be borderline at the time I'm trying to find out whether the glass is pretty full. If we have to pick a different sentence each time we want to explain why someone can't figure out whether the glass is pretty full then the explanation loses its generality.

Furthermore, had we used *S* in a precise way we wouldn't be in a better epistemic situation regarding Harry's head. We may therefore be able to explain why no one *in fact* knows whether the glass is pretty full by appealing to a selected sentence's borderlineness, but in order to explain why we *couldn't* have known whether the glass is pretty full, even if *S* had been used in a precise way, we need also to relativize to worlds. This opens up the possibility that we can't know that the glass is pretty full because the sentence 'John is tall' is used in a vague way at another world to mean that the glass is pretty full.

Let me end by mentioning one more point which, although not a fully fleshed out objection, is something I find worrisome. In many ways explaining IGNORANCE is one of the easier jobs for a linguistic theorist. Other facts are harder to explain. Suppose that Harry is as before a borderline case of baldness, and that we rationally believe this. Then it seems that:

AGNOSTICISM: It would be irrational to believe (given what we know) that the glass is pretty full, and it would be irrational to believe that the glass isn't pretty full.

This fact is prima facie quite puzzling. After all, you know that either Harry is bald or he isn't, so at least one of the two beliefs above is true. Furthermore you have all the evidence you could possibly have available. It feels like there should be some general fact about vagueness-related phenomena that explains this.

Since there appear to be plenty of things we do not *know* which are rational to believe, it seems, therefore, that there is a separate and harder problem of explaining what feature of linguistically vague sentences prevents us from rationally believing certain propositions. This seems a lot harder to do—for example, in a discussion of Williamson's epistemicism, Horwich writes: 'the ignorance due to vagueness is attributed to a special form of unreliability—an external failure—and not, as it should be, to the internal difficulty in making a judgement' (Horwich [71]). But surely any explanation of our ignorance that appeals to the use of a term in a public language is going to be an external one.[6] Even if we bracket the problems surrounding the explanation of IGNORANCE, I am much less confident that an explanation of AGNOSTICISM in terms of language use can be given.

5.2.1 Explaining ignorance via metalinguistic safety principles

Let me now turn to a specific attempt to explain IGNORANCE within a linguistic theory. This attempt arises in the context of the epistemicist account of vagueness defended by Williamson [156]. Williamson's view, as I am characterizing it, is linguistic—vague sentences are *semantically plastic*: slight variations in the use of language will result in that sentence expressing a slightly different proposition. A sentence is borderline relative to a linguistic community C and parameters p iff the following holds: there are close worlds, with respect to how that language is used by C, where the sentence says something false relative to p, and similarly there are close worlds where it says something true. It earns the name 'epistemicism' as it allegedly *entails* the ignorance thesis. If this is true it is a significant benefit of the view over other linguistic theories. That said, even if one rejects Williamson's analysis of vagueness in terms of semantic plasticity, one might still think that his explanation of ignorance in terms of semantic plasticity is fundamentally sound by maintaining that vagueness and semantic plasticity, although not identical, come hand in hand. Thus Williamson's explanation of vagueness-related ignorance has interest that is independent of the success of his brand of epistemicism.

The crux of Williamson's explanation is a controversial principle that has come to be known as the 'metalinguistic safety principle'. In order to understand the principle we firstly need to introduce some definitions. Say that A's belief that P is *safe* iff it couldn't easily have been the case that (i) A believes that P and (ii) it is not the case that P. A's belief that P is *metalinguistically safe* iff it couldn't easily have been the case that (i) A produces the belief token that actually resulted in a belief that P and (ii) that belief token expresses a false proposition (let us say that a belief token 'expresses' a proposition P if it constitutes a belief that P). The expression 'it couldn't easily have

---

[6] Horwich's own view may be an exception: Horwich, unlike most linguistic theorists, takes the language in question to be an internal language of thought instead of a public language. This exception aside, the problem of explaining AGNOSTICISM seems harder for most linguistic theories.

been the case that P' is a term of art, and means something roughly like P isn't true in any nearby world, where what is 'nearby' is determined by some epistemically significant measure of similarity. It is unclear whether one can get an understanding of the relevant notion of similarity without already having a grip on the concept of knowledge, but however this question turns out, the notion of a safe belief is still of interest and can be used to put some important structural constraints on knowledge that would be hard to motivate without invoking the notion.

Although it is by no means uncontroversial, a sizeable number of philosophers take the notion of a safe belief to have some epistemic force. In particular, these philosophers subscribe to something like the following safety principle:[7]

SAFETY: One knows that P only if one does so via a belief that is safe.

Roughly, the thought is if you could easily have falsely believed that P, then even if you were in fact correct about P you were lucky to be correct, and so your belief could not constitute knowledge. It is important to contrast this principle with its metalinguistic variant:

METALINGUISTIC SAFETY: One knows that P only if one does so via a belief that is metalinguistically safe.[8]

The safety principle says that if you know that P then you couldn't easily have been wrong about P. The metalinguistic safety principle entails no such thing: being easily wrong about P is neither necessary nor sufficient for having a metalinguistically unsafe belief—all you need to do to be metalinguistically unsafe is to have a false belief in some proposition in a nearby world; although not necessarily a false belief that P, it need only belong to a belief-type whose tokens are beliefs that P in the actual world.

It should be clear that the ordinary safety principle cannot explain IGNORANCE. For example, in the present example, given my knowledge, the glass is 70% full in all worlds that count as nearby for me, and since whether the glass is pretty full supervenes on how full it is, it is either pretty full in all nearby worlds or not pretty full in all nearby worlds. Thus for all the safety principle says, a belief that the glass was pretty full could constitute knowledge provided it was in fact a true belief. Metalinguistic safety does better in this regard. One could make a reasonable

---

[7] The principle as stated below will probably need some refinement. For example, it is common to additionally stipulate that a belief that P is only safe if it couldn't easily have been the case that the agent falsely believed that P by the same *method* she actually came to believe that P.

[8] Let me mention a variant metalinguistic safety principle that I find more plausible, although I won't discuss it in detail since it involves accepting one of the views I have set aside for section 5.3. The variant says: an agent knows that P relative to a mode of presentation m only if she couldn't have easily had a false belief via the same mode of presentation. This principle can certainly be invoked to explain *something*, but it's unclear whether it explains why we cannot know whether the glass is pretty full or merely the fact that we cannot know whether the glass is pretty full via certain modes of presentation. The principle leaves it open, for example, that one could come to find out whether the glass is pretty full by accessing it via a mode of presentation that isn't semantically plastic.

case that the semantic properties of belief tokens are correlated to the semantic properties of corresponding public language sentence tokens. On that hypothesis, we can attribute a similar degree of semantic plasticity to vague beliefs corresponding to vague public language sentences, and we can also attribute to them the feature Williamson identifies with borderlineness: in some nearby worlds they express a true belief and in others they express a false belief. If I form a belief that the glass is pretty full by forming a belief that is borderline in this sense, then this belief is not metalinguistically safe: although I couldn't easily have been wrong about whether the glass is pretty full, I could easily have been wrong about another proposition—one that, in that same nearby world, would have been expressed by the same belief token.

Let me firstly note, along with many others, my reservations about the metalinguistic safety principle.[9] Suppose that, unbeknownst to me, my fellow English speakers had decided to start using the numeral '1' to mean 100. Grant also the assumption, needed in the explanation above, that the semantic properties of beliefs and their corresponding public language sentences are correlated. Since the actual world certainly counts as nearby, it appears as though there are nearby worlds where a belief token corresponding to '1+1=2' expresses the false proposition that 100+100=2; in other words my belief that 1+1=2 is metalinguistically unsafe. Yet, I hope, it should be clear that this form of unsafety does not in any way undermine my knowledge that 1+1=2, a fact which I can verify by performing a simple calculation.

Although the metalinguistic safety principle seems somewhat suspect, notice that the ordinary safety principle, which is on much firmer footing, does allow us to explain a related fact: that we cannot know whether the sentence 'this glass is pretty full' is true or not in English relative to the context and other parameters described. If that sentence is borderline in that context then it expresses a false proposition in a nearby world, and so, we are to suppose, the belief corresponding to this sentence is not metalinguistically safe.

Recall, however, that I took pains to distinguish IGNORANCE from LINGUISTIC IGNORANCE—the latter claim only states that it is impossible to know whether the sentence 'this glass is pretty full' is true in English relative to the relevant parameters. The fact that there are close worlds where we use 'this glass is pretty full' differently may explain, by *ordinary* safety principles, why we cannot know that this sentence is true in English. But whether the glass is pretty full or not is not unstable in the same way: it is just as full as it actually is at all the relevantly close worlds. So there is also the further fact that, given the status of the glass, we cannot know whether it is pretty full, and there is no obvious way to infer ignorance of the latter fact from ignorance of the former linguistic fact.

One way to bridge this gap would be to invoke knowledge of the disquotational schema. Suppose I know that if this glass is pretty full then the sentence 'this glass is

---

[9] For more criticisms of the metalinguistic safety principle, see Kearns and Magidor [77]. See also Caie [22], Hawthorne [68], Mahtani [99], and Sennet [130] for related discussion.

pretty full' is true in English, at the present context, etc. However, since we have just shown that I do not know the consequent on the basis of an orthodox safety principle, one can infer that I do not know that the glass is pretty full, assuming a small amount of closure. A parallel argument could be made to show that I do not know that the glass is not pretty full either. Of course, this argument still suffers from the defect that it is not completely general. To explain why a monolingual Spanish speaker does not know whether the glass is pretty full we'd appeal to knowledge of a principle that does not take the form of a disquotational principle when it is stated in English: if Lucía knows that this glass is pretty full then she knows that 'este vaso está bastante lleno' is a true sentence of Spanish in this context. The fact that neither I nor Lucía know whether the glass is full seems to call out for a general explanation—something holding both the cases together—which the explanation I have just given does not seem to provide. The explanation we gave relied on specific knowledge I had about English, and Lucía had about Spanish—knowledge relating the truth of sentences in these languages to a fact about the glass—which is not present in all cases where there is ignorance about whether the glass is pretty full.

Generality aside, there are some important limits to when knowledge of disquotational reasoning can be appealed to, even among people who speak the relevant languages. When words change their meanings, or we are no longer sure of a word's meaning, we are generally not in a position to know instances of disquotational principles involving those words. To demonstrate the general point, let us suppose that Alice is looking at Madagascar for the first time, and correctly concludes that it is an island. However, Alice is living during the time in which the word 'Madagascar' was being used to refer to a portion of mainland Somalia. So although Alice does in fact have a correct belief that Madagascar is an island, and in fact she knows it is an island, she also knows that the sentence 'Madagascar is an island' is not a true sentence of English. Similarly, cases where a word's meaning is unknown give rise to cases where the relevant disquotational principles are unknown. If Alice is observing a particular cow chewing grass, then it's plausible that she knows that the cow is masticating, although if she doesn't know that 'masticating' means chewing, she might not know that the sentence 'the cow is masticating' is true in English in her context.

The critical point here is that if the argument from semantic plasticity is to work, then there are nearby worlds in which the borderline sentence does not mean what it actually means. Assuming the ordinary safety principle, it follows that unless our beliefs about meanings are extraordinarily sensitive to very slight differences in meaning due to slight differences in use, we simply do not know what borderline sentences mean. Thus it follows that the above argument for IGNORANCE appeals to exactly the kind of knowledge of disquotational principles that we have seen to be suspect.[10] Indeed we can effectively prove that this instance of the disquotational

---

[10] Hawthorne [68], notes that there is a case to be made that the T-schema '"p" is true if and only if p' is true in all nearby worlds, due to penumbral connections between the word 'true' and the vague words

principle is unknown from SAFETY. That is, we can argue that although *the sentence* '"Harry is bald" is true if and only if Harry is bald' expresses a truth at all nearby worlds (even if it is a different truth at each world), the regular safety principle predicts that, were there nearby worlds where 'Harry is bald' is true even though Harry isn't bald (for example), we would not count as knowing that 'Harry is bald' is true iff Harry is bald. Let us suppose for the sake of argument that Harry is not bald (a symmetrical argument can be made if he is) and that he has the same number of hairs at every nearby world (let's suppose I know how many hairs he has). So there are no nearby worlds at which Harry is bald, since baldness supervenes on hair number. Yet by hypothesis the sentence is borderline, so there are nearby worlds where 'Harry is bald' expresses a truth and nearby worlds where it expresses a falsehood. So there are nearby worlds where 'Harry is bald' is true even though Harry is not bald, so by the safety principle it follows that I do not know that 'Harry is bald' is true if and only if Harry is bald.

The epistemicist might insist at this juncture that although there are nearby worlds where Harry is bald, even though 'Harry is bald' isn't true in English (because it means something else), our beliefs are surprisingly sensitive to the differences in meaning of 'Harry is bald' between these different worlds. Sensitive enough that we are able to have the belief that Harry is bald iff 'Harry is bald' is true in English only at the worlds where 'Harry is bald' does indeed mean something that's true iff Harry is bald. On its face this suggestion looks absurd: according to the picture described, the meaning of 'Harry is bald' in English depends on very particular features of usage, and can change on the basis of slight differences of use that are clearly far beyond the knowledge of ordinary humans. On appearances this sounds like the type of view where we cannot know what vague sentences mean—the best we can hope for is to know the range of meanings a word could have.

The epistemicist might try to defend the position by noting that at the nearby worlds where 'Harry is bald' means something else, so does the corresponding disquotational principle: '"Harry is bald" means that Harry is bald in English'. Indeed, whatever 'Harry is bald' means at this world—p, let's say—the instance of the disquotational principle will mean something true, namely that 'Harry is bald' means that p, and not the false claim that 'Harry is bald' means that Harry is bald. The thought, then, is that we get to know what 'Harry is bald' means simply by tokening some mentalese equivalent of the sentence '"Harry is bald" means that Harry is bald', which is guaranteed to express a truth whatever it means.

It is hard not to think there is something extremely fishy about this way of acquiring knowledge of something. Consider again Alice, who does not know what 'masticate'

---

appearing in 'p'. However, this is still not sufficient to guarantee that one knows that 'p' is true if and only if p, unless one has knowledge of the more complicated instance of the T-schema: '"p" is true if and only if p' is true if and only if 'p' is true if and only if p. But knowledge of this is suspect for the same reason that knowledge of the initial instance is suspect.

means, but can see, and therefore knows, that a particular cow is masticating, even though she has no idea whether the *sentence* 'the cow is masticating' is true in English. I take it that there is something obviously wrong about inferring from her non-linguistic knowledge about the cows eating behaviour, that a particular English sentence, whose meaning she is completely unsure about, is true. To conclude that 'the cow is masticating' is true in English in Alice's position seems just as obviously wrong even once we have pointed out that if she were to form a belief of the form 'if the cow is masticating then "the cow is masticating" is true in English' her belief would most likely express a truth, even if she does not know which truth it is.

5.3 Denying Ignorance about the Vague

The arguments in section 5.2.1 rested on a couple of assumptions. The first was a relatively fine-grained theory of propositions: one in which there are two kinds of proposition, the vague and the precise. The second assumption was that the former kind of proposition is distinguished from the latter by being a distinctive source of ignorance. On these assumptions we saw that linguistic accounts of vagueness had a hard time explaining this kind of ignorance.

However, one could imagine denying that vagueness involves ignorance by denying the first assumption: opting for a coarse-grained theory of propositions in which the proposition that Harry is bald is just identical to a precise proposition you know. One could also directly deny that vagueness involves ignorance, even in the context of a fine-grained theory of vague propositions. If vagueness is not a distinctive source of ignorance then there is no special phenomenon that a linguistic theorist is hard pushed to explain.

Here we will explore the idea that vagueness does not involve ignorance in the context of both fine-grained and coarse-grained theories of propositions.

5.3.1 The fine-grained no-ignorance view

Failures of principles like IGNORANCE should strike us as extremely surprising. To deny such principles would be to take seriously the idea that, for example, not only is there a nanosecond at which I stopped being a child, but that we in fact typically can, and often do, *know* which nanosecond this is! The view seems to be open to a simple refutation—as a matter of sociological fact, no one, not even those who reject the ignorance thesis, will ever go so far as to try to answer a question like 'what is the length of my childhood in nanoseconds?', even after I've filled them in with all the relevant details of my development. Yet their inability to answer these questions appears to be in tension with their claim to know how long my childhood is. It seems uncomfortable for someone to endorse this view, whilst at the same time being unable to answer these questions which, by their own lights, they know the answer to.

While claims like IGNORANCE are widely acknowledged, their acceptance isn't universal. Both David Barnett [12] and Cian Dorr [34] argue that when people are knowledgeable about the relevant precise facts, people often do have knowledge of borderline cases.[11],[12]

Why is it, then, that we don't say 'yes' or 'no' to borderline questions whose answers we know? Let's consider a specific example: Suppose that Harry is a borderline case of the predicate 'bald'. By the law of excluded middle, an assumption Barnett and Dorr accept, Harry is either bald or he isn't. Let us suppose for the sake of argument that Harry is bald; so by the no-ignorance view I know that Harry is bald, assuming I have the relevant precise facts to hand (hair number, distribution, and so on). Why is it, then, that I don't simply say 'yes' when someone asks me the question 'Is Harry bald?'?

There is an important choice to be made at this juncture regarding what one asserts when one utters a vague sentence. Let us begin by assuming a relatively fine-grained theory of propositions: one in which there is a vague proposition, the proposition that Harry is bald, distinct from the precise proposition that Harry has at most n hairs, for each choice of n. This view is naturally paired with the thesis that a vague sentence (uniquely) expresses a particular vague proposition. In particular:

The sentence 'Harry is bald' expresses a unique vague proposition, namely the proposition that Harry is bald. The belief that Harry is bald can thus be communicated uniquely by uttering the sentence 'Harry is bald'.

While Dorr's theory, to be discussed in section 5.3.2, is consistent with a coarse-grained theory of propositions, Barnett treats indeterminacy as an operator and is therefore committed to the existence of vague propositions. It is thus extremely natural to pair Barnett with the above thesis about the proposition expressed by the sentence 'Harry is bald'.

Given this picture the problem of explaining why we are reluctant to assert 'Harry is bald' or 'Harry isn't bald' is extremely pressing. If I know that Harry is bald, and I can easily communicate that knowledge by simply uttering the sentence 'Harry is bald', why don't I? In order to explain this, Barnett appeals to the principle:

One should aim to clearly satisfy the rule: (M) assert p only if p.

According to Barnett there will be cases where you know whether p, but you are not allowed to assert p (or $\neg p$) because it would result in its being unclear whether you've satisfied (M). That is, *even though you know you've satisfied (M)*, i.e. even though you

---

[11] Barnett [12] argues for the mere possibility of knowing borderline truths, although doesn't conclude that we actually know any borderline truths. In Barnett [10] the more radical view that there are *in fact* borderline truths that we should believe is defended, although he doesn't talk about whether we would know them. It would, however, be puzzling to think that we're not in a position to know what we should believe in these cases. In what follows I shall continue to talk about knowledge although I think a similar discussion could be run if I replaced 'knows' with 'ought to believe'.

[12] Crispin Wright [167] also denies principles like IGNORANCE, although his theory seems to require that one accept an intuitionistic logic. Jeremy Goodman [65] has also questioned whether vagueness in general precludes knowledge, although his arguments do not cast doubt on the majority of instances of the general principle, including IGNORANCE.

know you've asserted p only if p, you would not have *clearly* satisfied (M) and this in itself is supposed to be a bad thing. No explanation for this rule is given; it is just a primitive norm of assertion.

There's a natural analogy to be drawn here between Barnett's rule and the claim that one shouldn't assert things that are rude, or mean, or that would break a promise to keep something secret. Even when you know that p, it can bad to assert p if it would violate these principles. But even so, one might think these types of norms can sometimes be trumped: maybe by other norms of communication, or by moral norms, or by something else. For example, if some knowledge you possess could somehow save someone's life, then one should provide it even if it would mean breaking a promise to keep it secret.

Now, if we are to believe the kinds of claims made in Barnett [10] and Barnett [12], Barnett knows, or at least is in a position to know, which the last small number is.[13] Yet due to this primitive norm of assertion, he's not in normal circumstances permitted to tell us which it is. In the particular case at hand, however, I think the rule (M) *is* trumped by other considerations. If Barnett really does know which the last small number is, it would be of great value to the rest of the philosophical community if he were to tell us. After all, there has been a long-standing debate amongst classical and non-classical logicians about whether there even is one; if Barnett could settle the matter constructively, that would be significant progress. Therefore, given the great benefits, it seems like it would easily be worth the cost of flouting the rule and telling us which the last small number is. Yet he has not told us. A much more appealing explanation of Barnett's silence on this issue is not that he is afraid of violating the rule (M), but that he simply doesn't know which number the last small number is.[14]

5.3.2 The coarse-grained no-ignorance view

Unlike Barnett, Dorr's version of the No-Ignorance view is one in which all propositions are precise. Dorr tells a very different story about assertion:

When one asserts a vague sentence one asserts a large number of very similar precise propositions. When the sentence is borderline, some of those propositions are true and others false.

Thus, for example, when I utter the sentence 'Harry is bald' I assert, for each n in an admissible range, the proposition that Harry has at most n hairs. When Harry has a borderline number of hairs—suppose that he has exactly k hairs—then some of these asserted propositions are true and some of them are false: the proposition that he has at most $k + 1$ hairs is asserted and true, and the proposition that he has at most $k - 1$ hairs is asserted but is false.

---

[13] Providing, of course, that his cognitive conditions are reasonably ideal, and so on and so forth. See, for example, p. 28 of Barnett [12].

[14] Barnett might perhaps play down the value of knowledge to the philosophical community when it is not clear knowledge. Yet I for one would value knowledge of the last small number, clear or not, so there is certainly reason to break the norm (M) if only to satisfy my own idiosyncratic preferences.

What happened to the proposition that Harry is bald? Since all propositions are precise on this view, the proposition that Harry is bald is just identical to the proposition that Harry has at most N hairs, where N is the cutoff point for baldness. Thus when one asserts 'Harry is bald' one asserts the proposition that Harry is bald (i.e. the proposition that Harry has at most N hairs), along with further propositions.[15]

Dorr has a straightforward explanation for why we do not assert the sentence 'Harry is bald': if I were to assertively utter this sentence I would increase my audience's confidence in a number of false propositions. Thus when forced to answer the question 'Is Harry bald?' neither the answer 'yes' nor 'no' is appropriate: it is as though I have been asked a number of precise questions of the form 'Does Harry have less than n hairs?' for lots of different n, and been forced to either answer 'yes' to all of them at once, or 'no' to all of them at once.

This is not to say, however, that I can't communicate my knowledge that Harry is bald: we have only demonstrated that I can't communicate this knowledge using the sentence 'Harry is bald'. In fact, I *can* communicate my knowledge that Harry is bald: I can do this by uttering the sentence 'Harry has at most N hairs', since this sentence expresses the proposition that Harry has at most N hairs, and the proposition that Harry is bald is identical to the proposition that Harry has at most N hairs. It follows that the sort of *ad hominem* charge we levelled against Barnett cannot be extended to someone theorizing in terms of coarse-grained contents.

It should be evident that disquotational principles for dissent and assent cannot be a part of this view. Take for example:

DISSENT: Don't dissent to 'A' if you know that A.

By 'dissent' I mean to include a range of linguistic responses that include flat out denial as well as a principled and persistent refusal to assent. According to Dorr, one should dissent from 'Harry is bald' in some cases, even when one knows that Harry is bald.

Now of course the scope of DISSENT needs to be limited somewhat if it is to remain at all plausible. It doesn't apply if verbally dissenting to 'A' will have other bad consequences—you might, for example, wake the baby (if you dissent too loudly), or perhaps you're in a case where lying is morally justified, or 'A' has a presupposition that you want to avoid.[16] However, it seems like a prima facie cost if one cannot accept some suitably limited principle of this form—such principles seem to be integral to the way that we learn what people believe on the basis of their linguistic behaviour: if I look all puzzled and refuse to say 'yes' or 'no' when you ask me a question like 'is Harry bald?' English speakers typically conclude, mistakenly according to Dorr, that I don't know whether Harry is bald.

---

[15] Although note that the sentence 'the proposition that Harry is bald is identical to the proposition that Harry has at most N hairs' will be unassertable for the kinds of reasons 'Harry is bald' is unassertable, since the descriptions flanking each side of the identity do not determinately refer to a single proposition.

[16] I suspect one may also have to restrict DISSENT in light of liar-like paradoxes involving the notion of dissent.

Here is another worry. Supposing that I know the number of hairs on Harry's head, then even though I know that Harry is bald I won't assert the sentence 'Harry is bald', for that would be to assert (or at least, raise my audiences confidence in) a number of similar but false propositions. Surely, one might think, if I know that Harry is bald it should be simple for me to introspect on that fact, and come to know that I know that Harry is bald. Why then couldn't I just assert the sentence 'I know that Harry is bald' and allow my audience to conclude that Harry is bald from that? According to Dorr it is for exactly the same reasons I cannot assert the sentence 'Harry is bald'—the knowledge ascription is also borderline. Generally, when S is borderline 'I know that S' is also borderline provided I'm in possession of the relevant precise facts.

Note that there are two aspects of our linguistic behaviour that need explaining. One is our refusal to accept a sentence or its negation when it is a simple borderline sentence that does not involve attitude reports, such as 'Harry is bald'. Dorr's explanation seems to account for this adequately. It is not clear, however, whether Dorr can accommodate our linguistic behaviour regarding borderline sentences involving knowledge and belief ascriptions like 'Alice knows that Harry is bald'. Unlike the simple borderline sentence 'Harry is bald', where we are not inclined to assert it or its negation, we are inclined to outright deny things like 'Alice knows that Harry is bald' and to even assert its negation. Yet according to Dorr, this sentence is also borderline, and we should not be asserting either it or its negation.

Not only are utterances of 'Alice knows that Harry is bald' utterances of borderline sentences, according to Dorr, but one would expect these utterances to be untrue in cases where the knower is in possession of the relevant precise facts. It seems, then, that this theory must endorse some kind of error theory regarding attitude reports: speakers frequently assent to the false sentence 'Alice doesn't believe/know that Harry is bald'.

Things get worse. The sentence 'Alice believes that Harry is bald', let's suppose, is borderline. Now, according to Dorr, if I'm in possession of the relevant precise facts (perhaps the fact that Alice believes that Harry has exactly n hairs) then I know whether Alice believes that Harry is bald. Thus I am in an even worse position than someone who goes about assertively uttering false sentences: I am going about asserting 'Alice does not believe that Harry is bald' *when I in fact know* that Alice believes that Harry is bald.

5.3.3 Non-linguistic behaviour

Setting aside the troubles with attitude reports, an adequate theory ought to do more than just account for linguistic behaviour that is associated with vagueness. It seems to be extremely hard to eradicate vagueness-related uncertainty from explanations of our non-linguistic behaviour as well. It is therefore hard to see, even in principle, how a view in which vagueness is a public language phenomenon could accommodate this behaviour.

For example: I may know exactly how much cheese I have, but still be unsure about whether I have enough to make a cheese sandwich that tastes reasonably good—that, for example, doesn't have too high a ratio of bread to cheese. It is natural to attribute this uncertainty to the fact that I have a borderline amount of cheese. There is no precise fact that I am unsure of, but yet I still hesitate: it would be a waste of bread if I don't have enough cheese, but it would satisfy my need for food if I do. How is this to be explained if all ignorance is ignorance about the precise? You might try to attribute the hesitation to uncertainty about some precise fact about what the resulting sandwich *would* taste like. Let me clarify then: I know exactly what it would taste like, I just don't know whether that taste is tasty—with that cheese to bread ratio, the result would be borderline. The hesitation is not like the refusal to assert or deny a sentence—I needn't even speak a language to be in this situation. The problem is completely internal to me: I don't know whether the sandwich would be tasty, and so my actions will depend on how confident I am about its tastiness.[17] Phenomenal sorites show that these kinds of scenarios are ubiquitous—the states you are in when something looks red to you, when you feel cold or feel hungry, and so on, are all soritesable, and these states all seem to play an important role in guiding our non-linguistic behaviour in ways that couldn't be explained if we were always certain about whether we were cold or hungry.

Another example in which vagueness makes its way into our thought, although plausibly not via a public language, is when we acquire evidence through imperfect perceptual faculties. If I see a tree in the distance I learn some things about its height. It seems implausible that my total evidence, after seeing the tree, is that the tree is between x and y centimetres tall, or any precise proposition of this type—my credences will presumably fit some kind of smooth curve over the possible precise heights, but no credence that is gotten by conditioning on a precise proposition would have this smooth shape. It is natural to think that in this case my total evidence is vague; yet the visual experience and my resulting epistemic state had nothing to do with my ability to speak a language. Of course, there is much more to be said about this argument, and we'll return to that in chapter 6, but on the face of it, vagueness is pervasive in our non-linguistic mental lives.

5.3.4 The contextualist no-ignorance view

One of the main problems with the no-ignorance account described above is its treatment of attitude reports. Indeed, one can see this as an instance of a more general problem of trying to account for propositional attitude ascriptions in a theory where the objects of attitudes are reasonably coarse-grained. If the proposition that Harry is bald just is the proposition that Harry has less than n hairs, for some n, then the fact

---

[17] Whether it's rational to care intrinsically about whether the sandwich is tasty, as opposed to the particular tastes it could have, is another question; but at any rate, some people do care, and this fact about them explains what they do.

that we are disinclined to say that someone believes that Harry is bald even if we are inclined to say that they believe that Harry has less than n hairs might be subsumed by a more general theory for dealing with Frege puzzles.

Of course, one solution is to adopt a fine-grained account of the things appearing in the complements of attitude ascriptions (i.e. 'propositions', in my terminology). This would allow you to accept the conjunction of IGNORANCE and KNOWLEDGE, and falls under the fine-grained views we have already considered in earlier sections. If we want to maintain a coarse-grained account of propositions it seems as though we'd either have to bite the bullet and accept the problematic attitude reports, or adopt a contextualist view in which the word 'knows' expresses a different relation between people and propositions in different contexts.[18] On this view it is consistent to say that the expressions 'that Harry has less than n hairs' and 'that Harry is bald' denote the same proposition, p, but still maintain that utterances of 'Jane knows that Harry has less than n hairs' and 'Jane knows that Harry is bald' can have different truth values provided that they are made in different contexts. In one context 'knows' must express a relation that Jane bears to p, whilst in the other context it must express a relation which she doesn't bear to p.

Let me suggest one way to cash this idea out a bit further. When a person has a belief-like relation to some proposition p they typically do so by being in a particular kind of mental state. Naturally lots of different mental states can result in the same proposition being believed: I might mentally token something like the sentence 'Hesperus is Hesperus' in believing the necessary proposition, or I might mentally token something like the sentence 'Hesperus is Phosphorus'. Presumably it would be fairly easy to come to know the necessary proposition by believing it the first way, and not so easy to come to know it by believing it the second way. For convenience, call these different ways of believing a proposition 'modes of presentation'. Presumably before the discovery that Hesperus was Phosphorus, nobody knew the necessary proposition via the second mode of presentation, although everyone knew it via the first mode of presentation.

So far I have been using locutions like 'S knows/believes p via m'. Let us grant the empirical assumption that this is really what is going on psychologically when we believe things—that this expression picks out a natural three-place relation between propositions, people, and modes of presentation. How does this technical ternary relation relate to the binary ordinary language expressions 'knows' and 'believes' and so on? These are the terms that we have been theorizing with, and are the terms relevant to the puzzles we have been developing. According to contextualism, there

---

[18] See Crimmins and Perry [30] and Richard [120] for views of this kind in the context of attitude reports involving proper names. I do not know of a fully worked out defence of this view in the context of vagueness in print; however, both John Hawthorne and Tim Williamson have suggested this kind of view to me in personal communications, and it is certainly a natural way to avoid some of the puzzles I have been pushing in this chapter.

is no one relation that these expressions pick out: in contexts where a mode of presentation m is salient, 'knows' picks out the relation of knowing via m, and in other contexts knowing via m'. A context, then, can be thought of as providing a way of matching up coarse-grained propositions with modes of presentation.

It's natural to think that the that-clause we use when making an attitude ascription brings to salience certain modes of presentations and not others. Thus when I say 'Alice doesn't know whether Harry is bald' I bring to salience a vague mode of presentation corresponding to the vague sentence 'Harry is bald', and when I say 'Alice does know whether Harry has less than n hairs' I bring to salience a precise mode of presentation, even if, in both cases, I am just ascribing some relation between Alice and one and the same proposition. Note that the view in question is importantly different from the view which treats propositions, the denotations of that-clauses, to be ordered pairs of sets of worlds (or some other coarse-grained entity) and modes of presentations. There are affinities, but this view is a version of the fine-graining strategy: that view can straightforwardly make sense of the conjunction of IGNORANCE and KNOWLEDGE, and is therefore covered by the criticisms in section 5.3.3.

The contextualist cannot accept the conjunction of IGNORANCE and KNOWLEDGE—at least, not unless the context changes mid-sentence.[19] For whatever mode of presentation is salient, since we are relating the agent to one and the same propositions, it cannot be both known and not known relative to that mode of presentation.

At this point it is worth drawing the analogy between this type of view and a similar view in the philosophy of modality. For example, when railing against the third grade of modal involvement, Quine writes 'being necessarily or possibly thus and so is in general not a trait of the object concerned, but depends on the manner of referring to the object' [114]. According to Quine an object is only necessarily F relative to some linguistic 'mode of presentation' of that object: the number nine is necessarily composite relative to the mode of presentation 'the square of three' but not 'the number of planets'. It's natural for the contextualist to make an analogous move here concerning determinacy operators. Rather than taking the relation 'S is borderline in L relative to parameters $\bar{p}$' as primitive (as suggested in section 4.2), this theorist might theorize instead with a slightly more complex expression 'It's borderline whether p relative to the mode of presentation S in L and parameters $\bar{p}$'.

The view does not completely solve all the problems. It predicts, for example, the existence of a puzzling context in which it is okay to assert 'Alice knows whether Harry is bald'. Moreover, there will be no context in which it is okay to assert 'John knows that Jane's height is less than $x$cm but does not know whether she's tall': whatever

---

[19] One could just insist that the contextually salient mode of presentation changes to match the embedded sentence with which attitude ascription is made, but this view seems hard to distinguish from the fine-grained view where one takes propositions to be ordered pairs of sets of worlds and modes of presentations. One could try to pin the difference on some important line between pragmatics and semantics, but it seems unlikely to me that this would draw a substantive difference between the two views.

mapping from propositions to modes of presentation the context provides, if the proposition that Jane is less than $x$cm is identical to the proposition that she's tall, it will be assigned the same mode of presentation by the context and John must either know the proposition or not know it relative to that mode of presentation.

5.3.5 More on non-linguistic behaviour

The puzzles raised about the role of vague beliefs in guiding our non-linguistic behaviour in section 5.3.3 seem to be just as pertinent here. If it is modes of presentations that are vague, and these are vague in virtue of their relation to public language sentences (perhaps they are in some sense synonymous with vague public language sentences), it is hard to see how they play a role in behaviour that doesn't seem to involve language.

Moreover, most of our vague beliefs aren't articulable, either in a public language or even in a private language of thought. If I've been blindfolded and rolled down a hill, I have lots of beliefs about which direction is roughly up, which I acquire through some form of proprioception. I certainly have these beliefs, since they affect the actions that I make, but I couldn't articulate these beliefs in English, or even mentally. They are just beliefs that I have, they do not appear to be beliefs about any precise fact, and they do not appear to be linguistic. Such facts are puzzling for a view on which it is only linguistic items that are vague.

Consider the following scenario: Alice ran because she believed that Jack the Ripper was following her—not because she believed that Harmless Harry was following her, even though Jack the Ripper and Harmless Harry are the very same person. In order to explain why Alice acted in the way that she did we must appeal to a potentially hyperintensional distinction in her attitudes. The way we behave depends on what we believe and desire, and any decent theory of propositional attitudes ought to be able to accommodate explanations like this one. It is crucial to note that hyperintensional differences in a person's beliefs do not just generate differences in *linguistic* behaviour—the contextualist would not be hard pressed to accommodate differences in linguistic behaviour within her theory of modes of presentation—they also generate differences in non-linguistic behaviour as well. Alice's behavioural profile could be instantiated by someone who didn't speak any languages whatsoever.

Decision theory is an extremely general and powerful framework for representing a rational agent's decisions, beliefs, and desires. The intuition behind decision theory is very straightforward: to find out how good things are conditional on some supposition, A, first of all assume A, and calculate a weighted average of how good the remaining epistemic possibilities are, with each possibility weighted by its probability given A.

A standard way to formalize this intuitive idea, due to Richard Jeffrey [73], represents propositions by sets of indices. In Jeffrey's presentation the indices are in fact possible worlds, but nothing turns on this—they could be something more fine-grained like epistemically possible worlds. A *utility function* is a function mapping each index to a real value, representing how good that index represents things to be, and one's degrees of belief are presented by a probability function on the set of propositions (i.e. the sets of indices) satisfying the usual axioms of probability.[20] The *expected utility*, or *news value*, of a proposition is then given by summing (or integrating if necessary), over each index, the result of multiplying the utility of that index with the probability of that index conditional on that proposition. According to the orthodox interpretation of this formalism, the indices correspond to epistemic possibilities, sets of them to propositions, and the probability and utility function correspond to graded attitudes equivalent in nature to the ungraded attitudes of belief and desire. (For a more thorough presentation of decision theory, see chapter 9.)

Decision theory is well established. There should, I think, be a default obligation on those who wish to abandon it to provide some reasonable alternative. Now, a crucial feature of this framework is that it is a theory of rational action that operates directly on the *contents* of the agent's beliefs and desires. The different ways of having beliefs with those contents, the modes of presentation, play no role whatsoever in its formulation—perhaps such things are needed in the correct semantics of belief reports, but for orthodox decision theory they remain completely idle.[21] Conclusions that can be drawn about fineness of grain in this framework, then, cannot be explained away by standard contextualist manoeuvres invoking modes of presentation.

According to orthodox decision theory, then, whether an agent's action is rational or not supervenes on the contents of her beliefs and desires (construed broadly to include assignments of credence and utilities):

CONTENT TO ACTION: If Alice and Bob's beliefs and desires have the same contents, then they will act the same way if they are rational and have the same actions available to them.

CONTENT TO ACTION allows us to draw conclusions about how fine-grained contents are in a way that are impervious to the standard contextualist responses. For example, it entails that if two rational people can behave differently whilst having only beliefs and desires with necessarily equivalent contents then contents are more fine-grained than sets of worlds.

Let us demonstrate this strategy with an utterly trivial example. If we were willing to relativize all operators to guises, in the way the contextualist does to attitudinal operators, then it is a live option that there are only two propositions: the true and the false. However, even if we can account for attitude reports and other non-truth

---

[20] I should note here that many people feel that Jeffrey's specific theory delivers the wrong verdict on Newcomb's paradox. However, even these dissenters accept the features I have listed here—my discussion extends *mutatis mutandis* to these theories.

[21] Modes of presentation might contribute to individuating the indices *themselves*. This might occur, say, on a view in which propositions are identified with ordered pairs of sets of worlds and modes of presentation—but it is important to remember that this is a view in which propositions are not sets of worlds and is thus importantly different from the contextualist we are considering.

functional operators using the contextualist apparatus there is a distinct problem deriving from decision theory. Given CONTENT TO ACTION there are only sixteen types of people, even including probabilistically incoherent people, differing only in the combinations of beliefs and desires they hold towards the two propositions. It is clear, however, that there are more than sixteen different rational ways of behaving.

This is, of course, a very boring version of the argument; however, you might think a similar argument can also be applied to show that propositions are more fine-grained than sets of worlds too. The basic point here is that the argument from attitude reports and the argument from decision theory are two distinct problems that both need addressing.

One can see how these considerations might also be relevant to the case of vagueness: almost all of our beliefs and desires, involving 'medium sized dry goods', are vague. Although I concede that one could in principle have a perfectly precise belief or desire, perhaps about some aspect of logic or mathematics, we would not be equipped to deal with ordinary day-to-day life if all our beliefs and desires were like this. I will not pursue the application of these ideas to vagueness here, however, as I will resume discussion of these issues in chapter 10 which is devoted to this topic.

Returning to our initial line of thought, it should be clear, I think, that a contextualist possible worlds theorist can at least in principle accommodate our intuitions about the two sentences beginning with 'Alice ran because . . . ' by postulating a change in context. The problem is not that the contextualist cannot explain particular utterances that purport to explain actions in terms of beliefs and desires. The problem is the rather more theoretical one of integrating particular explanations of this sort into a general theory of rational action. To my knowledge, no theorist of this stripe has ever developed anything looking like a half decent decision theory involving modes of presentations that avoids CONTENT TO ACTION.[22] To do this, one would have to replace orthodox decision theory with a theory where the rationality of an action depends not only on the contents of your beliefs and desires, but on how you have them, or in other words, that depends on the modes of presentation under which you assign credences and utilities.

I think the technical obstacles to this project would be immense. Here are just few problems I can foresee, although I doubt these will exhaust the difficulties. Firstly, in order for credence to depend on modes of presentation one needs to relativize credences to modes of presentations. I can see what it might mean to have a credence of 0.6 in the necessary proposition via a mode of presentation corresponding to 'Hesperus is Phosphorus', but then I presumably won't have a credence of any defined value in the proposition that snow is white relative to this mode of presentation.

---

[22] Note that you can treat the objects of thought as coarse-grained proposition/modes of presentation pairs. This kind of theory has the ability to accept IGNORANCE and KNOWLEDGE simultaneously and is the kind of view discussed in the first half of this chapter. This section is aimed at theories that identify propositions (objects of thought/denotations of that-clauses) with the coarse-grained propositions.

This makes it unclear how to even formulate probabilistic axioms for mode-of-presentation-relative-credences so understood. Secondly, recall that the basic thesis of decision theory is that the expected value of a proposition p is the sum, weighted according to probabilities conditional on p, of the utilities of a partition of epistemic possibilities. On the present view people only assign probabilities and utilities to epistemic possibilities relative to modes of presentation, and there is no canonical way of pairing epistemic possibilities with modes of presentations of the possibilities. There is therefore no reason to expect that one can assign a unique expected utility to p relative to a mode of presentation for p, since one has a choice about how to pair the epistemic possibilities with modes of presentation. Even worse, I think, is that there is no guarantee that the probabilities defined relative to some pairing will sum to one, leaving it open whether the resulting notion of expected utility will conform to the ordinary constraints on rational preferences.

Until basic problems like these are solved, it's totally unclear how to go about formulating a decision theory that takes into account modes of presentation. The simplest way to avoid these troubles would be to treat propositions as ordered pairs of sets of worlds and modes of presentation, and to treat negation, disjunction, and so on as operations on both coordinates of these things simultaneously. This is, of course, just a version of the fine-graining strategy I have been urging for: a view in which the objects of our attitudes are more fine-grained than sets of worlds.

6

Vagueness and Evidence

In chapter 5, I argued against linguistic accounts of vagueness on the grounds that if it is true that we cannot know whether Harry is bald, then this is not because of the way the English sentence 'Harry is bald' happens to be used. Whether a belief that Harry is bald constitutes knowledge or not, I suggested, is just not sensitive to facts about your linguistic environment.

Whether or not vagueness is linguistic, there is a quite general puzzle as to how we acquire vague beliefs at all, and as to what role, if any, language plays in the acquisition of these beliefs. A natural, and reasonably common thought among linguistic theorists is the following: in order to have a belief that Harry is bald you need to be able to internally token a certain kind of sentence in your mental language; one that is synonymous with, or has the same conceptual role as the English sentence 'Harry is bald'.

Could a mentalese sentence have this kind of role without the thinker standing in some kind of important relation to a public language? If vagueness is a linguistic phenomenon, it is hard to see how this story could work unless the thinker did stand in such a relation. According to the analysis of vagueness in terms of semantic indecision, for example, the feature that distinguishes sentences like 'Harry is bald' from sentences like 'Harry has at most n hairs' is to be spelled out in terms of the different ways in which these sentences are used to coordinate beliefs between members of a linguistic community. For this talk of coordination to even make much sense, we must be talking about a community that contains at least two members.[1] Thus whether one has vague beliefs at all, according to this proposal, is dependent on one speaking a public language. Indeed this kind upshot is already partially anticipated by conclusions of Burge [21], that purportedly show that whether one has beliefs involving arthritis and other 'deferential' concepts can be sensitive to your linguistic environment. While these arguments have enjoyed a moderate amount of acceptance among philosophers, the present proposal is far more radical and far reaching. Even semantic externalists wouldn't argue that it is impossible to have beliefs about arthritis without speaking a public language. More importantly, pretty much all of our beliefs are vague: the exceptions seem to be beliefs about logic and mathematics.

---

[1] See the discussion of this point in Dorr [34].

An argument that purports to show that one cannot have vague beliefs unless one speaks a public language would threaten to show that we could have barely any beliefs at all unless we spoke a public language.

There are plenty of other things to say about this proposal. My purpose in this chapter is rather to highlight other mechanisms by which we acquire vague beliefs which the linguistic theorist cannot account for. I shall argue that most of our beliefs do not require any familiarity with a public language; most of our vague beliefs are acquired via our sensory faculties, vision, smell, proprioception, or other non-linguistic faculties such as memory. Sometimes public language sentences express vague propositions and these sentences can be used to communicate vague beliefs. But the class of vague propositions which satisfy the evidential profiles outlined vastly outstrips the class of propositions expressed in any given language and, moreover, the way in which language features in our acquisition of vague beliefs acquired this way is minimal.

In section 6.1, I introduce the idea of one's evidence being 'inexact' and argue that there are important connections between these cases and cases in which one's evidence consists of vague propositions. More specifically, I argue that updating on vague evidence is a particular instance of a more general account of updating on inexact evidence: Jeffrey updating relative to a partition—in this case relative to a set of 'precisifications'. This observation leads to a couple of insights into the evidential relation between vague propositions and precise propositions that will form the basis for the account of vagueness I am advancing in this book. One of these, which I will develop in chapter 8, is that one's beliefs in the vague propositions are completely fixed (in a sense to be spelled out) by one's beliefs in the precise. Thus one's beliefs in a particular vague proposition can be uniquely determined by one's beliefs in the precise and a particular evidential relationship to the precise propositions. The way in which credences in a vague proposition depend on credences in the precise propositions will be called the *evidential role* of the vague proposition. The second insight, which I defend in this chapter, is a **Principle of Plenitude**: that for any possible evidential role in thought, there is some vague proposition which has that role.

6.1 Inexact Evidence

Suppose you are looking out of your window at a tree in the distance. In doing so you obtain some knowledge about the tree. However, your knowledge is not exact; for example, while you now know that the tree is larger than 10cm and less than 1000cm, based on what you can see the exact height of the tree in cm is still unknown to you.

In [155] Timothy Williamson introduced the term 'inexact knowledge' for the kind of epistemic state you would be in in situations like the one above. A parallel distinction arises with respect to your evidence in such situations. Given that you have only seen the tree from a distance, and your eyesight is not perfect, your evidence includes some propositions—that there is a tree, that it's larger than 10cm and smaller than 1000cm—but does not include the proposition stating the tree's exact height. We may call your evidence in such cases inexact evidence.

Being inexact is not the same as being inaccurate; if the tree appeared to be less than 500cm when it was in fact greater than 600cm, then your evidence, or apparent evidence, would be inaccurate, but in the case described in the opening paragraph your beliefs are not inaccurate. All of your evidence, or apparent evidence, is accurate. On the other hand, if you went out and measured the tree, the relevant source of inexactness in your evidence about the tree's height would be eliminated. But the phenomenon at hand is not ignorance or lack of evidence about the tree's exact height. If you had never seen the tree, but had been informed by a highly reliable person that the tree was less than 600cm, then your evidence would be exact, even though you still would not know the height of the tree. Nor is the relevant sense of exactness specificity: if you have been told that the tree is between 400cm and 500cm then your evidence is more specific than it would have been if you had only been told that the tree was between 300cm and 600cm, but the latter case is not closer in kind to the state of evidence you would find in the opening example.

I hope it is clear that the evidential status of someone who has seen a tree from a distance is relevantly different from any of the preceding examples. While I cannot give an uncontroversial definition of what must be involved in cases where one's evidence is inexact, I hope the preceding examples have elucidated it sufficiently to provide a good enough handle on the kind of situation I am interested in to be worth discussing further.

Evidence obtained via imperfect sensory faculties are paradigm examples of inexact evidence. However, it is not clear that it is always the case that when your evidence is inexact, some sensory faculty is at fault. Having taken a cursory look at my bookcase, I gain inexact evidence about the number of books I own; I have a general feel for how many books there are but I do not know exactly how many books there are without counting them. However, it is not completely obvious that my evidence would be significantly better if I had perfect vision. It seems more natural to say that it was not my visual experience, but rather my ability to process it, which was to blame.

I take it that most ways of obtaining evidence leave room for the possibility that evidence obtained in that way is inexact.[2] I also take it that most of our evidence *is* inexact. The main thesis of this chapter is the claim that inexact evidence is vague evidence; that when we find ourselves with inexact evidence the strongest proposition we ought to be certain of on the basis of this evidence is a vague proposition. This partly vindicates the claim that we do not need to be acquainted with a public language to have vague beliefs. Furthermore, since almost all of our evidence is inexact, it follows that vague propositions occupy a very distinctive evidential role in thought. They are usually the best pieces of information we have when we learn from

---

[2] One might argue that certain *a priori* methods, such as mathematical proof, never leave one with inexact evidence. I would dispute this, but we can set this case aside for the purposes of this chapter.

experience, they are what we usually reason from and to, and they are the objects of our desires upon which we act. If the evidential role of a proposition is partly constitutive of what it is to be that proposition, then our thesis also gives us some insight into what vagueness and vague propositions are.

The claim that inexact evidence is vague evidence requires some refining. If I have examined a man's head carefully and have determined that he has no hairs at all, then I have evidence that he is bald. In such a scenario my evidence about the man's hair number is exact, yet my evidence includes the vague proposition that the man is bald. Having evidence for a vague proposition is thus not sufficient for being in the kind of circumstance characteristic of inexact evidence. However, the proposition that the man is bald is not the strongest proposition I have evidence for—the strongest proposition is the proposition that the man has no hairs. While this proposition entails that the man is bald, it is not equivalent to it: a man with one short hair is bald, so one can be bald without having no hairs at all. The claim I am interested in is rather the claim:

VAGUE EVIDENCE: When your evidence about a subject matter is inexact, your total evidence about that subject matter is a vague proposition.[3]

It is important to note that I am operating here with the distinction between vagueness and precision, which must be sharply distinguished from the distinction between being borderline and determinate. The proposition that Patrick Stewart is bald is a vague proposition even though it is in fact determinately true. It is especially important in this context not to conflate borderlineness with vagueness: it is quite natural, given a factive conception of evidence, to think that evidence is never borderline, even if one's evidence is usually vague.

The basic argument for this claim (very roughly) is that when our evidence about, say, the height of a tree is inexact, the distribution of probabilities over possible tree heights forms a smooth curve. On the other hand, it is not possible to achieve this smooth curve by conditioning on a precise proposition that's about the tree (and not about, say, our experiences); therefore our total evidence cannot consist only of precise propositions. The primary goal of section 6.2 will be to finesse this argument.

Before I move on, however, it should be noted that there are a couple of questions that I shall not remain neutral on and that require some further remarks. The first of these is the thesis of probabilism: the view that ways of measuring the degree to which your evidence supports hypotheses—the credences one epistemically ought to have—are governed by the probability axioms. This is controversial in the present context

---

[3] I am operating under the assumption that *there is* such a thing as your total evidence: the conjunction of all propositions that are part of your evidence. Thus, I am assuming evidence is closed under conjunctions. It is possible that this assumption can be dispensed with in most of what follows: the conjunction of your evidence is well-defined, even if that conjunction is not itself part of your evidence. We may then instead consider the thesis that this conjunction is vague when your evidence is inexact.

because some philosophers (see, especially, Field [53] and Schiffer [126]) have argued that to include vague propositions within the remit of such theories would require relaxing the ordinary probabilistic axioms. Although I'll be assuming probabilism in this chapter, I shall return to that issue and give it a proper defence in chapter 7.

The second point requires a little more discussion. One might object that in the cases described my evidence is not propositional at all, but rather consists of an imprecise visual experience, a hazy memory, or some other non-propositional object. The thought that there is a proposition that summarizes everything that is learnt is what Jeffrey calls the 'empiricist myth of the sensuously given data proposition' ([74], p. 3). It is, in fact, possible to make VAGUE EVIDENCE nominally consistent with this view by replacing talk of evidence with talk of what your evidence fully supports: even if your evidence consists in visual experiences, memories, or what have you, what that evidence supports, presumably, is always a proposition. Thus VAGUE EVIDENCE can be thought of as just the claim that when your evidence is inexact the strongest proposition fully supported by your evidence is vague.

Although this formulation is neutral concerning whether evidence is propositional, there is a more substantial disagreement around the corner: many philosophers who reject propositional evidence also maintain that very little is maximally supported by your evidence, and is therefore not in a position to accept even the revised formulation (see, for example, Jeffrey [73]).[4] This is related to a second but distinct objection often levelled at views in which evidence is propositional. According to these views any proposition entailed by your evidence will be fully supported by your evidence and will be such that one ought to assign it maximal credence. This has surprising consequences about betting behaviour, assuming a standard connection between credences and betting: there will be contingent propositions which you should bet your entire life on for a penny. VAGUE EVIDENCE clearly does not speak to this objection, and it would take us too far afield to respond to it here. However, it is important to be clear about this odd consequence of accepting the view that evidence is propositional, and that it is separate from the other reasons that motivate one to deny that evidence is propositional.

At any rate, VAGUE EVIDENCE is supposed to be understood as an answer to Jeffrey's first challenge to produce a proposition that summarizes all and only the things learnt after obtaining inexact evidence. It is, however, important to separate this challenge from the more demanding one of producing a *sentence* that summarizes all and only the facts learnt. It is rarely ever possible to articulate what you've learnt. Belief, on one natural picture, is a process by which we rule out certain possibilities, and this ruling out process is characterized by its connection with rational action: it might not be a conscious thought, or the explicit articulation of a sentence in a mental language.

---

[4] Note, however, that the revised formulation of VAGUE EVIDENCE is compatible with a Jeffrey-style picture about how our credences in the precise are distributed: it could be that one can become fully confident in a vague proposition whilst remaining uncertain about almost all precise matters.

It might be that a belief that P coincides with the tokening of some sentence-like entity in some animals, such as humans; but the case of belief in animals surely demonstrates that the requirement that we be able to articulate our beliefs in some language (mental or otherwise) is too strong.[5]

There are a great many questions about the nature of evidence that I have left open: How exactly do we obtain evidence? What kinds of propositions are part of our evidence? and so on. For the most part, my discussion can remain neutral on these questions. However, to make the discussion more concrete it will help to have some particular answers to these questions on the table. Following Williamson [160], there are a number of natural propositional attitudes that seem to always confer the status of evidence to things they are held toward. Here are a few:

1. S saw that p.
2. S remembered that p.
3. S could hear that p.
4. S could feel that p.
5. S could see that p.

These are all paradigm cases of what I shall call 'evidential attitudes': attitudes one cannot have towards p without thereby having evidence that p.[6] In cases 3, 4, and 5 the auxiliary 'could' usually marks that the verb is being understood perceptually. Furthermore, evidential attitudes are characteristically factive, whereas 'heard that' and 'felt that', without the modal, are not factive or evidential. 'Jane heard that Hector is getting fired, although she didn't have any reason to believe that he was getting fired' seems to be a fine thing to say if, for example, Jane's source was known to be completely unreliable, whereas 'Jane could hear that Hector was getting fired, although she didn't have any reason to believe that he was getting fired' seems to be harder to maintain. The latter sentence implies that the actual firing is directly audible to Jane, whereas the former sentence implies that she has heard second-hand that Hector is being fired, but hearing in this way may or may not have any evidential import. On the other hand, both 1 and 5 are factive and evidential. The auxiliary in 5 can usually be inserted to force the reading where the evidence at hand is perceptual. This is not always the case, however, as seen by the sentences 'Hector saw that the theorem was true' and 'Hector could see that the theorem was true'—both seem to be okay things to say, yet in neither case is the kind of seeing a perceptual one. The fact remains, however, that whether perceptual or not, one cannot see that p without p becoming part of your evidence.

---

[5] Some authors conflate these challenges—Richard Bradley, for example, makes this stronger demand in [17]. One only needs to meet the weaker demand to respond to Jeffrey's objection.

[6] Here I borrow heavily from the discussion of 'factive mental state operators' in Williamson's [160]. Most of what I say in what follows is compatible with Williamson's view that evidence is just knowledge; however, it is also compatible with the weaker theory that evidence is just the disjunction of a more limited class of non-inferential factive mental state operators. For reasons to prefer this view over Williamson's, see Bacon [6].

It should be pointed out that one does not need to be particularly linguistically competent to have any of these attitudes towards a proposition. One could hear (see, remember, feel) that *p* without speaking any particular public language or standing in any relation to a representation—one only needs the relevant auditory (visual, recollection, sensory) capacities. It is perhaps slightly more controversial to say that one does not need a *private* language, or to stand in any kind of relation to a mental representation, in order to hear (see, remember, feel) that *p*. Although I would be willing to make this further claim, my discussion will not rely on this assumption.

We do not always have neat ways of expressing evidential attitudes in natural language. We generally have lots of propositional evidence about whether we are cold or thirsty, where our limbs are, whether we are moving, which way is up, and so on and so forth, which we have by our capacity for proprioception. In English, at least, we do not have any simple verbs for describing these states.

It should be clear that most evidence obtained in the ways detailed above is inexact. In the clearest cases the evidence is inexact because the subject's eyesight is poor, or she only caught a short glimpse of something, or it was in the periphery of her visual field, or her memories had faded, and so on. But even when our sensory organs are all in good condition our evidence will be inexact. The amount of computing power we would need to calculate the exact path a tennis ball would follow on the information that it has been hit a certain way is typically too high for most humans (even more so when the examples become more complicated, such as spilling a bag of rice). However, we are generally able to do rough-and-ready estimations, without any conscious calculation, which give us an approximate idea of where the ball will land. My best evidence is not that the ball will land exactly at location *l* or that it will land within the circular region *r* of diameter 3.8m or whatever; it will rather be inexact in the way characteristic of the examples we have been discussing. Most of our conscious and unconscious decisions are based on inexact information of this nature, and understanding this information is crucial if we are to apply these epistemological considerations beyond the contrived decision problems, typical in formal epistemology, to commonplace decisions such as whether to hit a tennis ball when it looks like it might be going out.

Let's now focus on a particular example. Suppose that, after seeing Harry for the first time, I gain some evidence about how much hair he has. I can see that he has *some* hair, but, as usual, my evidence is inexact. I claim that

For some, but not all *n*, I can see that Harry has less than *n* hairs.

As with all the evidential attitudes we have discussed, 'sees that' is a propositional attitude: syntactically *p* in '*S* sees that *p*' can be substituted for any grammatical sentence, and semantically we may view 'sees' as expressing a relation between a subject and a proposition. One slightly distracting feature of the above claim is that these perceptual reports often carry certain conversational implicatures: if I say that I can see that Harry has at most a million hairs it suggests that it's not the case that I can see that he has at most one. It is important, however, to resist the temptation to think that this presupposition is an entailment.

We may thus ask what the *strongest* proposition I stand in the seeing relation to is in this situation.[7] One may similarly ask what the *strongest* proposition that my evidence fully supports is in this situation. Assuming, for simplicity, that the only evidence I have in this case is my perceptual evidence we should expect the answers to both these questions to be the same.

I shall consider two answers to this question:

1. For some n, the strongest proposition about Harry's head that my evidence fully supports is the proposition that Harry has at most n hairs.
2. The strongest proposition about Harry's head that my evidence fully supports is a vague proposition about the number of hairs Harry has.

Before we tackle this question, a few clarifications are in order. By the strongest proposition about Harry's head that my evidence supports, I mean a proposition about Harry's head which my evidence fully supports and which entails all other propositions about Harry's head that my evidence fully supports. Assuming that one's evidence is closed under logical consequence, one can show that there always is a strongest proposition that is part of one's evidence: the conjunction of all the propositions supported by your evidence. Similarly (assuming that the propositions about Harry's head are closed under conjunctions), there will always be a strongest proposition about Harry's head that my evidence supports.

The notion of entailment between propositions cannot be taken to be strict implication if we are to include vague propositions in this algebra. I shall assume, as per usual, that whether someone is bald or not supervenes on the precise facts about hair number, distribution, and colour. To spare myself a few words, I shall simplify things further by assuming that whether someone is bald only depends on hair number. Under our simplification, it follows that for some n, it is necessary that Harry is bald if and only if Harry has less than n hairs—although it is vague for which n this holds. This notion of entailment is useless for epistemic matters such as evaluating your evidence, for it is exactly these necessities we cannot know because of vagueness. We shall need to avail ourselves of a broader notion of entailment, to be spelled out further in chapter 11, in which the claim that p entails q implies not only that p strictly implies q, but that it determinately implies q, and moreover, that the result of prefixing any string of Δs and $\Box$s to the material conditional $p \rightarrow q$ is true.

It seems clear that the propositions stating that Harry has at most n hairs, where n ranges over numbers, are linearly ordered by entailment: that Harry has at most n hairs entails that he has at most m hairs whenever $m \geq n$. Where does the proposition that

---

[7] In general, it is odd to say that we see, if we are rational, every logical consequence of what we see. But at least in this case it is natural to think that I've seen most of the relevant propositions about Harry's hair number that are logical consequences of the propositions I've seen.

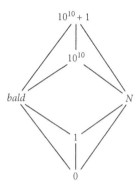

Figure 6.1. The proposition that Harry is bald (i) is entailed by the proposition that Harry has at most 1 hair, (ii) entails the proposition that Harry has at most 10^{10} hairs, and (iii) neither entails nor is entailed by the proposition that Harry has at most N hairs.

Harry is bald fall in this ordering? The answer, of course, is that it doesn't. It appears below (i.e. it entails) some of the precise propositions, including the proposition that Harry has at most 10^{10} hairs, and appears above (is entailed by) others, including the proposition that Harry has at most 0 hairs (Figure 6.1). But if it's borderline whether someone with N hairs is bald, and we know that Harry has exactly N hairs we cannot know whether Harry is bald. It's not determinately false (and thus epistemically possible) that Harry is not bald and has N hairs, so the proposition that Harry has at most N hairs does not entail the proposition that Harry is bald. Conversely, if I know Harry has $N + 1$ hairs I don't know whether he's bald for that would also be to know something borderline. So it would be not determinately false (and hence epistemically possible) that Harry is bald and has $N + 1$ hairs, establishing that the proposition that Harry is bald does not entail that Harry has at most N hairs.

Despite the fact that certain vague propositions are independent of the precise propositions entailment-wise, they are not probabilistically independent of one another. Suppose that the probabilities representing the degree to which my evidence supports various hypotheses about hair number are initially uniformly distributed over the possible numbers of hair Harry might have.[8]

What are the probabilities of these hair numbers given that Harry is bald? The precise propositions entailed by the proposition that Harry is bald now have probability 1 conditional on his baldness. But what about the precise propositions that neither entail nor are entailed by the proposition that Harry is bald? Deferring a rigorous justification for later, let us just present an intuitive picture. It is natural to think that

---

[8] Technically there are infinitely many possible hair numbers Harry could have. So we don't have to worry about infinities I shall just stipulate that we know that he has less than a billion hairs, but our credences are otherwise uniformly distributed over the remaining possibilities. This is an idealization—presumably the probabilities will trail off slowly—however, the substance of my discussion won't be affected by removing this idealization.

the probabilities of various hypotheses about hair number conditional on Harry's baldness will drop smoothly: for each n, if the probability that Harry has exactly n hairs is x then the probability that Harry has exactly $n + 1$ hairs will be just below (or the same as x once it levels out). Thus if N is the smallest number such that the proposition that Harry is bald entails that Harry has at most N hairs then, although the proposition that Harry has at most $N - 1$ hairs neither entails nor is entailed by the proposition that Harry is bald, it is probabilistically supported by it, in the sense that its relatively low initial probability will increase so that it is almost 1 on the supposition that Harry is bald. If I learn that Harry is bald, the probability in some of the propositions about hair number not entailed by the proposition that Harry is bald will increase, and the probability of others will decrease.

Formally, P is probabilistically independent of Q for S if $Pr(P \mid Q \wedge E) = Pr(P \mid E)$ where Pr is a rational ur-prior[9] and $Pr(\cdot \mid E)$ represents the degree to which something is supported by S's total evidence E—her credences if she is rational. Moreover, Q provides evidential support for P relative to background evidence E iff $Pr(P \mid Q \wedge E) > Pr(P \mid E)$. The foregoing demonstrates that the vague and precise propositions are not probabilistically independent for a rational agent with my evidence, and that being bald provides evidential support for certain hypotheses about hair number. In my view, this kind of probabilistic dependence is no accident: if you were to take any rational ur-prior and condition it on a maximally strong consistent precise proposition, there is a particular credence it should assign to the proposition that Harry is bald—this will be a small number if you condition on a proposition that entails that Harry has a large number of hairs, and a higher number if you pick maximally strong consistent precise propositions entailing he has lower numbers of hairs. To adopt a prior probability function that violates this constraint is to exhibit a kind of conceptual incoherence akin to believing that there are married bachelors or, perhaps more pertinently, bald people with millions of hairs. This fact explains why your credences about Harry's hair number and baldness are not probabilistically independent; if the only evidence you have is that described above, every rational ur-prior should agree that on that evidence hypotheses about baldness provide certain levels of evidential support for hypotheses about hair number and vice versa. This can be contrasted with a certain kind of Bayesian permissivism in which all probability functions represent a possible coherent assignment of prior degrees of belief. For this Bayesian there will be rational ur-priors that violate the probabilistic dependence constraints between the vague and the precise. So the thesis I have suggested is inconsistent with this kind of permissivism.

If one were to begin with evenly distributed credences and were to update on the proposition that Harry has less than n hairs, one would have a sharp probability function: $Cr(Hk) = \frac{1}{n}$ for $k \leq n$ and $Cr(Hk) = 0$ otherwise, where Hk is the proposition

---

[9] A credence it would be rational to have if you had no evidence at all.

Figure 6.2. A smooth curve and a sharp curve of n against credence that Harry has n hairs.

that Harry has k hairs. It would look something like the probability function depicted on the right in Figure 6.2, rather than the smooth one on the left. The smooth curve is intuitively what you'd expect to be the correct one, and, I would conjecture, is closer to the curve that corresponds to the credences people usually adopt after having learnt from inexact experience.

6.2 Updating on Vague Evidence

Let us now canvass some of the alternatives to VAGUE EVIDENCE. If our total evidence is not a vague proposition then either it is propositional and consists in a precise proposition, or it consists in a non-propositional piece of evidence such as an experience. Views in which evidence consists in a proposition are most naturally paired with the view that one should conditionalize your credences on your evidence: your credence in P after learning E should be the proportion of E that P takes up according to your credences prior to learning E. Your posterior credences can thus be calculated by the ration formula: $Cr(E \wedge P)/Cr(E)$, where Cr represents your prior credences. One can conditionalize on a proposition, but not on an experience: the latter view is therefore usually paired with another method of updating credences known as Jeffrey conditioning.

6.2.1 Conditioning on a precise proposition

Suppose that when I look at a tree in the distance I come to learn a precise proposition. What precise proposition do I learn? Two answers from the epistemological literature naturally present themselves. According to a more externalist conception of evidence, I simply learn something about the height of the tree. This view is in line with the picture outlined above that seeing, hearing, remembering, and so on, that p, all result in one's having evidence that p. According to a more internalist conception of evidence, I don't learn anything about the height of the tree (even though I have seen that it is above a certain height). The most I learn is something about my own experiences which makes various hypotheses about the height of the tree more or less probable.

The latter kind of view relies heavily on the agent having certain kinds of ur-priors. If prior to receiving any evidence, I was fairly confident that I was a brain in a vat, learning that I've had the experience e of seeing hands will not provide strong support for the hypothesis that I have hands. But it should be noted that even if my priors favour non-sceptical scenarios, I still can't rule out *a priori* that I'm a brain in a vat.

So even after having the experience e of seeing hands in front of me, and conditioning on the proposition that I've had this experience, I will still reserve a small amount of credence for the hypothesis that I'm a brain in a vat having this experience with no hands in front of me.

Unfortunately, this moderate kind of scepticism seems to undermine itself. If one should reserve some credence for the hypothesis that one does not have hands, even in the good case where you in fact can see that you have hands, it is hard to see why one shouldn't also reserve some credence for the hypothesis that you are not really experiencing e either. After all, you can be ignorant about what experiences you are having in some cases, such as when you are having one of several similar but indiscriminable experiences (see Williamson [155]), and it is not too much of a stretch to think that in other cases one can be completely mistaken about what experiences one is having. However, the view under consideration is one in which having the experience e confers certainty that you've had the experience e—your credences after experiencing e are the result of *conditioning* your prior credences on the proposition that you have experienced e—ruling out any kind of uncertainty about what experiences you've had.

A more pressing concern in the present context is that it is not obvious that the proposition that I've experienced e is a precise proposition. After all, one can easily imagine a situation in which one has clearly had the visual experience e of a red ball. But by slowly varying certain parameters—perhaps by considering agents with varying degrees of colour blindness, or by varying the apparent colour of the experienced ball—one can create a sorites sequence ending in a situation in which one has clearly not had experience e; in the cases towards the middle it will be, presumably, borderline whether one has had experience e.

Let us turn to the externalist view that one directly learns something about the height of the tree by looking at it, and that what one learns is precise. Which propositions, then, are both (i) precise and (ii) about the height of the tree? The maximally strong consistent propositions satisfying (i) and (ii) are propositions that state, for each choice of n, that the tree is exactly $n$cm tall, and an arbitrary proposition that's both precise and about the tree will be an arbitrary disjunction of these maximally specific propositions. A little reflection reveals that the only propositions like this that are in any way plausible candidates for my total evidence are propositions of the form: the tree is between $x$cm and $y$cm tall. For the sake of concreteness, suppose that I learn that the tree is between 300cm and 500cm. For simplicity suppose also that the tree can only have integer heights in cm and that your priors about the tree's height are approximately uniformly distributed over the heights less than 1000cm.[10] (Our conclusions will carry through with weaker assumptions, but for now we are just trying to get some basic features of the view on the table.)

---

[10] It is easy enough to dispense with this assumption and assume a continuous distribution. However, these technicalities are not really relevant to the present point.

Figure 6.3. The result of conditioning uniform priors on the proposition that the tree is between 300cm and 500cm high.

The result of conditioning on the proposition that the tree is between 300cm and 500cm tall can be plotted as a graph of our posterior credence in the possible heights against the possible heights. Plotted this way, the graph will represent a kind of rectangle: the credence I assign to the tree having a height less than 300cm or greater than 500cm is 0, and for each height between 300 and 500 I assign the same credence to the tree having that height (namely, 1/200). See Figure 6.3.

This sharp curve does not seem to be the credences we in fact have, or indeed ought to have, after seeing a tree in the distance. Assuming the tree is in fact about 400cm I should have comparatively high confidence that the tree is exactly 400cm, less confidence that the tree is exactly 350cm or 450cm, with my credence dropping off smoothly to 0 either side. See Figure 6.4 for a depiction of the intuitive curve.

We can dramatize this intuition further by looking at the kind of betting behaviour these distributions would permit. Consider three bets on the proposition that the tree is exactly 299cm, 300cm, and 301cm. If my evidence is the proposition that the tree is between 300cm and 500cm, then I treat the pair of bets on 299cm vs 300cm very differently than I treat the pair 300cm vs 301cm. I should be completely indifferent between the latter two bets, willing to sell and buy either at exactly the same odds, but with the former pair I should reject at any odds the proposition that the tree is 299cm, whilst accepting bets that the tree is 300cm. Our evidence in this scenario is inexact, and simply does not permit the betting behaviour that discriminates sharply between the first pair of bets and the latter. If our credences conformed to Figure 6.4, then our attitude toward neighbouring bets would be pretty similar.

For similar reasons, the view that we learn a precise proposition about the tree will not straightforwardly account for the following scenario. Suppose that Alice and Bob initially have identical evidential probabilities and that they both are looking at a tree in the distance, which, let us suppose, is 400cm high. As expected, their evidence is characteristically inexact. Suppose furthermore that in this situation they both have exactly the same precise evidence; the strongest precise proposition that is part of Alice and Bob's evidence is the proposition that the tree is between 300cm and 500cm (say). However, there is a difference: Alice's eyesight is better than Bob's. While neither Alice nor Bob can rule out that the tree is 301cm or 499cm, Alice's probability distribution forms a high peak at 400cm which quickly subsides remaining quite low at the edges (300cm and 500cm). Bob's probabilities are much more evenly distributed; they increase steadily from 300cm, peaking at 400cm, and decrease steadily until 500cm. This difference is easily explained by a difference in their propositional evidence on my view, but it cannot be straightforwardly explained by a difference in their precise evidence.

What this example shows is that one's evidential probability distribution does not supervene on one's precise evidence. Both Alice and Bob have the same precise evidence, yet they have different evidential probability distributions. In particular, this shows that evidential probability is not determined by conditioning on your precise evidence; thus either the agent's evidential probability is not determined by conditioning on her evidence, or her evidence does not consist only of precise propositions.

In light of these puzzles one might consider a hybrid of the two views I have described above. When one looks at a tree in the distance perhaps one learns both something about the tree and something about your experiences. Because conditioning on a conjunction is the same as conditioning on each conjunct in succession, this is equivalent to taking the credence represented by the rectangular distribution and conditioning it on the proposition that I have experienced e. Since the proposition that you've had e can make the various hypotheses about the height of the tree more or less probable, the resulting distribution will be no longer be rectangular. But because we've ruled out the hypotheses that the height of the tree is less than 300cm and greater than 500cm (hypotheses which aren't ruled out by the proposition that I've had experience e) the curve won't be smooth either: its derivatives will be discontinuous at 300cm and 500cm. In this regard it would be similar to a curve that drew out a semicircle between 300cm and 500cm, but was a straight line, with constant value 0, elsewhere: there would still be sharp corners at 300cm and 500cm. Although there isn't a clean argument against such a view from betting behaviour in this case, this is intuitively not the kind curve one expects here.[11]

6.2.2 Jeffrey conditioning

What if our evidence is not the proposition that we have experienced e, but the experience e itself? If our evidence is not propositional, it no longer makes sense to conditionalize on it. However, one might hope to apply a different method for revising your credences in response to non-propositional evidence, based on a generalization of conditionalization developed by Richard Jeffrey in Jeffrey [73]. We shall see, however, that Jeffrey's generalization of conditionalization does not tell us how we ought to respond to inexact evidence. We shall see, in fact, that the theory of inexact evidence presented here is not an alternative to Jeffrey's theory but a supplementation of it.

To illustrate Jeffrey's theory, suppose that prior to seeing the tree my credences are evenly distributed over the possible heights of the tree, and that after having an

---

[11] It is also worth noting that non-human animals can have inexact evidence. However, it is quite contentious whether animals have beliefs about their own experiences.

inexact experience of the tree my credences about the tree's height change as depicted in Figure 6.4. However, this inexact experience also has a bearing on many other propositions that are not directly about the height of the tree: for example, since I can only climb trees that are not too high, I may become more or less confident that I can climb the tree. How in general should we calculate the effect of an inexact experience on these other propositions?

Intuitively, if I thought it was x% likely that I could climb the tree conditional on the tree being exactly $n$cm high prior to having the experience, it should remain that likely conditional on it being $n$cm afterwards: my experience does not directly bear on my tree climbing ability. Thus, for example, if it's initially likely that I can climb the tree conditional on the tree being between n and $m$cm, and unlikely otherwise, and if it becomes more probable that the tree is between n and $m$cm after my experience, then I will become more confident that I can climb the tree after the experience.

According to Jeffrey, for each n, we should calculate how likely I thought it was that I'd climb the tree conditional on the tree being exactly $n$cm *before* having the experience, multiply that by the probability that the tree is $n$cm that we assign after having the experience, and add them up. More generally, let us suppose that the experience, e, effects a direct change in your credences about the propositions $E_1 \ldots E_n$, where $E_1 \ldots E_n$ are mutually exclusive and exhaustive propositions (that is, they form a *partition*). Then your posterior credence in some proposition P is given by

$$Cr'(P) = \sum_{i \leq n} Cr'(E_i)Cr(P \mid E_i)$$

where Cr' are your posterior credences, and Cr your credences before the observation. If the partition is just $E, \neg E$, and if after I have the experience I become certain in E and have zero credence is $\neg E$, Jeffrey conditioning says that my posterior credence in P should be $Cr'(E)Cr(P \mid E) = 1 \times Cr(P \mid E)$. This is just your conditional credence. Thus we can see that Jeffrey conditioning is a generalization of conditioning where your posteriors can be less than 1.

A helpful tool for visualizing Jeffrey conditioning is to picture logical space as a two-dimensional plane, with $E_1 \ldots E_n$ carving up logical space into non-overlapping cells, each with an area representing its probability. An arbitrary proposition will in general cut across the cells, and thus will take up a proportion of each cell. According to the Jeffrey formula, when an experience effects a change in the sizes of the cells, their contents will be stretched out uniformly, keeping the proportion of a cell that any proposition takes up the same. Thus the probability of a proposition that cuts across multiple cells will be stretched in a way that can be calculated in a patchwork fashion, by looking at how each of the cells it overlaps are stretched.

If E_i is the proposition that the tree is $n$cm then it is straightforward to find an instance of Jeffrey's equation which matches the sorts of smooth curves seen in Figures 6.2 and 6.4, provided none of the E_i have a prior credence of 0 (one simply sets the coefficients $Cr'(E_i)$ to the values of the graph). However, we must be clear from the start which problem Jeffrey's extension of conditionalization is supposed to solve. In Jeffrey's own words:

The problem is this. Given that a passage of experience has led the agent to change his degrees of beliefs in certain propositions $B_1, B_2, \ldots, B_n$ from their original values, $Cr(E_1), Cr(E_2), \ldots, Cr(E_n)$ to new values, $Cr'(E_1), Cr'(E_2), \ldots, Cr'(E_n)$ how should these changes be propagated over the rest of the structure of his beliefs? If the original probability measure was Cr, and the new one is Cr', and if P is a proposition in the agent's preference ranking but is not one of the n propositions whose probabilities were directly affected by the passage of experience, how should $Cr'(P)$ be determined? [Notation has been modified to fit current conventions.]   (Jeffrey [73])

It should be noted that formally any partition of propositions can be substituted for $E_1 \ldots E_n$ in this equation. For Jeffrey propositions $E_1, \ldots, E_n$ are those propositions 'whose probabilities [are] directly affected by the passage of experience'. What might this mean? A natural thought is that the propositions, $E_1 \ldots E_n$, are those propositions whose credences are rationally determined by your background credences and the fact that you've had a certain experience. However, if the posterior credences in any partition of propositions is rationally determined by your new experience and your prior credences, all partitions of propositions are. This is, after all, what Jeffrey conditionalization itself says: once your posteriors over a given partition are fixed, the rest of your credences ought to be determined from this by Jeffrey conditionalization.

If we assume that evidence is propositional, and that we update by conditioning on our evidence, then it is possible to state what contribution a piece of evidence should make to two people's doxastic state in a way that depends only on the kinds of prior beliefs they have in propositions that entail their evidence. Within Jeffrey's framework, in which it is often assumed that evidence is non-propositional, it is not so easy to separate out the contribution a piece of evidence makes from the contribution your prior credences make. Given a non-propositional piece of evidence and some prior credences, it would be nice to know what your posterior credences *should* look like. There are many posterior credences you could have after acquiring a non-propositional piece of evidence that seem completely unreasonable. Indeed, some people go further and suggest there's only one reasonable way to respond to a piece of evidence: if two people with identical priors receive the same evidence they should have the same resultant credences.[12] Unfortunately, for all Jeffrey conditioning says, two people with identical priors and identical evidence could respond in radically different ways. Jeffrey conditioning is in fact extremely permissive. To illustrate: suppose that prior to receiving visual experience e, resulting from observing a tree in the distance, the propositions about the height of the tree and the number of stars

---

[12] If one is taking evidence to be non-propositional then there is an issue as to what counts as two people having the 'same' evidence. A common, albeit deniable picture is that people receiving phenomenally identical visual experiences should respond in the same way.

are probabilistically independent. Suppose also that after experiencing e, I become certain that the number of stars is odd. It seems that this kind of transition of credences simply isn't supported by the kind of evidence that I've received.[13] Yet this change of credences is completely consistent with the constraints of updating by Jeffrey conditioning, since I can get it by choosing the right partition and coefficients. Even if we insist on using the 'intuitively correct' partition in this case, there are many permitted transitions that seem irrational: a visual experience showing a tree that's roughly 400cm ought to increase your credence that the tree is 400cm if they were initially uniform, but Jeffrey's condition doesn't rule out a response where you become very confident that the tree is 1000cm, or where you become certain it's an even number of cm tall. I shall refer to a transition between two credences that *is* permitted by some possible evidence as a *realizable* Jeffrey conditioning. It is perhaps possible that some experience could have the effects described, and so it's open whether the transitions described above are realizable, but intuitively they are not warranted by the experience described.

What we would like is some correlation between experiences and propositions telling you how much those experiences support those propositions, from which one can determine, given someone's background credences, how much posterior credence *they* should assign to that proposition—Jeffrey conditioning alone does not do this. This is called the 'input problem', and as yet there is no particularly satisfactory answer to it (see the exchange between Field [49] and Garber [62] for a failed solution to the input problem). Without a solution to the input problem, Jeffrey conditioning has little content. It just states a relation that must hold between your prior and posterior credences. Furthermore, that relation is very liberal: under simplifying assumptions, if we choose a fine enough partition it can be shown that any transition between two probability functions that preserves certainty is permitted by the constraints of Jeffrey conditioning. This is perhaps too liberal; one might think that there are some changes of credential state consistent with the Jeffrey conditioning constraint that would not be permitted by any possible evidence you could obtain.[14]

Unlike conditionalization, Jeffrey conditioning should not be read as telling you how to respond to your evidence, but should rather be read as telling you how, once you have changed your credence in E to some degree, the rest of your credences redistribute to accommodate this change.[15] As a theory of inexact evidence, Jeffrey

---

[13] It should be noted that as far as Jeffrey himself was concerned, this is a perfectly rational transition: how your credences respond to your experiences is a purely causal involuntary matter, about which rationality has nothing to say.

[14] One reason to think this is that the *order* in which you receive the evidence between two times shouldn't matter to your credence at the latter time. But if any Jeffrey conditioning counts as a rational transition, then there are pairs of Jeffrey conditionings which result in a different posterior depending on the order in which you update on them.

[15] Conditionalization can be read as telling you how to redistribute your credences once you have become certain in E, but it is most naturally read as telling you *to become* certain in E (and redistribute accordingly) when E is part of your evidence. There is no simple connection between your evidence and Jeffrey conditioning in the same way.

conditioning remains silent. I may increase my credence in P when I have no new evidence for p, or even if all the evidence is against P, and still comply with Jeffrey conditioning by redistributing my credences correctly. Similarly, it is in full compliance with Jeffrey conditioning to make no change at all to my credences in P even when there is strong evidence for P. We want to know: when *should* I change my credences and how. When I see a tree in the distance, for all Jeffrey conditioning says, I may just keep my credences about its height the same even though I have new information.

Jeffrey's theory is thus not really a theory about how you ought to revise your credences in response to inexact evidence after all, nor is it a theory of what inexact evidence is. In order to have a satisfactory theory of inexact evidence, Jeffrey conditioning must be supplemented. In section 6.1, I defended the claim that when your evidence is inexact, your evidence consists of a vague proposition, and that your credences ought to be the result of conditioning your priors on this proposition. This, I claim, is just the sort of supplementation Jeffrey conditioning requires. The following is thus another feature of the role in thought that vague propositions have:

THE EVIDENTIAL ROLE OF VAGUENESS: With respect to the precise propositions, conditioning on a vague proposition has the effect of a realizable Jeffrey conditioning on a partition of maximally strong consistent precise propositions.[16]

Those who like the ideology of precisifications could state this principle in terms of Jeffrey conditioning relative to each of a number of precisifications. However, for reasons that will become clear in chapter 12, I prefer this more general formulation. Above, I speak of the partition of maximally specific consistent precise propositions; however, in practical contexts, we may simply use a coarser partition. For example, if we are talking about conditioning on the proposition that Harry is bald, we can think of that as Jeffrey conditioning on the partition consisting of each of the propositions that Harry has exactly n hairs, for each n. To determine the coefficients of this Jeffrey update we look at the conditional probabilities we appealed to earlier: the conditional probability of Harry being bald conditional on each proposition about hair number.[17] (Indeed, the conditional probability of a vague proposition on each maximally strong consistent precise proposition represents a theoretically important aspect of that vague role of propositions in thought, and will be a central concept in what follows.) So our principle states that conditioning on the proposition that Harry is bald is equivalent to Jeffrey conditioning over the partition of propositions containing, for each n, the proposition that Harry has exactly n hairs.

---

[16] It should be noted that the converse to our claim—that every realizable Jeffrey conditioning over the space of precise propositions is the result of conditioning on a vague proposition—is not being considered here. Formally, one can Jeffrey condition over a set of maximally strong consistent precise propositions without becoming certain in a vague proposition; however, one could in principle make the case that such Jeffrey conditionings would never count as realizable in the sense outlined earlier.

[17] See section 18.2 for a precise account of the relation between the Jeffrey coefficients and the conditional probabilities of Harry being bald on his hair number.

6.2.3 Conditioning on a vague proposition

Assume that my credences are initially evenly distributed over the possible tree heights. How confident should I be about the possible tree heights after learning the following vague proposition?

The tree is about 400cm tall.

Answering this question is instructive for the view that when your evidence in a subject matter is inexact it consists in a vague proposition. While it is unlikely that the proposition one learns after seeing a tree in the distance is easily expressed by an English sentence, such as the one above, the effect the above proposition has on your credences is representative of the effect a vague proposition of the kind you might learn will have.

We are primarily interested in the shape of the curve that plots the various height hypotheses against our credence in those hypotheses. Our strategy will be as follows: note that by Bayes' theorem the probability that the tree is exactly Ncm tall conditional on it being about 400cm is proportional to the converse conditional probability of it being about 400cm conditional on it being exactly Ncm tall. The constant of proportionality is the prior probability that the tree is Ncm, which by assumption is the same for each N. Thus the curve plotting N against the number Cr(exactly Ncm | about 400) is proportional to, and thus has the same shape as the curve plotting N against Cr(about 400| exactly Ncm).

Now we proceed in three stages. For stage one, consider a sequence of 1,000 trees of increasing heights starting with a tree that is 0cm and ending with a tree that is 1000cm. What rational credence should we have that the Nth tree is about 400cm? Presumably our credence that the 400cm tree is about 400cm should be 1, and our credence that the 0cm tree and the 1000cm tree are about 400cm should be 0. What about the tree that's 425cm? It is borderline whether this tree is about 400cm, thus we should be uncertain: our credence must lie strictly between 1 and 0. Indeed, intuitively as the distance between N and 400 increases, the credence that the Nth tree is about 400cm should get less and less, in a smooth way much like that depicted in Figure 6.4.

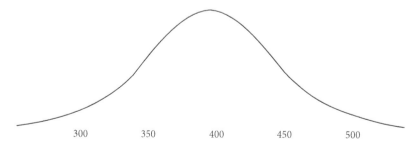

Figure 6.4. A graph of n against the proportion of epistemic states where the tree's height is $n$cm where it's also about 400cm.

Now instead of 1,000 trees, imagine we just have one tree whose height is unknown to us, and we are instead considering whether the tree is about 400cm conditional on each hypothesis about its height. Conditional on the hypothesis that the tree is Ncm our credence that the tree is about 400cm ought to be the same as our credence that the Nth tree in the last paragraph is about 400cm. After all, if I were to learn that our tree was Ncm, I'd be in the same epistemic situation as I am towards the Nth tree from the last paragraph. Thus our conditional credences in the proposition that the tree is about 400cm is the curve in Figure 6.4.

Finally, by Bayes theorem (and the fact that we were initially uniformly distributed over possible heights) we can conclude that our credences, after conditioning on the proposition that the tree is about 400cm, is proportional to the curve in Figure 6.4.

Let us try and construct a simple model of this. Note that propositions cannot simply be modelled by sets of possible worlds, for otherwise (given a natural superve-nience assumption) the proposition that the tree is about 400cm will be identical to the proposition that the tree is in the range 400cm $\pm x$ for some x. In order to represent this situation, then, we shall model propositions as sets of epistemic possibilities. These can be thought of as certain kinds of maximally strong consistent propositions; all that really matters is that since we do not know whether a tree that is 420cm is about 400cm (because, let's suppose, it is borderline) we require there to be an epistemic state in which the tree is 420cm and is about 400cm, and a state in which the tree is 420cm and it's not about 400cm. For every possible height of the tree, h, and for every epistemically possible cutoff point for 'being about 400cm', $\pm c$cm, there will be an epistemic state where the tree's height is h, and the cutoff point for being about 400cm is being within $c$cm of 400cm. The relevant epistemic states can thus be represented by conjunctions of propositions taken from the following two sets: {the proposition that the tree is $h$cm | $h \in \mathbb{R}$} and {the proposition that, necessarily, a tree is about 400cm iff it's between $400 - c$cm and $400 + c$cm| $c \in [0, 400]$}. We can thus represent the epistemically possible worlds as ordered pairs (h, c) where h is the height of the tree, and $400 \pm c$ represents the range between which a tree counts as being about 400cm according to that world. There are therefore many more states than there would be if we had only countenanced precise possible worlds.

The proposition that the tree is about 400cm tall is thus the set of epistemically pos-sible worlds where the tree's height is between the cutoff point for being about 400cm:

$$\{(h, c) \mid 400 - c \leq h \leq 400 + c\}$$

Suppose, for mathematical simplicity, that we restrict the possible values c and h may take to some large but finite set, and that my credences are initially uniform over this set: I assign the same credence to the proposition that any given c in this set is the cutoff for being about 400cm, and the same credence to any proposition that a height in this set is the height of the tree. After updating on the proposition that the tree is about 400cm the possible heights of the tree graphed against my posterior credences over the possible precise heights of the tree would form a triangular shape with a point at 400cm. In reality, however, my credences would not be uniform: I take it that the cutoff point $c = 1cm$ is very unlikely, as is the cutoff point $c = 400cm$. In reality we should therefore expect my credences to conform roughly to Figure 6.4.

6.2.4 Evidence for the whereabouts of cutoff points

We observed in chapter 4 that considerations of vagueness-related ignorance indicate that vagueness is a distinctive source of fineness of grain. Thus many theorists are committed to the existence of vague propositions, in this very minimal sense. Once we have accepted the existence of these propositions it is also extremely natural to think that we can sometimes learn them. A standard thought about testimony, for example, entails that when you hear a trustworthy person assertively utter a sentence you often get to know and have as part of your evidence the proposition that that sentence semantically expresses. Thus if a trustworthy person utters the sentence 'that tree is around 400cm', in normal cases, we get to add a vague proposition to our evidence.[18]

The view that we sometimes update on vague propositions therefore seems quite plausible independently of the theory of inexact evidence defended here. However, there is a somewhat surprising consequence for any view of this kind.[19] According to any view that maintains both that we are ignorant about the vague and the thesis of probabilism, vague propositions can provide one with evidence for the whereabouts of cutoff points. Suppose that I already know that the tree is between 380cm and 420cm, but I am uniformly distributed over those possible heights. If I then learn that the tree is around 400cm in height, then I rule out states where, for some c, the cutoff point for *being about 400cm* is being within a margin of $c$cm of 400cm and in which the tree is not within that margin of $c$cm of 400. Let us suppose the largest candidate for c is 20. It follows that out of the states in which the cutoff is a margin of 20cm, I do not rule out any of the possible heights of the tree: I already knew the tree was within 20cm of 400cm. Out of the states where the cutoff is 10cm I rule out the states in which the tree is less than 390cm and over 410cm—i.e. roughly half of those states. Finally, out the states where the cutoff is 0cm or close to 0cm I rule out basically all of the heights except for 400cm. In summary, since we are initially uniformly distributed over the possible heights of the tree, this means we end up ruling out far more states in which the cutoff points are low, e.g. the ±1cm cutoff points, than states in which they are high. We should therefore increase our confidence that the cutoff point for *being about 400cm* is a large margin around 400cm.

---

[18] Of course, when we hear someone utter the sentence 'that tree is around 400cm' we learn lots of things: implicatures, facts about their accent, the pitch of their voice, and so on. We are looking for an example in which the strongest thing learnt is vague, but for all I've said the conjunction of things I learn is precise. This worry is easily set to rest: since in normal cases where the proposition semantically expressed is vague and not about the pitch of your voice or other such things, it will be epistemically independent of the other things I learn. Thus their conjunction will also be vague even if the other things are all precise.

[19] Thanks to Cian Dorr here for pointing this consequence out to me.

The thought behind this argument can be illustrated quite simply using an analogy. Suppose that, instead of being ignorant about cutoff points, someone has rolled a twenty-sided die which has landed on some number, X, whose value you are ignorant about. Suppose also that as before you know the tree is between 380cm and 420cm and each of the forty possible heights is supported equally by your evidence. If someone tells you that the tree is within the unknown Xcm of 400cm it is clear that you should become both more confident that the tree is closer to 400cm and that the die landed on a higher number. You can make the point even more vivid if you imagine that there are 100 trees you know to have a random height between 380cm and 420cm and that the height of each is independent of the height of any other. If someone told you that all of the trees were within the unknown margin Xcm of 400cm you can become pretty confident that X is large, since it is antecedently very unlikely that all the trees would be bunched tightly around the 400cm mark.

Now anyone who thinks that we are straightforwardly ignorant about the vague and accepts probabilism must accept the analogous argument that shows that vague propositions provide evidential support for certain hypotheses about the locations of cutoff points.[20] Although there are disanalogies between ignorance about cutoff points and ignorance about dice rolls, the only analogies we need to run this argument are being granted.

It should be stressed that although these experiences provide confirmation for the hypothesis that the cutoff for being around 400cm is on the wider side of things, the change of credence being recommended is not necessarily bringing our credences closer to the truth about the location of the cutoff point. Indeed, this partial confirmation that the cutoff is a large margin on either side of 400cm is only a temporary one, and further precise evidence about the exact height of the tree will bring my credences closer to my prior beliefs about the cutoff points. In addition, since one's evidence can never be borderline, once I have determined all the precise facts I will be as uncertain about the locations of the cutoff points as I was initially, and there will be no way to improve my epistemic position since I already know all the non-borderline facts.

6.3 A Principle of Plenitude for Vague Propositions

According to linguistic theories of vagueness, vague propositions have a derivative status. Usually there are no vague propositions or there are and they are parasitic on the vocabulary of the language. In either case, there are at most countably many vague propositions entailing any particular maximally strong consistent precise proposition

---

[20] Whether one can get around these issues by rejecting ignorance about the vague, or rejecting probabilism, bears further investigation. For more discussion of no-ignorance views, see chapter 5, and for a discussion of probabilism, see chapter 7.

corresponding to the distinctions one can make in the language.[21] This dependence on the language seems to be particularly odd, and in chapter 4, I suggested that there were vague propositions not expressed by any sentence.

An alternative to the modal way of individuating propositions, in which they are identified with sets of possible worlds, would be to individuate them by their attitudinal roles and to identify them with sets of maximally specific epistemic (bouletic/doxastic/etc.) possibilities. How should we express the idea that propositions exist independently of language and thought?

I propose the following **Principle of Plenitude** for propositions:

> For every possible evidential role in thought, there is some proposition with that evidential role.

Intuitively the evidential role of a proposition is a profile of the strength of evidential support that that proposition receives from each maximally specific precise description of the world. We may state the **Principle of Plenitude** rigorously as follows. Let P denote the set of maximally strong consistent precise propositions: the set of propositions that are (i) precise and (ii) entail any other precise proposition that they are consistent with. Then, an evidential role will be represented by a function $E : P \rightarrow [0, 1]$, telling you what your ur-priors should look like conditional on each maximally strong consistent precise proposition:

> **The Principle of Plenitude.**   Let E be an evidential role. Then there is some proposition, p, such that for every coherent ur-prior Pr and maximally strong consistent precise proposition $w \in P$, $Pr(p \mid w) = E(w)$.

As an example, the **Principle of Plenitude** entails that there is a proposition p such that conditional on the proposition that Harry has N hairs, everyone's ur-prior in p ought to be 0.798 (set $E(w) = 0.798$ whenever w entails that Harry has N hairs). The Principle of Plenitude entails the existence of an abundance of such propositions. Note that while the evidential roles generate propositions, they do not individuate them. For example, by **Plenitude** there is a proposition, p, such that one is required to have credence $\frac{1}{2}$ in it conditional on any maximally specific precise proposition. It follows by probability theory that $\neg p$ has the same evidential role as p, yet these propositions are always distinct in a Boolean algebra.

Roughly, the **Principle of Plenitude** guarantees that we have the following kind of picture: the space of epistemic possibilities will be divided up into non-overlapping cells, representing the maximally strong consistent precise propositions. Furthermore if for each cell I pick a proportion of that cell that I want to 'fill', plenitude guarantees

---

[21] In principle there could be $2^{\aleph_0}$ maximally *logically* consistent sets of sentences above a given set of precise sentences; however, it is not clear that they all correspond to conceptually consistent propositions. (For example, there are $2^{\aleph_0}$ maximally consistent sets extending Peano arithmetic, but most of these are not genuinely conceptually consistent because they are ω-inconsistent.)

there will be some vague proposition that fills each cell to that proportion. The notion of filling a cell by some proportion simply means having that probability conditional on the cell according to every ur-prior. (Notice that a proposition generated by the **Principle of Plenitude** is such that every ur-prior agrees about what proportion that proposition takes up of each cell, even if the ur-priors disagree about the size of the cells themselves. This is an important feature, although I shall defer its discussion until chapter 8.)

Although plenitude will be enough for most purposes, an even stronger principle is motivated by the picture I have been sketching. Roughly, if we cut up each cell (i.e. each maximally strong consistent precise proposition) into smaller non-overlapping cells, and assign proportions to each of these to be filled, then there is some vague proposition that fills each of the smaller cells to those proportions:

> Let P' be any subpartition of P and let E a function from P' into $[0, 1]$. Then there is some proposition, p, such that for every coherent ur-prior Pr and every proposition $x \in P'$, $Pr(p \mid x) = E(x)$.

This principle allows us to generate vague propositions from evidential relationships to other vague propositions that have already been generated.[22] In appendix 18.2 it is shown that both principles are consistent.

It is tempting to ask why one ought to have a particular credence, 0.798 say, in the proposition that Harry is bald given one knows the relevant facts about his head. This is simply not a question with a reasonable answer on this way of individuating propositions. Part of what it is to be the proposition that Harry is bald is to have the specific epistemic profile it in fact has, which includes this conceptual requirement. It is like asking, on the modal way of individuating propositions, why the proposition that Harry is bald corresponds to one set of worlds and not another.

However, the thought which I take it this question is trying to latch on to might be better expressed by the question: why does the sentence 'Harry is bald' denote a proposition with one set of conceptual requirements rather than another? This is presumably a question of metasemantics and a question whose answer is vague. The appearance of precision is really an illusion. A similar problem arises for the modal way of individuating propositions. One might ask why the proposition that Harry is bald contains worlds where Harry has 784 hairs but not worlds where he has 785 hairs. If propositions are just sets of worlds, this seems like a misguided question. The sensible question in the vicinity is presumably a metalinguistic one: why does the sentence 'Harry is bald' express this proposition and not another, or why are we

---

[22] For example, suppose that we know which exact shade a ball is, and that that shade is such that it's borderline whether it's turquoise and blue, just turquoise, or just blue. Although having further evidence that it was turquoise would be impossible, we can still ask if the proposition that the ball is turquoise is evidentially relevant by considering the conditional probabilities. Intuitively the probability that it's blue decreases on the hypothesis that the ball is turquoise since we have to rule out the blue and not turquoise possibilities, and there are more just turquoise than turquoise and blue possibilities.

justified in calling this set of worlds 'the proposition that Harry is bald'? Both are metasemantic questions, and one should expect there to be vagueness concerning which propositions are picked out by which sentences.

Another issue that might initially seem problematic for the **Principle of Plenitude**, is the existence of higher-order vagueness. Given that it is vague which propositions are precise, it's going to be vague which partition of logical space into non-overlapping cells corresponds to the partition of logical space into maximally strong consistent precise propositions.[23] Note, however, that this is not inconsistent with the **Principle of Plenitude** as stated, or even the claim that the **Principle of Plenitude** is determinate. The determinacy of plenitude guarantees that determinately, for any evidential role on the partition of maximally strong consistent precise propositions, there is some proposition satisfying that role. There is vagueness concerning which partition to use, and thus vagueness concerning which functions are evidential roles. But so long as whatever the partition and evidential roles are, there are vague propositions and ur-priors which accord with that role conditional on each member of the partition, the **Principle of Plenitude** will come out determinate. What one shouldn't expect, however, is that a proposition have its role determinately, or that it be completely determinate which class of probability functions are the rational ur-priors. A proposition p might have probability $\frac{1}{2}$ conditional on each cell, but there might be things that are not determinately not cells (i.e. not determinately not members of the partition of maximally strong consistent precise propositions) such that p does not have probability $\frac{1}{2}$ conditional on these 'would be' cells: p has the role of being $\frac{1}{2}$ on each cell, but not determinately so. Differences regarding the evidential role of a proposition could potentially force differences in which probability functions are rational. Neither of these things should be too surprising, since neither the notion of an evidential role or of a rational ur-prior, was introduced in a completely precise way. We shall have more to say about this in chapter 7.

Having developed a plenitudinous theory of vague propositions, we are now in a position to see how to give a theory of inexact evidence based on vague propositions. Recall the example of the tree seen from a distance. Assuming that my credences are initially uniform over the possible heights less than 1000cm, it seemed in that case that after the evidence is in my credences ought to conform to the curve displayed in Figure 6.4. What proposition would have the effect that updating this way would cause this change in credence? Here the **Principle of Plenitude** kicks in. According to the principle there will be a proposition that has the desired effect on my credences. The curve depicted can be represented by a function, E, mapping each possible exact height of the tree to a number. Suppose for simplicity we identify the cells of our partition with the partition of propositions saying that the tree is exactly $n$cm tall (this

---

[23] Were it a precise matter which partition this was, it would be a precise matter which propositions are the precise ones, since precise propositions are simply disjunctions of members of the partition of maximally strong consistent precise propositions.

simplification is harmless), then it follows that E is effectively an evidential role of the kind we were looking for: something which maps each member of this partition to a number between 0 and 1. The **Principle of Plenitude** entails there is a proposition, p, with evidential role E. In particular, if Cr is a coherent prior and e is the agent's total evidence, so that $Cr(\cdot \mid e)$ is my credence before the observing and is thus uniform over the partition p_n for $n \leq 1000$, then conditioning on p will result in a curve like that in Figure 6.3: $Cr(p_n \mid e \wedge p) = E(p_n)$.

6.4 Evidential Roles and Degrees of Truth

In many ways, maximally strong consistent precise propositions in this framework are approximations of possible worlds in an intensional framework. Assuming this analogy, vague propositions then determine a function from 'possible worlds' to real values in $[0, 1]$: the credence any conceptually coherent agent ought to have if she learnt that she was in that possible world.

There is a striking resemblance between this formalism and approaches to vagueness based on fuzzy logic. Propositions, according to the fuzzy logician, are not simply true or false: there are many truth values between the two, which we can represent on a scale including all the real numbers between 0 and 1, with 0 representing outright falsehoods and 1 representing truths. A borderline proposition, on this view, has a truth value somewhere strictly between 1 and 0. A proposition, then, determines a mapping from possible worlds to numbers in the interval $[0, 1]$: the mapping that takes a world w, to the truth value of that proposition at w.

These theorists typically work with a non-classical logic, due to the way in which the truth values at a world interact with the connectives. However, some theorists retain classical logic and consequently exhibit some strong parallels with the view I've been defending here. Edgington [39], for example, espouses a view in which the truth values behave like probability functions (see also Kamp [76], Lewis [89], and Williams [152], for versions and developments of the view in a supervaluationist setting). Given the analogies, one might wonder if the view I have defended here is just a version of the fuzzy view defended by Edgington.

Rational degrees of belief and degrees of truth might seem like different concepts on the surface: the former seem subjective and the latter objective. Unfortunately, this appearance is quite fragile: under the assumption of classical logic, sentences like 'Harry is bald even though the degree of truth that Harry is bald is $\frac{1}{2}$' are consistent, for effectively the same reason the 'Harry is bald even though he's borderline bald' is consistent in classical logic (see chapter 2).[24] In this respect, degrees of truth behave more like degrees of belief; it is certainly consistent that Harry is bald even though

---

[24] Suppose that Harry is bald to degree $\frac{1}{2}$. By LEM and classical reasoning it follows that either Harry is bald but that this has degree of truth $\frac{1}{2}$, or that he isn't and this has degree of truth $1 - \frac{1}{2}$ (i.e. $\frac{1}{2}$). Either way we have a truth of the form 'P but the degree of truth of P is $\frac{1}{2}$'.

you are uncertain. This also demonstrates that degrees of truth do not satisfy the disquotational portion of the truth role (see section 2.3): if degrees of truth deserve to be called an aletheic notion at all, they'd better play some other aletheic role. A natural candidate derives from the idea that it is generally better to have degrees of belief that are closer to the truth (see Joyce [75], for example): if you know all the precise facts, but are still unsure whether Harry is bald, then the best credence to have in his baldness is the degree to which it is true. Degrees of truth, if understood solely by this role, would just be evidential roles as I've defined them.[25]

Edgington, however, thinks that there are other important differences between truth values (which she calls 'verities') and degrees of belief. In Edgington [39] she writes:

Some philosophers (e.g. Williamson 1994) hold that vagueness is a species of epistemic uncertainty: there is a precise line which divides the red from the non-red, etc., but it is epistemically inaccessible to us. Were this true, verity would be a kind of credence: the credence that a person with no relevant ignorance other than about the precise line, would give to a statement like 'that's red'. If a is redder than b, but neither is clearly (certainly) red, then one must, if rational, be more confident that a is red than that b is: a is more likely to be above the mystery line than b is. The nearer to clearly red is the nearer to certainly red. Credence and verity, I argued, have the same logical structure. This could be interpreted as grist for the epistemicist's mill. What better explanation of the analogy I have developed, than that verity is credence, and so vagueness a kind of epistemic uncertainty?

The view I am defending is not epistemicism, but it agrees with the epistemicist in respects important to this discussion. According to Edgington degrees of truth and credences play different roles. Credences inform a rational person's actions, whereas degrees of truth do not. In support of this she notes the distinction between the following kinds of justifications:

I prefer A to B by a long way. Therefore I prefer A with certainty to A or B happening with equal uncertainty, i.e. probability $\frac{1}{2}$, and I should prefer A or B happening with equal uncertainty to B with certainty.

I prefer A to B by a long way. Therefore I prefer A with truth value 1 to A or B each with equal intermediate truth value, $\frac{1}{2}$, and I should prefer A or B with equal truth value to B with truth value 1.

The former principle is a valid principle in most decision theories, yet Edgington gives an example, in which whether A or B is borderline figures in my preferences, to demonstrate that the second is not a universally correct principle of decision making: I might have a definite preference for drinks that are definitely tea, or definitely coffee, but strongly dislike things that are borderline between the two.

---

[25] I am indebted to Robbie Williams for pressing me on this objection.

Other than Edgington's argument above (which I find convincing), I think there are two further points to be made. The first is that the truth norm articulated above is not entirely uncontentious. It seems in tension with the idea that one ought to proportion one's credence to the evidence: if one had strong evidence against a proposition that is nonetheless true, it's unclear that there's any good sense in which it's better to be more confident in that truth. Even putting this worry to one side, the motivations for the truth norm seem to also motivate a much stronger thesis. An ultra-objective norm on belief would require that one have credence 1 that Harry is bald if he's bald and 0 otherwise. Given excluded middle this entails that all my credences ought to either be 1 or 0. This version of the norm seems more principled, but leaves little room for degrees of truth to play.

The second point I want to make involves an analogy with chances. Note the similarity between the following two claims

> For each proposition, p there is a function, E, such that for each maximally strong consistent precise proposition w, and rational prior credence function Cr, $Cr(p \mid w) = E(w)$.

> For each proposition, p there is a function, F, such that for each maximally strong consistent piece of admissible evidence at t, w, and rational prior credence function Cr, $Cr(p \mid w) = F(w)$.

In both cases there is a special partition of the space of propositions such that every rational prior agrees with every other prior conditional on any element of that partition. The former principle is a consequence of our theory of propositions, the latter a consequence of the principal principle (see Lewis [91]). In the latter case $F(w)$ simply represents the chance that p at time t. In whatever sense E plays the truth role for the proposition p (relative to the partition of maximally specific precise propositions) so do chances (relative to the partition of worlds with the same admissible evidence at t). Since we are not at all tempted to call chances truth values on the basis of truth norms alone, I think that we should not be tempted to call credences truth values.

7

Probabilism, Assertion, and Higher-Order Vagueness

The theory of vague propositions outlined in chapter 6 appealed freely to the orthodox Bayesian theory of credences and learning, which assumes, among other things, that the correct theory of rational credence is one governed by the classical axioms of probability theory. I assumed this theory when I argued that one's credences after conditioning on a vague proposition conform to the kinds of smooth curves depicted in Figure 6.4.

However, the classical theory of probability is not entirely uncontroversial, especially when it is applied to vague propositions. For one thing, it takes sides on the question of what kind of doxastic attitude we should take towards P when we know it is borderline. This question turns out to be surprisingly central to the study of vagueness, and if one accepts the probability calculus, the answer is that one should be uncertain. That is, one should have the same credal attitude concerning Harry's baldness as one should have about the outcome of a coin flip, for example. For according to the probability calculus, the only alternative to having middling degrees of belief about a borderline proposition is to either assign a credence of 1 to that proposition, or to its negation—and this seems to be absurd: I shouldn't be certain, for example, that any number is the last small number.

This dichotomy, however, has been challenged in recent years by Hartry Field [53], who has given a precise formulation of the idea that we shouldn't be uncertain about the borderline.[1] Crucially, this idea doesn't collapse into the view that one should be certain about the locations of cutoff points, because it relaxes the probabilistic axioms governing rational degrees of belief.

Field's theory is one of the most influential alternatives to the standard probabilistic picture. However, it also serves as a good springboard for a discussion of the present theory of vagueness-related uncertainty, since it articulates in a particularly precise way, positions on two of the most central questions concerning this issue—it tells us what kind of attitude we ought to have toward a proposition we know to be borderline,

---

[1] Field's [53] paper assumes classical logic; in later work Field extends some of these ideas to a non-classical setting. Field's theory is closely related to some ideas of Stephen Schiffer's (another opponent of probabilism): see Schiffer [126]. See also Dietz [32], MacFarlane [97], Smith [134], and Williams [152].

and it tells us how this attitude ought to be extended to propositions we know to be higher-order borderline. This is summarized by the following two principles, which I'll state informally for now:

ITERATION: The credal attitude one ought to have towards A when one is certain that A is higher-order borderline is the same as the attitude one ought to have when one is certain that A is borderline at the first order.

REJECTION: The credal attitude one ought to have towards A when one is certain that A is borderline is simply that of fully rejecting A and fully rejecting $\neg A$: one's credence in both A and $\neg A$ must be 0.

Although Field's theory of credences will be a useful springboard for this discussion, many of the points I shall be considering here generalize to other questions involving attitudes like knowledge and assertion.

7.1 Field's Theory

Let us begin by outlining a fairly pared-down presentation of Field's theory. This version is governed by the following two axioms, where Q represents a rational credence function mapping propositions into the interval $[0, 1]$:

F1. $Q(\Delta A \vee \Delta B) = Q(A) + Q(B) - Q(A \wedge B)$

F2. Q respects the modal logic KT. This just means that Q assigns 1 to its theorems, 0 to the negations of its theorems and assigns A no more probability than B when B is provable from A.

The first axiom quite clearly bears a close resemblance to the additivity law from classical probability theory, except that the usual sum is being used to calculate the value of $\Delta A \vee \Delta B$ rather than the usual disjunction $A \vee B$ (the standard additivity law states that $Q(A \vee B) = Q(A) + Q(B) - Q(A \wedge B)$). The second axiom effectively amounts to the requirement that rational degrees of belief respect logical laws, where these laws are understood broadly enough to include the laws governing the determinacy operator.

The above theory is in fact strictly weaker than Field's theory. According to Field, Q must additionally respect the logic of determinacy S4, which controversially states that if something is determinate, it's determinately determinate.[2] However, the theory can be developed perfectly consistently without making this assumption. None of the problems I shall raise here will depend on this unnecessarily controversial assumption, so I shall simply put it to one side for now.

---

[2] Adding S4 suffices to derive the crucial formula that relates Field's theory to Dempster-Shafer theory: $Q(\bigvee_{i=1}^{n} \Delta A_i) = \sum_I (-1)^{|I|+1} Q(\bigwedge_{j \in I} A_j)$ where I ranges over subsets of $\{1, \ldots, n\}$. To prove this, it helps to note that $\Delta A \vee \Delta B$ is logically equivalent to $\Delta(\Delta A \vee \Delta B)$ in S4.

Two important consequences of these axioms are the following, which we already stated informally in the last section:

ITERATION: For any rational credence function Q, $Q(A) = Q(\Delta A)$

REJECTION: For any rational credence function Q, if $Q(\nabla A) = 1$ then $Q(A) = Q(\neg A) = 0$.

ITERATION follows from the first axiom by setting A equal B and then applying some fairly trivial logic using F2. REJECTION then follows straightforwardly from ITERATION given the definition of ∇ from Δ.

It is worth stressing, however, that the principle REJECTION, even in conjunction with the axiom F2, does not imply ITERATION. In fact, although there is an entailment in one direction, ITERATION and REJECTION correspond to two very different ideas that ought to be kept theoretically separate.

REJECTION allows for a particularly striking characterization of vagueness-related uncertainty. The standard probability calculus only allows you to be in one of two states regarding p: (i) be certain whether p: either having a credence of 1 in p and 0 in its negation, or vice versa, or (ii) be uncertain whether p: both p and its negation get a credence strictly between 1 and 0. However, in Field's calculus there is a third option: one assigns both p and its negation a credence of 0. Your credences are not intermediate, as they would be if you were uncertain, nor do they assign 1 to p or its negation, as they would if you were certain whether p. Whereas the epistemicist, for example, would treat uncertainty about whether Harry is bald just as she would treat uncertainty about, say, the mass of the moon, someone adopting Field's calculus can say something distinctive about the former case: vagueness does not merely amount to a kind of uncertainty, for one's credences in p aren't intermediate between 1 and 0. Yet this does not commit us to being certain about p either, since we do not assign 1 to p or its negation.

ITERATION entails something much stronger: it entails that anyone who is certain that A is borderline at the second order or higher, must also assign A and its negation credence 0. To illustrate, let us suppose that I am certain that A is second-order borderline: i.e. that $Q(\nabla\nabla A) = 1$. It follows that my credence in A and $\neg A$ must both be 0: if my credence in A were non-zero, then I would assign some credence to A being determinately determinate (applying ITERATION twice to my non-zero credence in A), and this is simply impossible if I'm certain that A is second-order indeterminate. A parallel argument establishes that my credence in $\neg A$ is 0. It is important to stress how this result is not entailed by REJECTION alone: if I'm certain that A is second-order borderline, it follows by REJECTION that my credence that it's borderline whether A and my credence that it's not borderline whether A must both be 0. But unfortunately I can't infer anything about my credence in A and its negation from this. If my credence in the borderlineness of A were 1 instead of 0, I could apply REJECTION again to conclude that my credence in A and its negation had to be 0. But this is not how it is on Field's theory: when you are certain that A is borderline borderline, your credence that A is borderline must be 0, not 1.

It is worth stressing that the thought behind these two ideas is quite different. The first tells us that we shouldn't be uncertain—in the sense of having middling credences—about the borderline. The second idea is that we should extend this state to the things we believe to be higher-order borderline. It turns out that neither of these theses is particularly friendly to the theory we developed in chapter 6.

7.2 Uncertainty in the Face of Higher-Order Vagueness

ITERATION encodes an assumption that is pretty pervasive in the philosophy of vagueness. In Field's case, it amounts to a position concerning what credal stance one ought to take toward the higher-order vague: credally, at least, one ought to respond to higher-order vagueness in exactly the same way one responds to first-order vagueness. Positions like this concerning credences, and other attitudes, are implicitly assumed in much of the literature on vagueness, although rarely acknowledged explicitly. Consider, for example, the case of knowledge: one might think that just as it's impossible to know the borderline, it is impossible to know the borderline borderline, or the borderline borderline borderline, and so on. According to this natural and pervasive thought, higher-order borderlineness is just as much a barrier to knowledge as borderlineness: something is known only if it is determinate at all orders. One natural motivation for such a view is there seems to be something distinctly odd about asserting P whilst simultaneously and deliberately withholding assent from ΔP.

These ideas have knock-on effects. For example, consider two popular, albeit competing, theories of assertion: that one ought to assert only what one knows, and that one ought to assert only what one believes or has sufficient credence in. The theses about knowledge and credence above, therefore, both constrain the relation between higher orders of vagueness and what one is in a position to assert. According to either view, one is not in a position to assert that A if A is (or is believed to be) higher-order borderline. Kit Fine, for example, articulates the connection between assertability and determinacy as follows: 'In asserting some propositions P1, P2, . . . , one is committed to more than their actual content, one is also committed to their being definitely the case, definitely definitely the case, definitely definitely definitely the case, and so on' (Fine [58], p. 114).

Another way in which this idea gets presupposed in the philosophy of vagueness is through the rule sometimes called 'Δ introduction' endorsed by many classical theories of vagueness: the rule that allows you to infer that A is determinate from A. In formalism: $A \vdash \Delta A$.[3] This rule encodes the legitimacy of a certain way of reasoning: if you believe that A, one can legitimately go on to infer that it's determinate

---

[3] See Fine [56] for one of the first articulations of this rule. The rule is often stated as a metarule: if $\Gamma \vdash A$, then infer that $\Gamma \vdash \Delta A$. This is equivalent to the rule $A \vdash \Delta A$. One direction follows straightforwardly from the fact that $A \vdash A$, so by the first version of the rule we can infer that $A \vdash \Delta A$. For the other direction suppose that $\Gamma \vdash A$. By the second version of the rule we have $A \vdash \Delta A$, so by the transitivity of $\vdash$, we have $\Gamma \vdash \Delta A$.

that A. Of course, note that this does not commit us to the idea that every truth is a determinate truth, or to the validity of the conditional $A \rightarrow \Delta A$—either of these would be disastrous, since they would entail that every proposition is determinate.[4] However, the rule of determinacy introduction does seem to commit us to the idea that when a belief that A is in good epistemic standing (in some sense or other), so is the belief that it's determinate that A. For otherwise it would just be a bad idea to reason with that rule: to continue calling it 'logically valid', whilst being unwilling to reason with it, would be to pay lip service to the rule in some sense, but would deny it any important theoretical status.

The rule of Δ introduction, then, suggests a more general thesis about committal attitudes: that when one has the attitude toward A, one also ought to have it toward the claim that it's determinate that A.

DETERMINACY INTRODUCTION (as a constraint on attitudes):

(DI1)   If you rationally believe (or have sufficient credence) that A, then you are in a position to rationally believe (or have sufficient credence) that it's determinate that A.

(DI2)   If you are in a position to know that A, you are in a position to know that it's determinate that A.

(DI3)   If you are in a position to assert that A, you are in a position to assert that it's determinate that A.

As I mentioned above, the rule of Δ introduction seems to be underpinned by the idea that higher-order vagueness imposes the same barriers to rational belief and knowledge as first-order vagueness. To see the connection in full, let us start by considering the cases of knowledge and assertion. The thesis we want to relate says:

(HO)   One is in a position to assert or know that A (if and) only if A is determinate at all orders.

Note here that I parenthetically include one half of the biconditional. The parenthetical direction only holds if we are in 'epistemically ideal circumstances with respect to A'—that we are not ignorant about some of the relevant precise facts, for example. I shall not attempt to spell out what this means precisely, but the idea is clear enough in specific examples: I'm in epistemically ideal circumstances regarding whether Harry is bald if I know how many hairs he has, their colour, distribution, and so on. The other direction is our primary concern and on the picture we are considering, holds in full generality.

---

[4] Formally, this move requires restricting the rule of conditional proof: the fact that $A \vdash \Delta A$ does not imply that $\vdash A \rightarrow \Delta A$. Modelled theoretically, there is a very straightforward way of making sense of this, by understanding a valid argument as one where the conclusion is true at every point in the model if the premises are. Validity does not imply that truth is preserved at any particular point in the model.

We can see how this principle is connected to the cluster of principles associated with DETERMINACY INTRODUCTION as follows. Starting with (DI2), suppose that I'm in a position to know that A. Then by applying (DI2) repeatedly we can infer that I'm in a position to know that it's determinate^n that A for any n. So by the factivity of 'in a position to know that' we can infer that A is determinate at all orders. So (DI2) entails that higher-order borderlineness precludes knowledge. Conversely, suppose that I'm in a position to know that A. So by (HO) it's determinate at all orders that A. But A is determinate at all orders only if the proposition that A is determinate is determinate at all orders.[5] So, at least in epistemically ideal circumstances,[6] I'm also in a position to know that A is determinate. A similar argument can be given relating the above principle for assertion to (DI3), although it is slightly more contentious: the analogous argument relies on the factivity of 'in a position to assert that' (some philosophers maintain that one can be in a position to assert a falsehood if one has strong enough evidence in its favour).

The principle (DI1) also bears a close connection to Field's principle ITERATION. For clearly if your credence in A is sufficiently high, your credence in ΔA is sufficiently high, since it is simply identical to your credence in A if we accept ITERATION. What is perhaps most surprising about the credal version of DETERMINACY INTRODUCTION is that it appears to entail the falsity of probabilism. For example, if you think that logical entailments do not permit a drop in rational credence, then the rule of Δ introduction entails ITERATION: your credence in A ought to be identical to your credence in ΔA, since (by factivity and Δ introduction) they entail one another. But as we demonstrated above, ITERATION requires us to relinquish probabilism since it entails REJECTION, which contradicts finite additivity. If we conceive of DETERMINACY INTRODUCTION as imposing a constraint on credences, we not only get a connection between our doxastic attitudes toward the vague and the higher-order vague, but we get Field's distinctive non-probabilistic account of those doxastic attitudes.

Although the credal version is more controversial, this account of assertion and knowledge is presupposed in much of the literature on vagueness. It might be tempting to think that we can simply derive this view from the principle that you cannot know or assert things that are borderline: the thought being that higher-order borderlineness is simply a type of borderlineness, and therefore precludes knowledge and assertability because ordinary borderlineness precludes these things. However, this crude way of motivating the view rests on a conflation: borderline borderline does not entail being borderline. Recall from chapter 2 that one of the distinctive features of a classical theory of vagueness is that a non-bald person can be still be borderline bald.

---

[5] The claim that A is determinate at all orders can be captured by the infinite conjunction $A \wedge \Delta A \wedge \Delta\Delta A \wedge \ldots$. Clearly the infinite conjunction $A \wedge \Delta A \wedge \Delta\Delta A \wedge \ldots$ entails $\Delta A \wedge \Delta\Delta A \wedge \ldots$ by conjunction elimination, and the latter just is the claim that ΔA is determinate at all orders.

[6] Which I am presumably in with respect to ΔA, since I know A.

The same therefore goes for borderline borderlineness: a determinate proposition can still be borderline determinate (and thus borderline borderline if it is borderline whether it is determinate or borderline). The idea that borderline propositions have some 'bad' assertion and knowledge precluding status therefore does not, on its own, entail that borderline borderline propositions have to have this status as well. This point is worth stressing, for it highlights the fact that we are making an assumption when we maintain that second-order borderline propositions are unassertable, and this assumption needs to be examined.

The alternative view, which I'll ultimately suggest is superior, is that borderlineness is the only barrier to knowledge and assertion—when you are in epistemically ideal circumstances with respect to A:[7]

(FO)   One is in a position to assert/know that A (if and) only if it's determinate that A.

(Again, the parenthetical direction applies only when we have access to all the evidence we could hope to have that's relevant to the truth of A.) So that we have something to call these two views, let's call the above principle 'first-orderism', and the principle that one is in a position to assert A only if it's determinate at all orders 'higher-orderism'. The credal and doxastic variants of these views correspond to an analogous principle about evidence: must one's evidence be determinate at all orders or can it merely be determinate? And of course, Field's theory of credences is a credal version of higher-orderism.

To illustrate the alternative, we shall focus on an instance where the two views deliver different verdicts. Consider a sorites sequence of people, starting off with people who are determinately bald, moving on to people who are borderline bald, and eventually people who are determinately not bald. By familiar classical reasoning, there's going to be a point in the sequence at which the people stop being determinately bald and start being borderline bald. At the boundary, it will be borderline whether the people are determinately bald or borderline bald. Let us suppose that Harry is one of these people: it is borderline whether Harry is determinately bald or borderline bald. Let us also suppose we know everything there is to know about Harry's head—how many hairs he has, their colour, and distribution, and so on. Because it is borderline whether Harry is determinately bald, for all we know Harry is determinately bald. So let us suppose, for the sake of argument, that he is: Harry is determinately bald, but it's borderline whether Harry is determinately bald.

Now our two views deliver differing verdicts. According to higher-orderism, the kind of view articulated by Fine above, we are neither in a position to know whether Harry is bald or to assert that he is bald since it is second-order borderline whether Harry is bald. First-orderism delivers the opposing verdict: we are in a

---

[7] See also Bacon [7] where I defend this view in the context of indeterminacy and the liar paradox.

position to know and assert that Harry is bald, because we have all the relevant information about his hairline and it is determinate that Harry is bald. Thus I am in a position to assert that Harry is bald even though he is borderline borderline bald because, as it happens, he's not borderline bald he's determinately bald.

It is worth noting that even though I'm in a position to know and assert that Harry is bald according to first-orderism, it doesn't follow that it's determinate that I'm in a position to assert that Harry is bald. Indeed we can give an argument that according to first-orderism, in this particular case, it's borderline whether I'm in a position to assert that Harry is bald. Since we are endorsing (FO), it is safe to assume this biconditional is determinate. By the set-up, it's borderline whether Harry is determinately bald, so the right-hand side of this biconditional is borderline. But then by some fairly uncontroversial reasoning—that if one side of a determinate biconditional is borderline, so is the other[8]—it follows that it's borderline whether I'm in a position to know and assert that Harry is bald.

7.2.1 Vagueness and assertion

This consequence of first-orderism—that it's often borderline whether one ought to assert—might at first look like a decisive consideration in favour of higher-orderism. Let's consider one type of reason that one might be uncomfortable with this consequence. It seems to imply that we will often find ourselves in tricky normative situations: cases where it's definite that I ought to do one of two or more actions (in our case, assert or refrain from asserting), but indeterminate which of those actions I should execute. When it comes to deciding what you should do, a verdict of 'it's indeterminate' would be exasperating. You have to do something, assert or refrain; there ought to always be an answer to the question of what you should do.

I think that there is something important to this objection. Note, however, that technically speaking there is an answer to what you should do: by the law of excluded middle, either I'm in a position to assert or I'm not, so I should assert in the former case and refrain in the latter. The problem isn't that there isn't an answer to the question of what I should do, it's that it is borderline which of these two cases obtain, and so it's impossible to know which the right thing to do is. I suspect that the real source of discomfort here stems from the idea that the conditions for proper assertion should be always knowable to the asserter. Surely, the objection might go, the conditions for proper assertion should be transparent in the sense that if the conditions for proper assertion obtain, the asserter is in a position to know that they do, and if the conditions do not obtain, the asserter is in a position to know that they don't. Of course, if the

---

[8] This follows using the K principle for determinacy, which says: if it's determinate that if P then Q, and it's determinate that P then it's determinate that Q. If it's determinate that P if and only if Q, then the following two conditionals are determinate: it's determinate that if P then Q, and it's determinate that if $\neg P$ then $\neg Q$. It follows by the K principle that if Q is not determinate, then P is not determinate, and that if $\neg Q$ is not determinate, then $\neg P$ is not determinate. Thus if Q is borderline—that is, that neither Q nor $\neg Q$ is determinate—then P is also borderline.

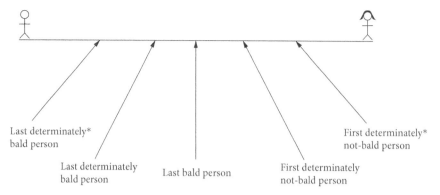

Last determinately*
bald person

Last determinately
bald person

Last bald person

First determinately
not-bald person

First determinately*
not-bald person

Figure 7.1. The cutoff points for baldness, determinate baldness, and determinate* baldness.

proper conditions for assertion are that you know the proposition you are asserting then this seems like too much to ask for; Williamson [155], for example, famously argues that we are not always in a position to know that we know when we know. But more generally, whatever the proper conditions for assertion, it seems unlikely that they would be transparent in the way required.[9]

Rather than defending this claim straight on, let me instead provide an argument that *whatever* your account of knowledge, assertion, or belief is, there will be cases completely analogous to the kinds we have been worrying about: cases where it is borderline what one is in a position to assert, know, or believe. I'll focus once again on the case of assertion. Consider once more the sorites for the property of being bald. As noted above this is not only a sorites for the property of being bald, it is also a sorites for the property of being definitely bald: the sequence begins with people who are definitely bald, and at some point it switches to people who aren't definitely bald, and the boundary is just as vague as the boundary between the bald and not bald. Arguably, this sequence is also a sorites for the property of being definitely bald at all orders (being bald, definitely bald, definitely definitely bald, and so on).[10] In Figure 7.1, moving from left to right, we see that the cutoff for being definitely bald at all orders occurs first, then the cutoff for being definitely bald, and finally the cutoff for being bald.

Now let us suppose that Alice has examined each individual in this sorites, documenting all the facts relevant to whether they are bald, such as hair number and so on. It is clear, I hope, that after acquiring all this information, Alice ought to be in a position to assert that the first member of the sorites sequence is bald. Of course, when we get to the individuals that are borderline bald, and beyond, she

---

[9] See, for example, Williamson [163].

[10] This latter assertion is more controversial because some maintain that nobody is definitely bald at all orders.

is no longer in a position to assert of these individuals that they are bald. As we have seen, it is a matter of controversy where the cutoff point is regarding which individual's baldness Alice is in a position to affirm: perhaps the cutoff lines up with the cutoff for determinate baldness, as suggested by (FO), or perhaps it lines up with the cutoff for being determinately bald at all orders, as (HO) contends, or perhaps it's neither of these two options. What I hope should be clear is that irrespective of which theory we endorse, the cutoff regarding when one is in a position to assert is not a determinate matter. The sorites we have described is a sorites for the (somewhat convoluted) property of being an individual whose baldness Alice is in a position to assert, and there is nothing particularly exceptional about this sorites: presumably the cases surrounding the boundary will be borderline cases. Of course, it is worth noting that not every sequence of small incremental changes constitutes a sorites sequence: if we keep adding grains of sand to a scale it will eventually tip, but this is not a sorites because it's a completely determinate matter at which point the scale tips. When a cutoff point in a sequence is determinate, we can usually find out where the cutoff is. The point along our sorites at which Alice is no longer in a position to assert does not seem to be like the case of the grains of sand on a scale—we cannot find out where the cutoff point is in the same way we can when the cutoff is precise.

It seems, then, that we have completely general reasons for thinking that it's sometimes borderline whether one is in a position to assert. Indeed it should not be particularly surprising that this property is vague, for vagueness is so pervasive—it would in fact be astonishing if the property *being an individual whose baldness Alice is in a position to assert* turned out to be precise.

It follows that whatever theory we adopt there will be cases where it is borderline whether we are in a position to assert; the view that assertion is indirectly regulated by what's determinate, rather than what's determinate at all orders, is therefore not alone in having this prediction.[11] It should be noted that similar arguments extend to knowledge, belief, and rational credence; it is presumably vague at which point along this sorites sequence Alice stops being in a position to know or rationally believe, and at which point her evidence stops supporting propositions about the baldness of the sorites members.

7.2.2 The role of borderlineness

A natural question to ask, once we've accepted that vagueness concerning what one is in a position to assert or know is inevitable, is what is left to recommend higher-orderism over first-orderism? In order to assess this question properly, we need to

---

[11] It is worth noting that some philosophers think that there are special logical reasons why the property of being determinately bald at all orders has no borderline cases. These reasons in fact rest on a controversial principle governing the way that determinacy iterates, the Brouwerian principle B. However, even if these considerations were correct, it is, I think, more natural to take this as evidence against higher-orderism rather than as evidence for the thesis that 'Alice is in a position to assert that …' has no borderline cases.

get a clearer handle on what borderlineness *is* in the first place, and how the notion fits into the study of vagueness. The easiest way to introduce the philosophical notion of vagueness is by looking at the kinds of puzzles in which the notion is invoked. Of course the paradigm puzzle is the sorites paradox: the sequence above, for example, begins with clear cases of baldness and ends with clear non-cases. The cases toward the middle, we notice, possess a distinctive feature that appears all over the place in other structurally similar sorites sequences. Philosophers of vagueness have rightly noticed that this feature, whatever it may be, is susceptible to a systematic theoretical treatment; for convenience let us temporarily label that feature 'X'. Upon further inspection, for example, we notice that whenever a person in the sorites possesses the feature X, the question of whether they are bald is somewhat elusive: if we know all the relevant facts about a person's hairline we usually are in a position to tell whether they're bald, yet in the cases where the feature X is present we seem to be ignorant, and this ignorance is not easy to eliminate. Indeed, we find that a whole cluster of phenomena are connected with this feature X, relating it to belief, assertion, and other rational attitudes.

On one way of conceiving of the study of vagueness, the phrase 'borderline case' is simply shorthand for whatever that elusive phenomena, X, is—the one that precludes knowledge, assertion, and certainty in that distinctive way. Of course, the distinction between the things that have the knowledge-precluding status, X, and those that don't, itself has borderline cases—this is hardly surprising, since very few distinctions are completely free from vagueness. At any rate, on this way of thinking about the role of borderlineness in the philosophy of vagueness, it is not at all strange that the borderline cases line up perfectly with the cases where the vagueness-related attitudes are appropriate: our handle on borderlineness was primarily given by the vagueness-related attitudes in the first place.

The higher-orderist must have a different conception of the study of vagueness. For according to them, borderlineness is not what causes that distinctive kind of ignorance, or that is associated with the other phenomena listed above. The source of this distinctive ignorance is actually a different and wider phenomena—this wider phenomena, as it happens, can be captured by infinitely iterating the borderlineness operator. It is the complex operator 'is borderline at some order or other', and not the borderlineness operator, that plays the interesting theoretical role we labelled 'X'. This leaves the actual role of the borderlineness operator quite mysterious; if its only purpose is to generate the interesting notion by iterating it, why don't we just introduce a name for the interesting iterated operator and do away with borderlineness altogether. The view that borderlineness and only borderlineness is responsible for the distinctive vagueness-related ignorance, I contend, gives a simpler explanation of the phenomenon of vagueness.

Moreover, simply stating that 'determinacy'/'borderlineness' is that operator which—when iterated—generates the theoretically central notion, the one we called 'X', doesn't narrow things down at all. For example, the determinacy operator Δ yields the notion of determinacy* when iterated, which is for the higher-orderist the more central notion. But so does the operator $\triangle\triangle$, $\triangle\triangle\triangle$, and so on. For if a proposition is *determinately determinate* at all orders, it is determinate at all orders, and conversely if it is determinate at all orders it is determinately determinate at all orders. That is to say, the conjunction $\bigwedge_n \triangle^n A$ is logically equivalent to $\bigwedge_n (\triangle\triangle)^n A$: every conjunct of the latter is a conjunct of the former, and every conjunct of the former is entailed by some conjunct of the latter (by the factivity of determinacy). So the former entails the latter, and the latter the former. Thus the role of determinacy, if given in terms of determinacy*, is radically undetermined.

This raises a serious question about what borderlineness is. The way that philosophers usually get a handle on it is by looking through paradigm cases, and realizing that there ought to be some general phenomena responsible for this distinctive incurable kind of ignorance that comes along with those cases. If this procedure singles out a different concept, some story about what borderlineness really is, and why it is even needed, is surely in order.

7.2.3 The forced march sorites

Let us now attempt to apply these conclusions to one of the most puzzling versions of the sorites paradox, the so-called 'forced march sorites'. This has the usual set-up: we begin with a sequence of increasingly hairy individuals. However, this time, you are going to be marched down the line of people and asked at each point whether you think the corresponding person is bald. At some point, you are going to have to switch from saying 'yes, this man is bald' to doing something else: you're unlikely to switch to saying 'no', but you'll at some point have to do something else such as saying 'it's borderline', 'I don't know', or just remaining silent.[12] The puzzle is that, assuming that it's definite at each stage whether you are affirming or not (you are not mumbling, for example), this procedure seems to allow us to associate a definite cutoff point of some sort to this sorites sequence. (Note that while classical logic guarantees the existence of cutoff points, it does not guarantee *definite* cutoff points, so the forced march on the face of it is more troubling than the ordinary sorites paradox).

On its own this is not much of a paradox—a madman might affirm at random, and even if the questionee is trying to be sincere, the cutoff might depend on something as irrelevant as their eyesight. The best version of the paradox arises when we focus on a 'perfect asserter': someone who has all the possible information about the hairlines of the sorites members at her disposal, and asserts only when she should. Note that we can give an argument that there could be such things as perfect asserters: imagine that we have a very large class of fully informed questionees, and that for each possible

---

[12] Presumably you cannot switch from saying 'He's bald' to 'It's borderline whether he's bald'. Assuming you can only assert something if it's determinate, this would commit you to the nth guy being determinately bald, and the $n+1$th guy being determinately borderline bald. This rules out certain combinations of higher-order vagueness you would have expected to be present in a standard sorites sequence like this one.

cutoff point there is someone in that class who stops affirming at that point. Because we know, by classical logic, that there's a cutoff point at which you ought to stop affirming, it follows that one of the questionees stops asserting exactly when they ought to. Of course, I've said nothing about why these people stop affirming when they do—they might stop affirming for all the wrong reasons—but asserting for the right reasons is not built into the definition of a perfect asserter: they need only to—in fact—assert when they should, and refrain when they shouldn't.

This version of the paradox, one might think, is particularly troublesome for the view I have endorsed here. Indeed, you might even think the forced march paradox provides a recipe for discovering the locations of seemingly undiscoverable cutoff points. For all you need to do is find a perfect asserter and simply observe the point along our sorites sequence at which she stops affirming. If I'm right, then this would allow us to discover the number of hairs (say) at which people stop being determinately bald: if the asserter is perfect, she is in a position to assert the baldness of all and only the determinately bald people.

This argument, however, does not really get any traction: in order for us to apply this procedure, we first need to find a perfect asserter, and this is a hard task—not because there aren't any, but because it's hard to know of someone that they're a perfect asserter. For example, suppose that as it happens the nth guy is the last determinately bald man in our sorites sequence, and suppose Alice stops asserting at the nth guy, and Bob stops asserting at the $n + 1$th guy. Even though, as it happens, either Alice or Bob is a perfect asserter, it's borderline which of the two is the perfect asserter. So even if I observe the points at which both Alice and Bob stop affirming, I still can't conclude anything about the location of the last determinately bald person, because I don't know which the perfect asserter is.

In short, the forced march paradox only gets off the ground if you restrict attention to idealized asserters. However, it is exactly because of the vagueness in the notion of 'being in a position to assert' that the forced march does not allow us to associate determinate cutoff points to a sorites sequence.

7.2.4 Paradoxes of higher-order vagueness

As we noted earlier, there is a tight connection between higher-orderism and the rule of DETERMINACY INTRODUCTION. Rules that preserve determinacy at all orders, like the rule of Δ introduction, are acceptable ways of reasoning because they preserve attitudes like rational belief and knowledge. This observation should give us reasons to be less than optimistic about higher-orderism, since, as we shall see in a moment, the rule of DETERMINACY INTRODUCTION is known to be susceptible to the paradoxes of higher-order vagueness. Delia Graff Fara has shown that it conflicts with some eminently plausible principles capturing the idea that we can't associate precise cutoff points with sorites sequences; Fara calls these principles the *gap principles* (see Fara [46]). Fara's argument explicitly appeals to the rule of DETERMINACY INTRODUCTION, which raises some questions about how to interpret rules of inference which I think are irrelevant to our present concerns. To circumvent these questions, I shall instead examine the principles about credence, assertion, and knowledge (DI1)–(DI3) directly, helping myself only to ordinary propositional logic. Indeed, we will be able to derive a direct contradiction between the hypothesis that we should be certain in the gap principles and several of the theses about belief, knowledge, and assertion we have considered so far, including Field's claim that for rational credences Q, $Q(A) = Q(\Delta A)$.

Gap principles effectively state that sorites sequences don't have sharp cutoff points at any orders. For example, if you had a sorites sequence of people starting with people with no hairs who are clearly bald and ending with people with many hairs who are clearly not bald, then the most basic gap principle amounts to the claim that there isn't a sharp cutoff between the bald and non-bald: if the nth guy is determinately bald, the $n + 1$th guy isn't determinately non-bald. In symbolism $\Delta A_i \to \neg\Delta\neg A_{i+1}$ where A_i represents the claim that the ith person is bald. To deny this would be to countenance a determinately bald person adjacent in the sequence to a determinately non-bald person, and this is just to say that it's not vague where the cutoff point for being bald is.

A sorites for the property of being bald is also a sorites for the property of being determinately bald. Although classical logic guarantees that there's a last determinately bald person in the sequence, it ought to be borderline which that last person is: so if it's determinate that a given person is determinately bald, it shouldn't be determinate that the adjacent person isn't determinately bald. This would be problematic for the same reasons that determinacy at the first order is problematic: if there were a single hair that could make the difference between being determinately bald and not determinately bald, and it weren't borderline where that cutoff is, why is it we can't determine where the cutoff point is? This kind of inability to say where the cutoff points are is exactly the kind of phenomenon that vagueness was supposed to explain, yet here we would have the phenomenon without the vagueness. Similar observations apply for the cutoff point for being determinately determinately bald, determinately determinately determinately bald, and so on. Thus, writing Δ^n for a sequence of n Δs, the general gap principle is:[13]

GAP PRINCIPLE: $\Delta\Delta^n A_i \to \neg\Delta\neg\Delta^n A_{i+1}$.

The informal considerations above, I hope, give us reason to accept the gap principles. Thus there ought to be rational credence functions, Q, that accept the gap principles by assigning them a probability of 1. Moreover, since it is practically a conceptual truth that someone with no hairs is bald and that someone with a million hairs isn't, this same rational credence function ought also accept A_0, the claim that the first guy is bald, and accept $\neg A_{1,000,000}$, the claim that the guy with a million hairs is not bald. (In fact, all we will need below is the assumption that $Q(A_0) > \frac{1}{2}$ and $Q(\neg A_{1,000,000}) > \frac{1}{2}$.)

---

[13] Note that if we assumed S4, as Field does, the gap principles are all equivalent to the $n = 1$ instance.

We can now turn these considerations into a formal inconsistency in Field's probability calculus. Below we use $\mathcal{M}$ to abbreviate $\neg\Delta\neg$:[14]

1. Q assigns the gap principle probability 1, so $Q(\Delta^m(\Delta^{n+1}A_i \to \mathcal{M}\Delta^nA_{i+1})) = 1$ applying ITERATION m times.
2. $Q(\mathcal{M}^m\Delta^{n+1}A_i \to \mathcal{M}^{m+1}\Delta^nA_{i+1}) = 1$ from 1, by the fact that $\Delta^m(A \to B)$ entails $(\mathcal{M}^mA \to \mathcal{M}^mB)$ in KT.
3. $Q(A_0) = 1$, so $Q(\Delta^{1,000,000}A_0) = 1$ by applying ITERATION a million times.
4. Then $Q(\mathcal{M}\Delta^{999,999}A_1) = 1$ from 3 by 2 with $n = 999,999$, $m = 0$, and $i = 0$.
5. Then $Q(\mathcal{M}\mathcal{M}\Delta^{999,998}A_2) = 1$ from 4 by 2 with $n = 999,998$, $m = 1$, and $i = 1$ ....
6. So $Q(\mathcal{M}^{1,000,000}A_{1,000,000}) = 1$.
7. But also $Q(\neg A_{1,000,000}) = 1$ so $Q(\Delta^{1,000,000}\neg A_{1,000,000}) = 1$. Thus Q assigns 1 to two claims that are inconsistent in KT (see 7 and 6), which contradicts the axiom F2.

Parallel arguments can be run for principles (DI1)–(DI3). For example, if I'm in a position to know that A_0, it follows by (DI2) that I'm in a position to know that $\Delta^{1,000,000}A_0$, and that if I'm in a position to know $\neg A_{1,000,000}$, I'm in a position to know $\Delta^{1,000,000}\neg A_{1,000,000}$. But if I'm also in a position to know the gap principles, and what I'm in a position to know is closed under logic, it follows that I'm in a position to know that $\mathcal{M}^{1,000,000}A_{1,000,000}$ by reasoning completely analogous to the above. Thus according to this argument I'm in a position to know two mutually inconsistent propositions: $\mathcal{M}^{1,000,000}A_{1,000,000}$ and $\Delta^{1,000,000}\neg A_{1,000,000}$.

First-orderism, by contrast, is not susceptible to the paradoxes of higher-order vagueness. It is sometimes suggested that the paradoxes of higher-order vagueness are paradoxes for everyone. For example, Zardini [169] formulates a version of Fara's paradox that does not rely on Δ introduction, but instead relies on the premises of Fara's argument being determinate at all orders. In particular, to get Zardini's argument off the ground, one must assume that not only are the gap principles true, but that they are determinate at all orders. According to the first-orderist, the gap principles may be true, and even assertable, but these concessions carry no commitment to the idea that they be determinate at all orders.

7.3 Should Our Credences in the Vague Obey the Probability Calculus?

In section 7.2.4 we suggested that Field's principle ITERATION, and the wider picture of the relation between attitudes and higher-order vagueness it is a part of, is not particularly attractive. But this position on the relation between credences and

---

[14] This argument is an adaptation of the argument presented in Fara [46].

higher-order vagueness was only one of the distinctive features of Field's theory. The other interesting feature of the theory was REJECTION: the view that if one is certain that A is borderline one should fully reject both it and its negation. In a probabilistic setting, this requires that one give up the axiom of finite additivity, which states that one's credence in a proposition and its negation add up to 1.

As we noted in section 7.2.4, ITERATION entails REJECTION. However, the converse is not true, and it is quite simple to construct models of REJECTION that do not validate ITERATION. (One can generate models of REJECTION as follows. Take any model of the determinacy operator—a reflexive Kripke frame—and take a classical probability function defined over the sets of indices in the model.[15] We can then identify the Fieldian credence in the proposition A as the classical probability assigned to the proposition that A is determinate in that model. These models won't in general satisfy ITERATION.[16]) One can, therefore, separate the problematic principle extending vagueness-related attitudes to the higher-order vague from the insight about what vagueness-related attitudes actually amount to, captured by REJECTION.

This all suggests that the kernel of Field's account of vagueness-related uncertainty is not susceptible on its own to the paradox in section 7.2.4, which depended essentially on ITERATION. One could get a reasonable theory of degrees of belief by looking at the weaker theory of probability gotten by looking at credence functions generated from Kripke frames in the way outlined above. This theory includes the axiom F2, REJECTION, and a principle of subadditivity stating that the probability of a disjunction of pairwise incompatible propositions is less than or equal to the sum of the probability of the disjuncts (see footnote 2).

To assess the prospects of the probability calculus, as an account of vagueness-related uncertainty, it is most natural to compare it with this weakened version of Field's theory that takes no stand on ITERATION.

7.3.1 Dutch book arguments

The standard way to get a handle on these kinds of questions is to look to rational betting behaviour to provide some constraints on what kinds of credences are acceptable. For example, it is common to assume that your credence in a proposition, p, can accurately be revealed by your dispositions to accept certain kinds of bets. Thus, for example, if an agent has a credence of x in a proposition p, then they would accept any bet that costs less than \$$x$ to buy and pays out \$1 if p and nothing otherwise. Given this connection between credences and betting behaviour, we can give arguments that certain structural constraints on credences are required by rationality: agents who

---

[15] Assume that there are only countably many worlds, so that we don't have to worry about unmeasurable sets.

[16] We can invalidate ITERATION as follows. Let $W = \{0, 1, 2\}$, Rxy iff $|x - y| \leq 1$, let Pr be the uniform probability distribution over W, and define $Q(A) := Pr(\Delta A) = Pr(\{x \mid y \in A \text{ whenever } Rxy\})$. Then REJECTION holds in this model, but $Q(\{0, 1\}) = \frac{1}{3}$ while $Q(\Delta\{0, 1\}) = 0$.

violate the constraints will buy bets that are guaranteed to lose them money no matter what happens—the classic 'Dutch book argument' for probabilism takes this form, for example. One might hope that such arguments could be extended to cases involving vague propositions and used either to vindicate probabilism or Field's theory, thereby settling our question.

Of course, the assumption that an agent's credences are revealed by their betting behaviour is not always reasonable. For example, you could imagine an extreme philanthropist who cares only about giving to others, but whose only means of giving is through losing bets. The philanthropist is not necessarily being incoherent when she accepts bets that are guaranteed to lose, nor is she even being irrational: she is achieving what she desires the most.

Another type of situation where the assumption isn't plausible: suppose your friend offers you a bet, to be paid out immediately, that we will have created strong artificial intelligence by the year 2214. Here I take it that your betting behaviour does not reveal your credences regarding the existence of AI by the year 2214, because you can be pretty sure that your friend doesn't know the answer and that neither of you will live to find out the answer. It follows that you cannot be certain that your friend will pay you if there will be AI by 2214. In what follows, it is important to bear in mind these exceptions to the credence–betting behaviour link.

On first glance, it looks as though actual betting behaviour closely matches the betting behaviour predicted as rational by Field's theory. When people are offered bets on a proposition they know to be borderline, it seems clear that one shouldn't accept favourable bets on that proposition, nor should they accept favourable bets on the negation of that proposition. If, for example, I know that it's borderline whether Harry is bald and someone offers me a bet that costs a cent and pays out $1 if Harry is bald it seems as though I should reject it, which by the betting–credence connection suggests my credence should be less than 0.01. Similarly, I should reject the symmetrical bet that costs a cent and pays out $1 if Harry is not bald. Thus, assuming the betting–credence link outlined above, my credence in the proposition that Harry is bald and in its negation is 0.01 in both cases, which is incompatible with probabilism but exactly the kind of situation predicted by Field's theory. One might hope to go one further and turn this into a Dutch book argument for Field's theory of probability, and in fact Richard Dietz does exactly this (see Dietz [32][17]).

However, on closer inspection it seems as though this argument involves exactly the kind of situation that we already set aside as being the kind of case where your betting behaviour does not reveal your credences. A bet on whether Harry is bald is like a bet on whether there will be AI by 2214 in the sense that in both cases we can be pretty confident that the bookie will never know the answer to these questions. You therefore cannot be sure that the bookie will pay you back in a way that corresponds

---

[17] Smith [135] also uses Dutch book considerations to argue for a non-standard theory of probability.

with the bet you accepted. Your dispositions regarding bets on the proposition that p only reveal your credence that p, if you are certain that you'll get the payoff if and only if p.

(Note, of course, that if Field is right, then we are not uncertain about whether the payout will correspond with the proposition we are betting on: we assign a credence of 0 both to the claim that it will correspond and to the claim that it won't. However, we are still ignorant in the sense that we know neither that the payout will correspond to p nor that it won't, and so it is like the kind of case we set aside in the respect that matters.)

There are ways to define what it means for something to count as a 'bet on p' in which it is plausible that you know the payoffs of a bet on a borderline proposition. For example, Dietz defines a bet on p to be something which is cancelled if p turns out to be neither true nor false. The bet pays out if p is true, you lose if p is false, and you get refunded otherwise. This manoeuvre merely disguises the fallacy and moves it elsewhere. When Dietz talks about truth, he is not using it in the disquotational way we have been using it—he rather means something like supertruth. The proposition that p and the proposition that it's supertrue that p are not identical—whenever p is supertrue, p, but not conversely. Consequently a bet that pays out if and only if p is not the same as a bet that pays out if and only if it's supertrue that p. It's natural to think that the agent's betting behaviour with respect to Dietz's bets might reveal something about her credence about whether it's supertrue that p. However, why think that such bets will reveal an agent's credences in p? If you thought that the agent's credence in the proposition that p and her credence in the proposition that it's supertrue that p had to be the same, then you could make a case that these two things amount to the same thing. But this is exactly Field's theory of probability, and so a Dutch book that assumes this theory from the get go cannot provide an independent argument for Field's theory, or an independent way of undermining probabilism.

Note that a similar move could be applied completely generally to bets made when the bookie is ignorant: bets on p could simply be cancelled when the bookie isn't in a position to verify whether p. Perhaps if this were the agreement, I'd be indifferent about making a bet with my friend about AI in 2214, but clearly the amount I'd be willing to spend on this bet wouldn't say anything about how likely I take it to be that there is artificial intelligence by 2214. If I were confident that the bet would be cancelled whatever happens, I could bet as I pleased, confident that I'll just get my money back with no loss or gain.

Is the idea of using betting behaviour to make some ground on the debate between probabilism and Field a completely lost cause? To actually get a connection between betting behaviour and credences in a borderline proposition, p, we'd have to somehow set things up so that the agent could be sure that she will get the payoff iff p.

This appears to be impossible when p is borderline; however, it is not entirely obvious that it can't happen. Consider the following scenario. Suppose that over a period of four years Fred gradually succumbs to madness. At the beginning of the four years he is perfectly sane, and by the end of the four years he is clearly insane. Let us suppose that the law works as follows: that at the time of a person's death, the beneficiaries of the most recent valid will automatically come to legally own any money and property left in that will, and that a will written by a madman is not valid. We may assume that madness is the only reason Fred's will might fail to be valid. Now suppose that at some point at which he is bordering on madness, Fred offers you a bet over whether he is currently mad. The bet costs $50 to participate in: if he is mad you'll get $100 and if not you'll get nothing. How can you possibly be certain that you'll receive the $100 just in case he's mad? It is simple—he writes it into his will. We may assume that when he dies it is indeterminate whether the $100 belongs to you, or to whomever it is promised in the other versions of his will. However, due to the nature of the situation, it is determinately true that you will legally own the $100 if and only if he was mad. Thus you may be certain that the bet is not defective; you are certain that you'll legally own the money if and only if he's mad.[18]

However, it is natural to think that this type of example involves the other kind of scenario where betting behaviour and credence come apart, discussed in relation to the philanthropist example. If the agent does not care about owning money, then we cannot conclude anything about her credences from her dispositions to accept bets with monetary payoffs. Now you might think that normal people only care about legally owning money because of the things they can buy with it, and thus would have little use for owning money that is only borderline owned. Since in the above example it will be at best borderline whether you own the money if you accept the bet, then it looks as though the assumption that our preferences correspond linearly with money breaks down.

One could respond by pointing out that even if normal people don't care intrinsically about owning money, you could imagine somebody who did—surely it's possible to care intrinsically about *anything*. In chapter 10, I will argue that even this response is limited—I'll argue that it's not rational to care intrinsically about the vague. For the time being, however, suffice it to note that to get a Dutch book for probabilism up and running is a tricky task that would have to be quite contrived: credences about borderline matters rarely ever reveal themselves through rational betting behaviour.

The Dutch book argument is hard to get going because it relies so heavily on using physical behaviour as a way of measuring an agent's credence. However, this does not mean that there are no good arguments in the vicinity. A better argument for probabilism might skip the connection between degrees of belief and physical behaviour, and work directly with the connection between degrees of belief and desire. Arguments that start with assumptions about what desires and preferences are rational, and from this determine that credences ought to be probabilistically coherent, are corollaries of the representation theorems for decision theory. Unfortunately, we do not have the

---

[18] Even though actual legal ownership does not work like this, this simplified example makes it plausible that a more realistic case could be constructed.

necessary background to give this argument here, but we shall return to this argument in section 9.4.

Although it does not seem as though Dutch book arguments are of much help for settling this question, these considerations do suggest a lacuna in the Fieldian account of probability. For although we can argue, from an ad hoc principle, that Field's theory predicts that someone may refuse bets with excellent odds on a proposition and with the same odds on its negation, if they are certain the proposition is borderline, it would be nice to derive this principle from a more general decision theory.

Unfortunately, it is very much an open question whether such a theory can be developed in sufficient generality. For example, standard decision theory works with the notion of the *expected utility* of an action, which has the property that for any proposition p that's independent of your actions, the expected utility of an action is identical to the expected utility of the action given p times the probability that p plus the expected utility of the action given that not p times the probability that not p. However, according to Field, this result would yield an expected utility of 0 whatever the action was, if we were to choose a p that we were certain was borderline: we would be multiplying by each summand by 0. This result would be devastating for Field, so whatever theory we use, it cannot be founded on the notion of expected utility as standardly conceived.

A natural way around this would be to look at classical probability functions related to Fieldian probability functions in suitable ways and use these probability functions to calculate the expected utilities. Since the probability of p and $\neg p$ can add up to less than 1 for Fieldian functions, Fieldian functions systematically underestimate probabilities relative to classical probabilities. However, for each Fieldian function there are probability functions that agree or assign higher probability than the Fieldian function for every proposition—indeed, there will typically be many such functions. For each choice of one of these functions, there will be a range of expected utilities we can assign to an action: there are thus a number of possibilities for defining preferences from these utilities. Perhaps the agent should prefer p to q if the lowest expected value p can take relative to one of these functions is greater than the greatest value q can take. This is a strong notion of preference: the range spanned by the possible utilities of p has to be completely above the range spanned by q. Other proposals in the ballpark are also possible: perhaps we should compare the lowest values of both p and q, or the two highest values, or the highest of p against the lowest of q.

I think none of these ideas accord with our intuitions particularly well. For example, let us suppose that I'm presented with a bag containing four balls. I know that three of them are clearly red, but one of them is a borderline red/orange ball. A ball has been selected at random; I'm pretty, but not totally, sure that the ball is red. Letting my Fieldian credence be Q and letting R be the proposition that the ball is red then $Q(R) = \frac{3}{4}$ and $Q(\neg R) = Q(\triangle \neg R) = 0$. Now I'm offered two bets: one pays out a dollar if the ball is red, the other pays out a dollar and twenty cents. Now of course, to get this to work we'd have to make the payouts penumbrally connected with the redness of the balls in contrived ways, but assuming that this can be done, it seems obvious that I should prefer the second bet to the first, and that this should be born out in my preferences over the proposition that I accept the first bet over the second. However, if we look at the first proposal we don't get this: the expected utilities of the former bet spans [0.75, 1] while the latter spans [0.9, 1.2], and the lowest value of the first bet is not greater than the highest value of the second. (Similar problems can be constructed if we instead compare the highest to the lowest value.)

Comparing the two lowest values doesn't do any better. If I'm certain that Harry is bald then my Fieldian credence in that and in its negation is 0. One can then construct classical probability functions related to your credences in the appropriate way that assign the conjunctive proposition that Harry is bald and I win the lottery the probability 0. Thus the proposal predicts that I should prefer it to be the case that I have won ten dollars than for it to be the case that I have won a million dollars and that Harry is bald. A similar problem can be constructed if we compare the two highest values (just replace winning the lottery with something very bad, and winning ten dollars with something bad but not very bad).

Lastly, rather than looking at the highest and lowest values that action can take, you might compare, for each choice of admissible probability function, the expected value of p to that of q. If each admissible probability function agrees about which is better, then one is better *simpliciter*. Unfortunately, if the functions don't all agree about which is better we get incomparable preferences: two options that aren't just as good as one another, but neither is one better than the other. There has been a long-standing puzzle concerning how we should relate such preferences to our decisions. We have to make a choice one way or the other when we're faced with a decision, so people typically just behave as though they did have a preference one way or another; therefore it's hard to say what the meaning of these incomparable preferences amount to. (See Elga [41] for further discussion of these problems.)[19]

The brief considerations above are quite clearly not supposed to be an exhaustive discussion of all the possible ways of developing a decision theory within Field's framework. It is merely a survey of the approaches that strike me as most promising; their shortcomings give us at least a good reason to suspect that an adequate theory will not be forthcoming.

7.3.2 Comparative probability judgements

As we noted in section 7.3.1, it is not obvious that we can settle the question of probabilism through a straightforward Dutch book argument: both the standard Dutch book argument for probabilism, or Dietz's non-standard Dutch book for Field's theory, require some implausible assumptions—bookies who know borderline propositions, people who care intrinsically about the vague, and so on.

---

[19] If we assume the principle of conditional excluded middle, it's also true that our unactualized dispositions to act also are determined by comparable preferences in the sense that either, if I were to be offered the choice between A and B, I'd choose A, or if I were to be offered that choice, I'd choose B.

Let us now turn to some other ways of settling this debate. A natural place to start looking is our pretheoretic intuitive judgements about probabilities. Needless to say, some of these judgements should be taken with a grain of salt. However, I think they do provide us with some kind of defeasible support for probabilism. I'll start by looking at a couple of intuitive probability judgements that cast doubt on Field's probability theory, and then I'll attempt to generalize this thought to give an argument for probabilism.

Consider the following example: imagine I'm about to roll a hundred-sided die, whose sides are labelled 0 to 99. The probability that the die will land on any particular number is 1%. So intuitively I should be fairly sure that the number it's going to land on won't be the last two digits of the age in nanoseconds at which I stopped being a child; which is, after all, a particular number. There are ninety-nine numbers which aren't this number, and only one that is, so there's a 99% chance it will land on one of the other numbers. So I should be 99% certain that it's not going to land on the last two digits of time in nanoseconds at which I stopped being a child. But I'm absolutely certain that whichever number it does land on, it'll be borderline whether it's the last two digits of the critical time at which I stopped being a child, because I'm certain there are at least 100 borderline cases (I was a borderline child for well over a second, so there will be thousands of borderline cases). Prima facie, this seems to be a case in which $Cr(\nabla p) = 1$ (or at least, high) and $Cr(p) = .99$, contradicting Field's prediction that $Cr(p) = 0$ (or low).

Field's proposal also runs afoul of our intuitions about comparisons of probabilities of borderline propositions. According to Field if you are certain that p and q are borderline, you cannot think that p is more likely than q, or that q is more likely than p. They both receive credence 0. But this, once again, does not accord with our intuitive judgements. Suppose you are looking at two glasses, one of which is 65% full and the other 67% full. Now you might be certain that it's borderline, in both cases, whether the glass is pretty full, but you should be *more* confident that the second is pretty full, since it is clearly more full than the former.

Let's see if we can generalize this thought. The framework that I shall adopt is Bruno de Finetti's theory of comparative probability whose only primitive is the notion of one proposition being at most as probable as another, written $A \leq B$. We can state what it means for A to be strictly less probable than B, $A < B$, with the definition $B \not\leq A$.[20] Given this interpretation the following principles (adapted from de Finetti's axiomatization) seem very natural, at least when restricted to the domain of precise propositions:

1. $\leq$ is a total ordering (it is reflexive, transitive and connected).
2. $\bot \leq A \leq \top, \bot < \top$.
3. $A \leq B$ if and only if $A \vee C \leq B \vee C$, whenever $AC \leq \bot$ and $BC \leq \bot$.

---

[20] The adequacy of the definition depends on the axioms governing $\leq$, and in particular, on the axiom that $\leq$ is connected.

One thing to observe about this axiomatization is that it makes very few logical assumptions. I have assumed that there is an inconsistent proposition, $\bot$, intuitively to be thought of as the conjunction of all propositions, and a tautologous proposition, $\top$, to be thought of as the disjunction of all propositions. The existence of these propositions are rarely, if ever, contested by deviant logicians and there is no straightforward reason why a deviant logician couldn't accept these principles. In de Finetti's original axiomatization, the claims '$AC \leq \bot$' and '$BC \leq \bot$' in the third principle—intuitively stating that AC and BC have no probability—are replaced by the statements that 'A and B are incompatible' and 'B and C are incompatible', where inconsistency in classical logic is taken to be sufficient for incompatibility. For this reason, de Finetti's axiomatization is not logic-neutral.

Are these principles any less attractive once we allow the domain of $\leq$ to include vague propositions? I take it that axiom 2 is practically definitional of $\bot$ and $\top$, and that the transitivity and reflexivity of $\leq$ are unassailable. This leaves us with axiom 3, which corresponds intuitively to finite additivity, and the connectedness of $\leq$.

To say that comparative likelihood is connected is just to say that for any pair of propositions one is at least as probable as the other. There are certainly reasons to worry about this principle that have nothing to do with vagueness. Here is an example from Elga [41]: 'A stranger approaches you on the street and starts pulling out objects from a bag. The first three objects he pulls out are a regular-sized tube of toothpaste, a live jellyfish, and a travel-sized tube of toothpaste.' It is very hard to imagine, in this scenario, that the proposition that the next object will be a jellyfish is either at least as probable or more probable than the proposition that it'll be a tube of toothpaste. You might feel that, intuitively, these propositions are *incomparable* because your evidence is just too unspecific to allow a determinate comparison of probability.

Elga goes on to describe a number of compelling arguments for the claim that, despite appearances, these propositions are comparable. However, the important point, as far as we're concerned, is that this phenomenon has nothing specifically to do with vagueness. Thus, for example, while it might be very hard to say whether or not it's more probable that Harry is bald than that Sally is old when they are both borderline, matters are no better in the precise case either. It is just as hard to imagine that it's more, less, or just as probable that I'll live until I'm eighty as that it'll rain tomorrow. Notice, however, that when the propositions are about similar subject matters, it becomes much easier to make comparative judgements in both the vague and in the precise cases. For example, it is much easier to say which of two people is more likely to live until they're eighty, assuming you know a little bit about their lifestyles. Similarly, you can quite easily make judgements about which of two people is more probably bald than the other, given you know roughly what their hairline looks like even when you know both people fall within the borderline region. It seems, then, that worries to do with comparability have nothing specifically to do with vagueness; it is more likely that they stem from the general difficulty of making probabilistic comparisons between different subject matters.

Of course, there are some who insist that probabilism is false for reasons having nothing to do with vagueness; such people are often motivated by the kinds of cases described above involving comparative probability judgements across subject matters. Let us set those sceptics aside for the time being. For the rest of us—who are fine with probabilism in the general case—the connectedness of comparative likelihood seems to be no more controversial in the presence of vagueness as elsewhere.

Axiom 3 states that if AC and BC are maximally improbable (are no more probable than the conjunction of every proposition) then A is no more probable than B if and only if $A \vee C$ is no more probable than $B \vee C$. In Field's theory this principle fails. For someone who knows that Harry is borderline bald, the proposition that Harry is bald is no more probable than the inconsistent proposition $\bot$, yet the proposition that either Harry is bald or he isn't is strictly *more* probable than the proposition that either the inconsistent proposition is true or Harry is not bald—the former has probability 1 and the latter 0.

Field's theory, however, collapses all distinctions of comparative probability between propositions we know to be borderline; as I said above, this strikes me as quite counterintuitive. Suppose, then, that we do acknowledge non-trivial comparisons of probability between propositions we know to be borderline. What kind of general principles governs these comparisons? Axiom 3, I think, is an extremely plausible candidate. Indeed, it seems to me like a fairly small step between acknowledging that non-trivial comparisons can be made between vague propositions to accepting the full generality of 3. If you judge A to be more probable that B, whether or not they are borderline, then you should judge $A \vee C$ to be more probable than $B \vee C$ whenever C is incompatible with both A and B.

Let us demonstrate the axiom with an example explicitly involving vagueness. Depending on the outcome of a coin flip, I am going to paint two clay pots, pot A and pot B. If the coin lands heads I'll paint them both the same shade of orange. If it lands tails, I'll paint both A and B two slightly different shades that are both borderline between green and blue. You do not know how the coin has landed, but you do know what shade each pot will be painted in each eventuality, and in particular you know that in fact, if the coin lands tails, A will be painted a shade that is closer to the green end of the spectrum than the shade that B is painted (and you know that both shades are borderline green). Presumably, even before you learn how the coin landed, and, thus, how the pots are painted, you should be more confident that A is green than that B is green. This is, at least, the intuition we began with. Once we have that premise, the following piece of reasoning seems to be valid:

1. You are more confident that A is green than that B is.
2. So, you should be more confident that either A is green or the coin will land heads than that B is green or the coin will land heads.

That is, if you accept the premise in the first place—you take the intuition that it's more probable that A is green than that B is at face value—the conclusion seems reasonable. This is exactly the kind of move that 3 permits.

Now what is quite surprising is that from these three minimal principles governing comparative probability, one is almost in a position to prove that your credences can be represented by a probability function. That is to say, given that we can make comparative judgements of probability, even when we know the things being compared are borderline, then it looks as though we have an argument for a probabilistic account of these judgements—one in which one's credences are given by probability functions.

Note that I say we are only 'almost' in a position to prove that our credences are represented by a probability function. Indeed, de Finetti conjectured exactly this: that for any ordering satisfying (1)–(3), it is possible to construct a classical finitely additive probability function which agrees with comparative ordering about the relative probabilities of each pair of propositions. Unfortunately, as Kraft et al. [83] showed the answer is 'no': there are some orderings over small finite algebras that satisfy de Finetti's axioms (and our variants) that are not representable by probability functions. However, the axioms do suffice for representability, provided the structure of propositions is sufficiently rich. To get around this, one can add some assumptions to ensure there are enough propositions, and thus enough comparisons floating about to generate a probability function. Here is one condition that suffices for:[21]

SUPPES' CONDITION: If $A \leq B$, then there is a C such that $A \vee C \approx B$.

Here $A \approx B$ simply means that $A \leq B$ and $B \leq A$. This principle isn't obviously purely about probability, but it ensures that there are enough propositions, and thus comparison facts, floating about to guarantee the existence of a probabilistic representation of the ordering facts.[22]

The upshot of all this, I think, is that unlike the Dutch book arguments, the argument from comparative probability judgements provides an argument for probabilism that has purchase even in the presence of vagueness.

7.3.3 Is there anything special about vagueness-related uncertainty?

I have argued for a broadly classical, Bayesian epistemology: a view in which rational credences are governed by the probability calculus, and in which rational credences are given by conditioning on the available evidence. Moreover, we have argued that these basic tenets of Bayesianism hold even when vague propositions are involved. By contrast, we have seen that one prominent alternative to the classical view requires one to adopt a deviant decision theory and conflicts with natural judgements about

---

[21] Suppes' condition entails that all atoms are equiprobable, which you might think is too strong a structure condition. LIER's CONDITION doesn't have this consequence and also suffices for representability: if A and B are atoms and $A < B$ then theres a C such that $A \vee C \approx B$. See Lier [96].

[22] Note that I have not included the Archimedian axiom which is often included to ensure that the entities that the probability function outputs will have the structure of the real numbers. Thus some of these comparative orderings won't literally be represented by probability functions; they will rather be represented by something pretty close to a probability function: either a function into the non-standard reals, or by a Popper function. However, these are close enough to probability functions to count the resultant ordering as probabilistic for my purposes. See Regoli [119].

the probabilities of vague propositions. In chapter 9, we will show that the present approach to propositional vagueness is also compatible with a completely classical decision theory as well.

The result of all this is that the kind of uncertainty that vagueness generates is not formally distinctive in the way that Field, and other authors, have argued. When you are uncertain about A because you know that it is borderline whether A, the psychological state you find yourself in is not fundamentally different from the psychological state you find yourself in when you are uncertain about where you've parked your car. Given this, it is natural to wonder whether there is *anything* special about vagueness-related uncertainty. We turn to this question now.

8

Vagueness and Uncertainty

Epistemicism is notorious for making a pair of counterintuitive claims.[1] According to epistemicism there was a nanosecond during which I stopped being a child: I was a child at the beginning of it, but not at the end. Moreover, according to the epistemicist, it is not possible to know when that critical nanosecond occurred. It is these consequences of epistemicism that regularly invite the incredulous stare and which have led people to try and find alternative theories that respect classical logic; the most famous alternative being supervaluationism.

Unfortunately for this project, it is possible to prove that there's a nanosecond at which I stopped being child, and that it is unknown which nanosecond it is, from a couple of eminently plausible premises. As we noted in chapter 1, apart from classical logic, all we need to derive the existence of a nanosecond at which I stopped being a child are the premises that I was a child after one nanosecond of my life had passed, but not after several billion had passed. But once we have accepted the existence of such a nanosecond it would be madness to suppose that we know which one it is. If you are not immediately convinced by this latter claim ask yourself which number it is: if you are unable to produce a satisfactory answer, I would suggest that this is because you do not know which number it is.

Thus the project of finding a more palatable alternative to epistemicism that accepts classical logic is pretty much a no-go if one takes these above consequences of epistemicism to be the source of the unpalatability. However, some theorists have suggested that the radical component of epistemicism is not the thesis that there is an unknown nanosecond at which I stopped being a child, but the claim that vagueness amounts to nothing more than this special kind of ignorance.[2] What explains these astonishing theses asserting the existence of unknown cutoff points, according to these theorists, is the thought that when there is vagueness about where a boundary lies, there is also no fact of the matter about where it lies.

---

[1] By 'epistemicism' I just mean the cluster of views often associated with Paul Horwich, Roy Sorenson, Tim Williamson, and a few others; a stricter definition would be unhelpful at this juncture.

[2] I personally find this response quite surprising—it is the former, not the latter claim that invites the incredulous stare. Once we have fully absorbed that there is an unknown nanosecond during which I stopped being a child, the correct response to these observations should be that epistemicism isn't quite as radical as it first seems.

Unfortunately, the locution 'there's no fact of the matter' is somewhat mysterious without further explanation. Since we are assuming classical logic, it follows that Harry is either bald or he isn't, even when there is no fact of the matter about whether Harry is bald. Thus one of the following must hold: either (i) Harry is bald but there's no fact of the matter about whether Harry is bald, or (ii) he isn't bald but there's no fact of the matter about whether he's bald. Both disjuncts are equivalent to something of the form 'P but there's no fact of the matter about whether P'. But this result does not seem to be consistent with our pretheoretic understanding of the locution 'there's no fact of the matter'. This suggests 'there's no fact of the matter' is at best a semi-technical notion in need of further explanation. At this juncture the epistemicist could triumphantly point out that 'P but it's impossible to know whether P' is perfectly consistent, and seems to play the right role in explanations as well— perhaps the semi-technical talk of there being no fact of the matter is just code for talk about things being unknowable for certain kinds of reasons. This would be an outright defeat for the classical theorist I am describing: the challenge, then, is to find some more technical notion that is consistent with the consequence that something can be true even when there's no fact of the matter, but excludes an epistemicist reading of it.[3]

As I have warned earlier, the task of attempting to find an uncontroversial classification of theories of vagueness as 'epistemicist' or 'not epistemicist' is not a productive one. That said, there is a set of questions that I think helps clarify the view that vagueness is *just* a matter of ignorance or uncertainty. If one can identify distinctive features of vagueness-related uncertainty that are dramatically different from ordinary uncertainty, it becomes harder to maintain that there's little more to vagueness than ignorance. For if uncertainty about whether Harry is bald is *just the same as* uncertainty about where I left my keys, for example, we shouldn't expect to see very dramatic differences. In recent years this strategy has been pursued by a number of authors.[4] The operative thought in all these cases is that when one is uncertain about something because it is believed it to be borderline one is in a very different kind of state than when one is uncertain about some ordinary matter such as where a set of keys has been left. However, these theorists typically end up rejecting the probability calculus; a thesis quite central to the present project (see chapter 7). In this chapter we shall explore, drawing on some analogies with expressivism about other subject matters, some distinctive features of beliefs about the vague that are predicted by the theory of vague propositions developed in chapter 6.

---

[3] This sort of challenge is also discussed extensively by Hartry Field. See, for example, Field [53].

[4] See, especially, Field [53], whose views we discussed extensively in chapter 7, and Schiffer [126]). See also MacFarlane[97], Smith [134], Williams [154], but also, in a certain sense, Barnett [12] and Dorr [34].

8.1 Expressivism about Vagueness

According to the theory of vague propositions in chapter 6, there's a more straightforward sense in which vagueness-related uncertainty differs from ordinary uncertainty, at least assuming a certain kind of permissive epistemology. In normal cases of uncertainty, two people can have exactly the same body of evidence but rationally respond to that evidence in very different ways: for example, Gideon Rosen [122] writes 'It should be obvious that reasonable people can disagree, even when confronted with the same body of evidence. When a jury or a court is divided in a difficult case, the mere fact of disagreement does not mean that someone is being unreasonable' (p. 71).[5]

Our uncertainty about the correct verdict in a court case seems to be in stark contrast with the kinds of vagueness-related uncertainty we have encountered so far. If we both have the evidence that a particular glass is two thirds full, and we additionally had all other relevant precise information about the glass, then I think it would be outright incoherent to have very high credence that the glass is pretty full. It would be similarly irrational to have very high credence that it isn't pretty full. In this case, it seems that the amount to which two people can have opposing opinions about whether the glass is pretty full, assuming they are both knowledgeable about the relevant precise matters, is extremely limited.[6] The theory of chapter 6 has an explanation for this limitation: there is some particular middling credence that one is supposed to have in the proposition that the glass is pretty full, given that you have all the relevant precise evidence.

The situation here bears a striking resemblance to a certain picture of disagreements about conditional matters. According to this view, if two people agree about the non-conditional 'categorical' matters—in particular, if they agree about the probability of $P \wedge Q$ and the probability of P—there's no room for them to disagree about the probability of the conditional proposition that if P then Q: it must be the ratio of the former probability to the latter (i.e. the conditional probability of Q on P). This view is typically associated with expressivism about conditionals: by asserting a conditional one merely expresses a *conditional attitude* towards a pair of ordinary non-conditional propositions. Once one has made one's mind up about the categorical, there is no further question of what your opinion in the conditional matters should be—there is nothing more to those opinions than your opinions in $P \wedge Q$ and in P. To the extent that there are conditional propositions at all, having a credence in one of them is just a matter of having your credences distributed in a certain way over non-conditional propositions. Similar ideas have been applied to epistemic modals: to accept 'might p'

---

[5] At the extreme, we have Bayesian permissivism which maintains that the only rational constraints on your ur-priors are that they be probabilistically coherent (and perhaps that they not rule out any contingent hypothesis *a priori*). In which case whenever E is consistent with both P and its negation, any credence in P is permissible, conditional on evidence E.

[6] Here and elsewhere I assume that you have all the relevant precise information if, upon conditioning on any other precise proposition, your credences would remain unchanged.

is to have non-zero credence in the unmodalized proposition p (Schulz [129]). For similar reasons, if two people agree about the non-modal, they must agree about the modal.

Expressivism has most famously been applied to moral talk, and this fits the same mould. According to a schematic and simplistic version of this view, when one utters a moral sentence, such as 'it is good that Alice is happy', one does not represent oneself as believing or knowing a proposition—the proposition that *it's good that Alice is happy*—one represents oneself as taking some kind of non-doxastic positive attitude towards Alice being happy. Generally, moral sentences are asserted to express a negative or positive non-doxastic attitude to a non-moral proposition, rather than a doxastic attitude to a moral one.

To illuminate my preferred theory of vagueness it will be instructive to compare it against a view one might call 'expressivism about vagueness' that is inspired by the expressivist views about conditionals and epistemic modals (and to a lesser extent, moral propositions) outlined above. Exploring various ways one might refine this theory gives us a natural route to the theory of vague propositions developed in chapter 6.

To fix ideas let us start with a simple conception of propositions as sets of possible worlds. Taking the lessons of chapter 5 to heart, such a view seems quite unfriendly to the existence of vague propositions, since one can be ignorant about whether Harry is bald even if you know which possible world obtains (i.e. you are not ignorant about any set-of-worlds proposition). Our expressivist thus maintains that vague sentences do not express propositions at all. When one utters a vague sentence, one does not represent oneself as having a high credence in a proposition: one merely represents oneself as having certain patterns of credences among the precise, sets-of-possible-worlds propositions.

To make this idea more rigorous, we may repurpose some of the formalism we introduced in chapter 6. Recall that an evidential role is a function, E, from maximally strong consistent precise propositions (possible worlds in this framework) to real numbers in $[0, 1]$. Let us suppose that each vague sentence for a language gets associated, via the conventional patterns of use for that language, not with a set of possible worlds but with an evidential role. To illustrate, the sentence 'Harry is bald' in English is associated with a role that maps worlds where Harry has no hairs to 1, worlds where he has lots of hair to 0, and worlds where his hair number lies in the border region to credences strictly between 0 and 1. Intuitively, the number $E(w)$ corresponds to the credence you would express by uttering 'Harry is bald' if you knew you were in world w.[7]

---

[7] There are some analogies to be drawn here between this idea and the theory developed in Horwich [72]. Horwich's theory provides an explanation of why we are disinclined to apply certain vague predicates in borderline cases. According to this theory, it is constitutive of the meaning of a vague word, such as 'bald', that to be competent with it you be disinclined to apply the word or its negation in cases were the subject is borderline bald. Of course, we are not disposed to apply the precise word 'has an even number of

Taking a cue from expressivism about conditionals, the idea is that when one utters a vague sentence, one does not express high credence in a vague proposition; rather one expresses that one has a certain pattern of credences in the precise. In particular, by uttering 'Harry is bald', associated with evidential role E, one expresses that one's credences are such that the sum $\sum_w E(w)Cr(w)$ is high: roughly, that one assigns low credences to worlds where $E(w)$ is low, and high credence to worlds where $E(w)$ is high. By analogy, according to the expressivist about conditionals, when one utters an indicative sentence, 'if P then Q' one's utterance is proper not if one has high credence in a conditional proposition, but rather if one has a certain distribution of credences over categorical propositions: in particular, if one's credence in $P \wedge Q$ and in P is such that the former divided by the latter is high.

Again, drawing on the parallel view about conditionals, one could introduce the notion of a credence in a vague sentence:

> EXPRESSIVISM ABOUT VAGUENESS: To have a credence of x in a vague sentence S is just for $\sum_w E(w)Cr(w) = x$ where E is the evidential role associated with S in the language, and Cr represent your distribution of credences over possible worlds.[8]

Note that expressivists about vagueness and conditionals alike can introduce, as we have done above, the technical notion of a credence in a conditional sentence or a credence in a vague sentence. However, since these sentences do not express propositions, they are not credences in any proposition; they are merely notions defined in terms of your credences in precise or categorical propositions. Note also that the initial credence one has in a sentence with evidential role E according to the expressivist theory is exactly the same as the initial credence one ought to assign to a vague proposition whose evidential role is E according to the theory of vague propositions in chapter 6: $Cr(p) = \sum_w Cr(w)Cr(p \mid w) = \sum_w Cr(w)E(w)$ since $E(w) = Cr(p \mid w)$ whenever E is the evidential role of p.

When I assert the sentence 'Harry is bald' and my audience takes me to be reliable, they ought to adjust their credences accordingly: they ought to become less confident that Harry has very high hair numbers, and more confident that he has lower hair numbers. If I had expressed a proposition, they could achieve this simply by conditioning their credences on the proposition I expressed. Since I haven't expressed

---

hairs' or its negation to people either. However, this pattern of linguistic inclinations obtains because we are ignorant of something whereas the application of 'bald' is precluded by our linguistic competence alone. Note, however, that this theory relies on a certain kind of 'meaning as use' theory of meaning whereas the present proposal does not. Moreover, it falls afoul of the problems we discussed in chapter 4. The theory can explain why competent English speakers are usually disinclined to utter the sentence 'Harry is bald' or 'Harry is not bald' when they know that Harry's hair number is in a certain range, but it doesn't explain why they don't know whether he's bald in those circumstances and this, I take it, is one of the most important jobs for a theory of vagueness.

[8] To cut down on formalism I have made the simplifying assumption there are a small (i.e. finite or countable) number of worlds.

a proposition they should condition their credences on the evidential role associated with the sentence. Formally, this is achieved by the following equation:[9]

$$\text{CONDITIONING ON AN EVIDENTIAL ROLE: } Cr(P \mid\mid E) = \frac{\sum_{w\in P} Cr(w)E(w)}{\sum_{w\in W} Cr(w)E(w)}.$$

Here W denotes the set of all worlds, and we write $Cr(w)$ for a world w to denote the credence assigned to w's singleton. I write $Cr(P \mid\mid E)$ to denote a function that takes a set of worlds and an evidential role to a real number, and reserve $Cr(P \mid Q)$ for the function that takes two sets of worlds and maps them to the conditional probability of one on the other. We can also see that each precise proposition—a set of worlds—corresponds to a special kind of evidential role: one that assigns either 1 or 0 to each world. If $E(w) = 1$ or 0 for each world w, then conditioning on E in the above sense is exactly equivalent to ordinary Bayesian conditioning on the set of worlds that E maps to 1.

The notion of conditioning on an evidential role is important for the theory of communication on this view. However, one could also apply this machinery to the view defended in chapter 6 that inexact experiences involve the acquisition of vague evidence. Indeed, it will deliver similar results to the theory of chapter 6 as witnessed by the following fact: for someone who doesn't already have any vague evidence, conditioning on an evidential role, E, has exactly the same effect as conditioning on a vague proposition which has that evidential role. However, because vague propositions are not always probabilistically independent of one another, the theory of chapter 6 will predict more interesting results for those who learn two vague propositions in succession.[10]

Expressivism about vagueness suffers some familiar problems. For starters, one might be worried by arguments from the apparent truth of quantified belief reports. I can say things like 'Alice is pretty sure that Harry is bald, Bob is unsure that Harry is bald, so there's something that Alice is pretty sure about which Bob isn't'. If the straightforward expressivist is correct, then this inference isn't in general good, because Alice and Bob don't really believe anything. Note, however, that we have introduced the notion of a credence (and thus a high credence) in an evidential role, and, thus, it would be easy to cash out this kind of quantification as quantification over

---

[9] For those wondering where the equation below comes from, note that it is the same equation you'd get from conditioning on the proposition that Harry is bald, *if there were such a proposition*. Suppose that there is a proposition, B, such that $Pr(B \mid w) = E(w)$ for all w. If P is a set of worlds, then note that $Pr(P \mid B) = \sum_{w\in P} Pr(w \mid B)$. Moreover, $Pr(w \mid B) = Pr(B \mid w)Pr(w)/Pr(B)$ by Bayes' law. The denominator is identical to $\sum_{w'\in W} Pr(B \mid w')$, so using the fact that $Pr(B \mid w) = E(w)$ we get that $Pr(w \mid B) = E(w)Pr(w)/(\sum_{w'\in W} E(w')Pr(w'))$. Finally $Pr(P \mid B) = \sum_{w\in P}(E(w)Pr(w)/(\sum_{w'\in W} E(w')Pr(w')))$. Although we have been working with a fictional proposition, B, note that the resulting formula in CONDITIONING ON AN EVIDENTIAL ROLE makes no reference to that proposition.

[10] It is well known that Jeffrey conditioning is not commutative. For these familiar reasons, condition-alization will not be commutative for the expressivist: conditioning on E and then E' may have a different effect from conditioning on E' and then E. This strikes me as one reason to prefer the theory of chapter 6 over this sort of expressivism.

evidential roles instead of quantification over sets of worlds. Since evidential roles can be learnt (we can condition on them), they're the objects of belief, and they are the meanings of simple vague sentences, it is already looking as though they play a large portion of the proposition role.

This form of expressivism also suffers from a Frege–Geach problem: although we have specified the conditions under which one can assert a simple vague sentence such as 'Harry is bald', we haven't explained how the assertability conditions of complex sentences get determined from their parts. The matter is simple for a negated sentence: if the evidential role of P maps w to $E(w)$, then the evidential role for $\neg P$ should map w to $1 - E(w)$, which guarantees that one's formal 'credence' in a negated vague sentence is 1 minus one's formal credence in the sentence. But it is very hard to say what happens to disjunctions and conjunctions. In the theory of vague propositions of chapter 6, each vague proposition determines a unique evidential role but not conversely: distinct propositions can have the same evidential role. At the extreme, take a vague proposition, P, that has a role that assigns every world the value $\frac{1}{2}$ (such a proposition must exist by the **Principle of Plenitude**). Then $\neg P$ has exactly the same role, but clearly $\neg P$ is not identical to P. One's credence in the disjunction $P \vee \neg P$ ought always be 1, while one's credence in $P \vee P$ should be one's credence in P, which is $\frac{1}{2}$. But if the only information the disjuncts contribute to a disjunction are their evidential role, then $P \vee \neg P$ and $P \vee P$ are constructed from the same evidential roles and thus whatever our rule for $\vee$ is, they must have the same output.

I do not know how to solve the Frege–Geach problem, but we may assume that it involves assigning some kind of technical device—a *gizmo*[11]—to each sentence of the language to account for its role in communication.[12] The problems gestured at above show that whatever these gizmos are, they have to be richer than evidential roles: in addition to an operation of negation, there must be operations of disjunction and conjunction that they are closed under. Moreover, we can also define what it is for a gizmo to be true or false using the now familiar disquotational schema: for example, the gizmo corresponding to 'Harry is bald' is true if Harry is bald, and false otherwise.

These gizmos, whatever they are, are beginning to look more and more like propositions, for we now know that they are closed under the Boolean operations of conjunction, disjunction, and negation, and there is a straightforward notion of truth that can be applied to them. If we add into the mix that almost all sentences of a natural language are vague, and therefore don't express sets of worlds, then it is hard to see this view as an alternative to the theory of vague propositions in chapter 6: it is

---

[11] The terminology of 'gizmos' for talking about expressivist semantics is from Cian Dorr.

[12] This conclusion is not *forced* on us. Often one can redescribe an expressivist theory that employs gizmos of some sort without employing them: for example, instead of introducing evidential roles, I could have just talked about the conditions on my credences over worlds under which certain sentences are assertable.

the gizmos that are playing the proposition role, not sets of worlds, and the gizmos have most of the distinctive features of the propositions of chapter 6.[13]

Indeed, some moral expressivists are willing to make exactly this concession.[14] For these expressivists, propositions—conceived merely as occupiers of the proposition role—are easy to come by. However, such theorists will typically also want to maintain that there is an important distinction between propositions to be made: some propositions correspond to the kinds of propositions we originally represented by sets of worlds, and others do not—the latter are, in some sense, 'metaphysically lightweight'.

Unfortunately, spelling out this notion of being 'metaphysically lightweight' has proved notoriously difficult, and it has long been an important challenge for such theories to offer some precise articulation of this idea. Indeed, without this distinction, the view effectively collapses into realism (see Dorr [33], Dreier [36], Field [50]). Since the problem of spelling out the difference between the metaphysically lightweight propositions bears a strong parallel with the problem of making sense of the notion of there being no fact of the matter, it will be worth taking a look at how some existing expressivists have attempted to articulate similar distinctions.

8.2 Disagreements about Morals, Conditionals, and Epistemic Modals

A distinguishing feature of the evidential roles introduced in section 8.1 was that one believes an evidential role only in virtue of having one's credences distributed over the precise (i.e. over the possible worlds) in a certain way. A trend among some more recent expressivists has been to cash out traditional expressivist slogans in terms of theses analogous to this that identify doxastic attitudes towards moral (conditional, epistemic) propositions with attitudes (possibly not doxastic) towards non-moral (conditional or epistemic) propositions (see, for example, Gibbard [64]). While the early views denied that there were beliefs in moral propositions and proffered other non-doxastic attitudes towards non-moral propositions as a replacement, these views freely acknowledge that there are moral propositions that serve as the objects of our belief, but instead *identify* these states of belief in moral propositions with other kinds of attitudes toward non-moral propositions.

In the three cases we mentioned earlier, it is natural to explore theses, such as those below, in which beliefs about certain suspect kinds of propositions are identified

---

[13] There is one possible difference. It should be noted that the formalism of evidential roles, conceived as functions from a determinate pre-existing set of worlds to the unit interval, is not very friendly to the possibility of higher-order vagueness. In the formalism of chapter 6, evidential roles are, instead, functions from maximally strong consistent precise propositions to the unit interval. Given higher-order vagueness it is natural to think it is vague which propositions are precise (and thus vague which functions satisfy my definition of an evidential role).

[14] See Schroeder [128] for an extended defence of the idea that expressivists ought to embrace propositions.

with an attitude towards an ordinary non-suspicious proposition. Such theses can be seen as ways of articulating (an important component of) the expressivist picture: attitudes towards the lightweight propositions are derivative on your attitudes towards the factual.

CONDITIONALS: To have a certain credence in a conditional proposition, that if p then q, is just to have certain credences in non-conditional propositions. In particular it is just for you to have credences in $p \wedge q$ and p that have that ratio.

(Adams [1], Edgington [38], and others)

EPISTEMIC MODALS: To have a certain credence in the proposition that it might be the case that p is just to have a certain credence in p. In particular, your credence in the proposition that might p is one if your credence in p is non-zero, and your credence is zero otherwise.

(Schulz [129])

MORAL PROPOSITIONS: To have a certain credence in a moral proposition, that it is good that p, is just to have a certain non-doxastic attitude towards the non-moral proposition that p. For example, it is for your credence that it is good that p to be identical to the degree to which you desire that p. (See the 'Desire as Belief' view criticized in Lewis [88]. A similar thesis identifying attitudes towards moral propositions is endorsed in Gibbard [64]).[15]

In each case, we have a thesis about the role that a certain type of proposition plays in thought replacing a metaphysical claim about the status of those truths. To believe that the cradle fell if the bough broke (to some degree, or fully) is to do nothing over and above believing, to certain degrees, non-conditional propositions about the bough breaking and the cradle falling.

At this juncture one might worry that the literal identifications of these attitudes are too strong. Take, for example, Gibbard's thesis that deciding what one ought to do is deciding what to do. It seems quite clear that one can decide that something is what one ought to do without deciding to do that thing. Perhaps a rational person will always make that last step, but it is a step that it seems as if one could fail to make. Similarly, it is natural to think that one could be fully confident that it's not raining, while also being confident that it might be raining. Of course, it would be completely irrational to do this, but it is not metaphysically impossible, just as it is irrational, but not metaphysically impossible, to believe that it's raining and it's not raining.

No doubt one could tell a special story about attitude reports that accommodates these intuitions. One could also weaken the principles to govern only rational credences. Whichever route one chooses, the principle that these identifications hold

---

[15] It is strictly speaking incorrect to talk of 'non-doxastic attitudes' since the view is one that sometimes identifies the purportedly non-doxastic attitudes with beliefs. Desire as belief theses aren't the only way to account for moral propositions—Gibbard maintains that deciding whether you ought to ϕ is just deciding whether to ϕ; thus having a certain practical attitude towards a moral proposition is simply having that attitude towards a non-moral proposition.

among *rational* people is granted by all parties, and I think that these restricted theses do a good job of elucidating some of these expressivist slogans.

The difference between moral, conditional, and epistemic matters now reduces to a thesis about *rational disagreements*. In general, two people can rationally agree about one subject matter whilst disagreeing about another. Arguably, such disagreements can persist even when the evidence is shared. On the other hand, the expressivist will maintain that this is not so when it comes to moral, conditional, and epistemic propositions: in such cases two subject matters can be intimately tied together. For example, suppose two people agree about the categorical matters—suppose, in particular, that they agree about how likely it is that a given coin was flipped, and about how likely it was that it was flipped and landed heads. Then according to the expressivist about conditionals, they cannot rationally disagree about the conditional matter of whether the coin landed heads if it was flipped. Note, however, that the proposition that the coin landed heads if it was flipped is logically independent of the propositions that it was flipped and the proposition that it was flipped and landed heads: indeed the conditional could be true or false consistently with the coin having not been flipped.[16] So agreement about the non-conditional forces agreement about conditional matters, even though such agreement is *not* forced by logic or conformity to the probability calculus. Similar agreement principles can be formulated for more familiar versions of expressivism. For example, a certain kind of moral expressivist might maintain that if two people agree about the non-moral propositions—they agree about how many lives each charity will save, say—and care about the same non-moral propositions—they care about saving lives—then they must agree about which charity they ought to donate to.

In each case, we have what I'll call the 'rational supervenience' of attitudes in one kind of proposition on another. For example, expressivists endorsing CONDITIONALS, *ipso facto*, endorse the thesis that, for rational people, your credences in the hypothetical propositions are completely determined by your credences in the categorical. Once you have distributed your credences over the categorical propositions you have no rational leeway regarding how you assign your credences over the remaining hypothetical propositions. That is, we have:

SUPERVENIENCE OF HYPOTHETICAL BELIEFS ON THE CATEGORICAL BELIEFS:
If two rational agents assign the same credence to every categorical proposition, they will assign the same credence to every hypothetical proposition.

This supervenience principle has a tight-knit connection to the principle CONDITION-ALS. CONDITIONALS tells us what it is to believe a conditional to a certain degree purely

---

[16] Of course, not everyone theorizing about conditionals thinks that a conditional is logically independent of its antecedent in this way. The idea that the falsity of the antecedent entails the truth of a conditional, along with modus ponens for the conditional, effectively entails that indicatives are material conditionals. The kind of expressivist we are interested in here cannot endorse the material account of the conditional.

in terms of our degrees of belief in non-conditional propositions—so credences in simple conditional propositions are straightforwardly determined by credences in the categorical. A shortcoming of CONDITIONALS, however, is that it only tells us what it is to believe a simple conditional proposition, and not an arbitrary hypothetical proposition. I take it that hypothetical propositions also include arbitrary disjunctions, conjunctions, and negations of conditional propositions, as well as conditional propositions with hypothetical antecedents and consequents. To get the full force of the supervenience principle we need principles that tell us what it is to have a credence in one of these extended hypothetical propositions purely in terms of our credences in categorical propositions.[17]

Regardless of the details, the takeaway message is that these versions of expressivism about each subject matter entail a kind of rational supervenience thesis. In the hypothetical case, for example, these effectively guarantee that all disagreements about hypothetical propositions derive from disagreements about the categorical.

Could these rational supervenience theses give precise cash value to the idea that hypothetical propositions are second-rate or derivative, while categorical propositions are not? The driving intuition here seems to be that while you can have genuine disagreements about categorical matters, all disagreements about the hypothetical boil down to disagreements about the first-rate, categorical facts. You might, therefore, think that the supervenience thesis goes a good way to elucidating some forms of non-factualism.

It is, however, unclear to me whether this strategy will succeed in *completely* capturing non-factualism about hypothetical facts. One reason for scepticism is that supervenience is not an asymmetric relation: it could turn out that categorical beliefs also supervene on the hypothetical in the sense that once you know what a rational agent's credences in the hypothetical propositions are you can work out their credences in the categorical propositions.[18] This symmetry problem is also a problem for the stronger thesis that simply identifies a credence in a hypothetical fact with a certain distribution of credences in categorical facts: the assertion of the identity of these states alone can't capture expressivism, because it could equally be taken to be a reduction of conditional categorical beliefs to unconditional hypothetical beliefs.

---

[17] This problem is quite technical, but the existence of a solution for the case of non-iterated conditionals is implied by the results in van Fraassen [147], and in the general case by the results in Bacon [6]. Using the results in [6] one could in principle state the assertability conditions for an arbitrary sentence purely in terms of complicated equations involving credences only in categorical propositions. Thus one could, in stating the metasemantic theory, dispense with hypothetical propositions altogether. This idea certainly seems to be more in line with writers working in the tradition of Adams [1].

[18] In evaluating this latter supervenience claim one has to be careful not to confuse a hypothetical proposition with a proposition that is expressed by a sentence containing conditionals: if A and B are categorical, the conjunction $A \wedge (A \rightarrow B)$ is logically equivalent to the categorical sentence $A \wedge B$, given some natural assumptions about the logic of conditionals, so the former sentence does not express a hypothetical proposition.

In light of all this, I'm not quite sure what to say on behalf of the expressivist. However, whether or not the supervenience theses (and the identity theses) fully capture the idea that certain facts are second-rate, they do help distinguish these views from a pretty stark kind of realism about conditional facts. According to the stark realist the truth of hypothetical propositions doesn't depend in any important way on the categorical facts. Even once you have formed opinions about all the categorical facts, there may still be a large number of opinions to be had about the hypothetical facts left open to you. If we are merely worried about distinguishing ourselves from the stark realist then the supervenience thesis does a good job of doing that.

8.3 Do All Rational Disagreements about the Vague Boil Down to Disagreements about the Precise?

Our primary concern in this chapter has been whether all classical accounts of vagueness collapse into views that, like epistemicism, accept a stark form of realism about vague matters. So a natural question to ask, given our previous discussion, is whether vague beliefs rationally supervene on precise beliefs in the way outlined above. If the supervenience thesis held, it could serve as a partial articulation of the idea that vague truths are not as substantial as precise truths.

The kind of supervenience thesis I prefer can be formulated as a constraint on which probability functions are 'conceptually coherent' where, very roughly, a probability function is conceptually coherent if adopting it doesn't involve making any conceptual confusions (a notion I'll elaborate on shortly). It is a straightforward variant of the corresponding supervenience principle for conditionals:

> **Rational Supervenience.**   For any pair of conceptually coherent ur-priors, Pr, Pr', if $Pr(p) = Pr'(p)$ for every precise proposition p, then $Pr(p) = Pr'(p)$ for every proposition p.

Given the assumption that the precise propositions form a complete atomic Boolean algebra—that every precise proposition can be expressed as a disjunction of *maximally strong consistent* precise propositions (see section 3.2)—we can reformulate this principle in a slightly more useful way. The reformulation states that all conceptually coherent priors agree with one another conditional on any maximally strong consistent precise proposition. In other words, if two people are knowledgeable about all precise matters, they must agree about everything:

> For any pair of conceptually coherent ur-priors, Pr, Pr', $Pr(p \mid w) = Pr'(p \mid w)$ for every maximally strong consistent precise proposition w, and proposition p.

Here is an informal way of visualizing the supervenience constraint. Think of logical space as being carved up into a number of non-overlapping cells representing the maximally strong consistent precise propositions: states within a

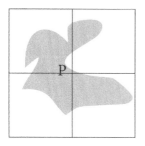 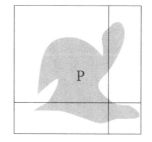 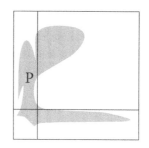

Figure 8.1. Three different priors assigning different probabilities to four maximally strong consistent precise propositions—'cells'—but agreeing on the proportion of each cell that the vague proposition P takes up. In this diagram, a proposition is represented by a subregion of each square, and the probability is represented by the magnitude of its area.

given cell will describe states of affairs that agree about all precise matters and differ from one another about things such as whether Harry is bald and the like. Different coherent priors will typically disagree with one another about the sizes of these cells— some will regard certain cells as probable and others as less so. (It helps to think in terms of a Venn diagram where the areas of the cells correspond to their probabilities: see Figure 8.1.) However, if the supervenience principle is true, then all priors must agree with one another about what *proportion* of each cell each proposition takes up. If the proposition p takes up three quarters of a cell according to at least one prior, it must take up three quarters of that cell according to all priors. Intuitively, although you can change the sizes of the cells as you move from prior to prior, these changes must only come about by 'stretching out' each cell uniformly, so that their contents scale proportionally. (To make the intuition even more vivid, you can think of each cell being like a separate sheet of rubber. A proposition is some portion of rubber, possibly spanning across multiple sheets. You can change the 'area' of a proposition, but only by making uniform stretches or shrinkages to each sheet individually.)

Say that a precise truth is relevant to p for an agent if their conditional credence in p on that truth is different from their unconditional credences. The supervenience constraint entails that two people who are knowledgeable about all the precise matters that are relevant to p must agree about p. For if they disagree about p, then since the supervenience principle states that they must agree conditional on the strongest precise truth, at least one of their credences must be different conditional on that truth and so at least one of them is not knowledgeable of all the relevant precise facts.

On the face of it, however, this principle falls short of a general principle stating that disagreements in the vague are rooted in disagreements about the precise. Indeed, this is no accident: it is at least theoretically possible that two people agreeing about the precise could have differing credences in the vague if (i) they have different evidence and (ii) at least one of them has evidence which is inexact—i.e. is such that their total evidence is vague. Since we argued in chapter 6 that one's total evidence could be vague, this is a live possibility. One must therefore be careful about how we formulate the disagreement thesis: in our framework no two people with the *same evidence* can agree about the precise and disagree about the vague. It should be noted that a similar qualification would need to be made for the other forms of expressivism we have considered if similar theses about evidence were accepted. Once the expressivist has acknowledged the existence of conditional propositions, for example, the question naturally arises as to whether one's total evidence could be a conditional proposition. If the answer is 'yes' then the principle that there can be no disagreements about the conditional without disagreements about the categorical must be similarly qualified.

Of course, in order for our thesis about disagreement to have any potency at all, we must accept a mild form of epistemic permissivism. According to the opposing view— an extreme form of anti-permissivism—there is only one correct doxastic attitude to make in response to any evidence, so that any two people with the same evidence must agree about everything. Under this assumption, the thesis that no two people with the same evidence can rationally agree about the precise but disagree about the vague would be true for uninteresting reasons, and would not succeed in singling out anything distinctive of vagueness.

In the Bayesian framework we have been using, the permissivist idea can be formulated as a question of whether there are many conceptually coherent ur-priors. According to the alternative 'neo-Carnapian' view there is only one true ur-prior (see Williamson [160]) and the supervenience principle is vacuously true—the ur-priors always agree with each other conditional on each maximally strong consistent precise proposition because there is only one ur-prior. If, however, the set of ur-priors is assumed to be sufficiently rich then it states a substantial thesis about the nature of vagueness and vagueness-related uncertainty. Indeed the supervenience thesis is consistent with a maximally permissive attitude towards the precise matters: one can have any initial credences one likes about the precise (provided you are probabilistically coherent); the only constraint is that once you have made your mind up about the precise, you distribute your credences over the vague according to the supervenience principle.

Even if one were tempted toward a neo-Carnapian theory, in which only one probability function satisfies all the constraints of all-things-considered rationality, there are weaker notions you might be interested in. For example, even if not all members of a jury respond in the maximally good way to the evidence in a complicated court case, their sin seems to be a lesser one than that of someone who remains pretty confident that a particular fox is male after learning that it's a vixen. In the latter case the agent seems to be conceptually confused. In **Rational Supervenience** I theorized in terms of the notion of a 'conceptually coherent prior': a class of initial credence functions that aren't conceptually confused in this kind of way—such confusions will include, but will not be limited to, failures to believe conceptual truths. On the picture I am endorsing, one can have pretty much any opinions about the precise you like without committing a conceptual confusion, although once those opinions about the precise are determined, deviation in your credences about the vague would involve a conceptual confusion of some sort: it would be confused, for example, to be almost certain that a glass is pretty full after learning it is two thirds full—as confused as, say, having some credence that Sally is a male cat after learning that she's a vixen.

The supervenience principle is distinctive to the kind of picture of vague propositions I have defended and, moreover, serves to distinguish that view from competing theories such as epistemicism. But is the principle true? To evaluate the principle, let us focus on the status of a particular collection of vague propositions in a situation in which all the precise truths that are relevant to those propositions are known. Imagine that we have a sorites sequence of individuals ranging from someone who has 0 hairs and ending with someone who has millions. Suppose also that the only precise truths that are probabilistically relevant to the baldness of these individuals are propositions about the hair number of the individuals, and suppose that we know all of the hair number facts. Assuming all this, what kinds of credences is it permissible—or, at least, not conceptually confused—to have about the baldness of each individual in our sorites? According to the supervenience principle, there isn't much leeway: if two people are knowledgeable about all of the relevant precise facts, then they must agree about the probability that each member of the sorites is bald; someone who denies the principle, by contrast, maintains that two people could be knowledgeable in this way and yet disagree.

Let us start by considering the first member of the sequence: the individual who has 0 hairs—call him John. Presumably it is a conceptual truth that if John has 0 hairs he is bald. This can be cashed out in the present framework in terms of conceptually coherent ur-priors: for every conceptually coherent probability function $Cr($John is bald | John has no hairs$) = 1$. This of course reflects the fact that the notion of conceptual coherence was introduced as a generalization of the idea that one should be certain in conceptual truths. For example, it would be conceptually confused to be less than certain that a thing is red conditional on it being scarlet, a fox conditional on it being a vixen, unmarried conditional on being a bachelor, and so on. Thus, at least in this special case, the supervenience principle delivers the correct results: any two people who know that John has 0 hairs ought to agree about whether John is bald. Indeed, they ought both to be certain he's bald—if there was any disagreement here, then at least one person has committed some kind of conceptual confusion. (Similar considerations presumably also apply to the individuals that have millions of hairs—in these cases everyone is required to have no credence that they are bald, in accordance with the supervenience principle.)

We know, by familiar soritical reasoning, that there is a last number N such that it's a conceptual truth that someone with N hairs is bald. Consider next the $N + 1$th person in the sorites—call him Bob. What kinds of credences is it permissible to have about whether Bob is bald, given that we know he has $N + 1$ hairs?

We have already considered the extreme anti-permissivist neo-Carnapian approach. Here it is natural to also compare our answer to a view at the other end of the spectrum: a common (extremely permissive) version of Bayesianism, in which any probability function whatsoever is a rational prior. Whereas the neo-Carnapian view makes the supervenience principle trivial, the Bayesian permissivist violates the supervenience principle. Bayesian permissivism requires certainty in a proposition p on the supposition of other propositions that logically or conceptually entail p. But when neither p nor $\neg p$ is logically or conceptually entailed by a supposition, then any credence in p on that supposition is permissible. It is here that the notion of conceptual coherence, as I am understanding it, is more general than the requirement that one be certain in conceptual truths: bearing in mind that all agree that it would be in some sense conceptually confused to be less than certain that the Nth person is bald (because it is a conceptual truth that people with N hairs are bald), it is similarly conceptually required that one have a high credence that Bob (the $N + 1$th person) is bald. Although it is permissible to be less than certain that the $N + 1$th person is bald, if I'm right, then having a very low credence in his baldness would be to commit a similar kind of conceptual confusion. Intuitively, it is not merely having a low credence that is incoherent: one should have a credence pretty close to your credence that the Nth guy is bald (namely, a credence of 1). So already you can see how the prediction that there is a particular credence that you ought to have that Bob is bald, given you are knowledgeable about all the relevant precise facts, seems to accord with intuition.

Finally, consider an individual towards the middle of the sequence that is borderline bald—call this guy Harry. Given that you know that Harry is in the borderline region for baldness, it would seem somewhat careless to be almost certain that he is bald, or almost certain that he's not bald: one should have a middling credence. This is in accordance with the supervenience principle, but contradicts a completely permissive version of Bayesianism. Again this brings out a quirk of this kind of Bayesian picture. To take an analogy, insofar as it is a conceptual truth, one is required to be certain that a particular ball is red conditional on it being scarlet (that all scarlet things are red is a paradigm example of a conceptual truth). However, it also seems that one is required for very similar reasons to have a middling credence (around $\frac{1}{2}$) that the ball is red conditional on it being auburn (or some other specific shade that puts the ball in the borderline region for redness). The standard way of articulating these constraints, using the probability calculus, treats these two cases very differently—the former is imposed by the probability calculus, assuming that the proposition that the ball is scarlet entails the proposition that the ball is red, whereas the latter is not.

Intuitively, if we were to plot rational credences in the proposition that the Nth person is bald against N, for a rational agent who knows all the relevant precise facts, we should get a smooth curve starting at 1 and smoothly decreasing until it hits 0 for high N. By contrast, the permissivist will require that for low N we assign credence 1 and for high N we assign credence 0 (because we must respect entailments), but will allow pretty much anything to happen in between: there could be a very steep and sudden drop between the 245th member of the sorites and the 246th, it could

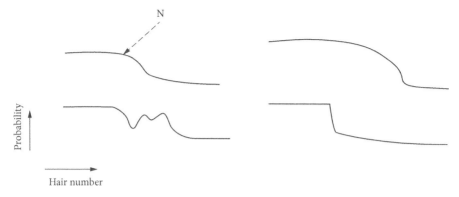

Figure 8.2. Possible credence distributions, given permissivism about ur-priors.

go up and down in a sine wave, and so on (Figure 8.2). So, at least in comparison to full-blown permissivism, the supervenience principle seems to get things right. One could go for a more moderate version of permissivism: one that requires that any two people who are knowledgeable about the relevant precise facts have a smooth distribution, with no steep drops, that decreases as N increases, but in which there are two or more distributions which are permitted. Note that to be smooth, decreasing, and to avoid steep drops, these distributions must, at best, be only slight variants of one another. The supervenience thesis is inconsistent with this kind of permissivism as well.

I do not have any direct arguments against this restricted form of permissivism, except to say that it is ad hoc: it feels like a doctored version of Bayesian permissivism, tailored specifically to account for very specific intuitions about the way our credences ought to be distributed over sorites sequences.[19] The supervenience thesis, by contrast, also gives us a natural restriction of full Bayesian permissivism, but is a simple powerful theory arising from more general considerations to do with vagueness.

It should be noted that the supervenience thesis says that there's a particular smooth distribution that all rational people are required to have when they know all the relevant precise facts. However, this does not entail that there is a particular smooth distribution that rational people are determinately required to have. The idea that one could discover that the credence that Harry is bald in the above scenario should be exactly 0.53184, for example, seems wild—it is not the kind of thing you could discover because it is itself, presumably, borderline. Thus it should be noted that insofar as the restricted permissivist is motivated by the idea that there aren't unique objective numbers like this which we can discover, nothing in that thought rules out the alternative view which accepts supervenience. Indeed my picture has

---

[19] Moreover, this watered-down version of permissivism faces awkward questions like 'how steep can a curve be before it becomes an impermissible way to distribute one's credences over a sorites?'.

much in common with the restricted permissivist: both agree that there are several distributions, all very slight variants of one another, which are not determinately irrational in this situation (the permissivist, in virtue of thinking, simply, that they are not irrational).

Vagueness concerning which credences we should have in this situation could have several sources. It could be due to vagueness concerning the notion of a rational or conceptually coherent credence. But it is also important not to discount the existence of vagueness that arises due to vagueness about which propositions are precise (i.e. higher-order vagueness). Given the existence of higher-order vagueness, it could be that w is a maximally strong consistent precise proposition but not determinately so. In which case one might in fact be required to have a certain credence in the proposition that Harry is bald conditional on w but, since it is borderline whether w is a maximally strong consistent precise proposition, it might be borderline whether we are required to have that credence in the proposition that Harry is bald. It's important to note that the existence of vagueness concerning which propositions are the maximally strong consistent precise propositions does not prevent **Rational Supervenience**, or indeed the principle **Plenitude** from chapter 6, from being determinately true. The **Rational Supervenience** entails that for every maximally strong consistent precise proposition, w, and vague proposition, p, there is some credence that conceptually coherent agents must assign to p conditional on w. Even if we assume that this principle is determinately true, we cannot infer from it that there is some credence that determinately all conceptually coherent agents must assign to p conditional on w, for that would be akin to inferring 'something is determinately F' from 'determinately, something is F'. This is a fallacy (for example, it's determinate that some number is the last small number, but no number is determinately the last small number).

Let us briefly relate **Rational Supervenience** back to the theory of vague propositions outlined in chapter 6. The **Principle of Plenitude** there told us that for every evidential role (roughly, an assignment of proportions to cells), there is a vague proposition that has that role (takes up the assigned proportion of each cell, according to each prior). Propositions generated by the **Principle of Plenitude** satisfy **Rational Supervenience**: they will take up the same proportion of each cell relative to any prior. What **Rational Supervenience** adds to this is a sort of completeness condition—*all* propositions have an evidential role, and so in a sense every proposition is generated by the **Principle of Plenitude**.

Although **Rational Supervenience** fits naturally with the theory of vague propositions endorsed in chapter 6, we have seen it to be a powerful principle in its own right. **Rational Supervenience** tells us that rational disagreements about the vague ultimately boil down to disagreements about the precise. For someone who wished to maintain that vagueness was *merely* a matter of ignorance or uncertainty, this sort of connection between the precise and the vague would be hard to explain. The theory is therefore more closely aligned with some of the ideas that have motivated recent expressivists. In this chapter we have primarily focused on the credal question—how vague beliefs are determined by precise beliefs—but there is a related question concerning desires: is agreement about how much you value the precise enough to ensure agreement in how much you value the vague? In order to answer this question we will need to develop a decision theory that can be applied in a setting with vague propositions. We turn to this now.

9

Vagueness and Decision

A compelling thought, no doubt inspired by the idea that vagueness is primarily linguistic, has it that beliefs and desires about vague matters are, in some sense, redundant. A naïve version of this idea might maintain that behaviour consists in precise bodily movements and that any explanation of this behaviour in terms of a person's beliefs and desires about vague matters could be equally or better put by appealing only to that person's beliefs and desires about precise matters. If you happened to know the precise propositions whose truth a person values, and to know how confident they are about all the precise eventualities, one would be able to determine how that person will behave, assuming they're rational.[1] This is the thesis I shall be examining:

> PRACTICAL IRRELEVANCE: The ranking of the available actions for a rational agent supervenes on the attitudes (typically beliefs and desires) the agent has towards precise propositions.[2]

It is completely consistent with this thesis that you have attitudes toward vague propositions. Indeed it is consistent that, for reasons of human psychology or convenience, most practical reasoning is conducted in vague terms. The thesis just states that these attitudes towards the vague are not necessary for the conclusion of this reasoning: the ranking, and thus the action that is most rational for an agent to perform, is determined completely by the attitudes a rational agent has to the precise.

---

[1] One might think that this all follows from a more general thesis that vagueness is causally inert. While one can certainly say that the rat scampered because he noticed the cat, even though talk of noticing cats is clearly vague, one might insist that there must always be some precise fact that was the basic cause of the rat scampering. There is a quite general puzzle about the connection between physical causation, relating fundamental physical (and maybe also precise) propositions, and the causal interactions we observe all the time, such as would feature in interactions between macroscopic objects, and, in particular, between our thoughts and the physical world. I shall take it for granted here that the everyday notion of causation is perfectly respectable, and that mental to physical causation is both possible and pervasive.

[2] Throughout this chapter I will be assuming Jeffrey's formulation of decision theory in which in general the action that is most rational for an agent to do supervenes on the agent's attitudes towards propositions (beliefs and desires). In Jeffrey's theory, actions, outcomes, and states of the world are all represented by propositions. The theory outlines rational constraints relating the propositions we ought to make true to the (graded) attitudes of belief and desire. By contrast in Savage's decision theory—the most prominent alternative to Jeffrey's—actions are physical events (like accepting a bet), and outcomes can be physical objects (such as a pot of gold).

The thought is best explained by examples. It seems clear that someone can care about being rich and consequently go about doing things that they believe will make them rich; this much is certainly consistent with PRACTICAL IRRELEVANCE. Someone subscribing to this thesis, however, will insist that what is really going on is that the person in question distributes their desires in such a way that they care about having certain precise amounts of money over others, with a preference towards larger amounts. This is what 'caring about being rich' amounts to, or at least supervenes on, and so our desire to be rich is in some sense derivative. A similar example can be run with belief: suppose that I can be described as believing that a certain glass is pretty full, and that consequently my credences concerning the various precise percentages that the glass is filled drop off smoothly. According to the above thesis, the practical significance of this belief is exhausted by the effects it has on my credences about the precise things—when it comes to making decisions it wouldn't matter whether I had the vague belief so long as my credences in the precise remained the same.

The idea that attitudes toward the vague are epiphenomenal is more of a slogan than a specific thesis, and the different precisifications of it bear no more than a family resemblance. While there is a version of the thesis that I like, I shall be giving reasons to resist the relatively striking PRACTICAL IRRELEVANCE principle.

For those who think that vagueness is just a matter of semantic indecision, or some other public language phenomenon, the idea that vagueness is practically irrelevant will seem extremely appealing. Presumably vagueness plays little role in our day-to-day non-verbal decision making: on one view the things we want, believe, and make happen are all precise—it is only when we want to latch on to these things using words that vagueness comes into play. If all propositions are precise then practical irrelevance straightforwardly falls out of standard versions of decision theory, in which the action that is most rational for an agent to perform is determined by the contents of the agent's beliefs and desires, and does not depend at all on the way, or mode of presentation, under which they have these beliefs and desires.[3]

Even those who admit the existence of vague propositions might be attracted to PRACTICAL IRRELEVANCE. For example, some versions of expressivism about vagueness described in chapter 8 admit the existence of vague propositions, but treat these as mere constructions out of precise things. There are also linguistic theorists who want to talk about vague propositions by constructing them out of precise things (e.g. as sets of world–precisification pairs). Again, on these views it is natural to think that the important psychological explanatory work can be done by appealing only to

---

[3] Indeed, for those who take belief and desire to be three-place relations between people, propositions, and modes of presentation, there is a non-trivial challenge in reformulating decision theory. See section 5.3.5.

our beliefs and desires about the way the world really is, with the aspects of content that derive from the way we talk playing little or no psychological role.[4]

It is natural for this kind of theorist to endorse a particularly strong precisification of the practical irrelevance thesis: if two people have exactly the same (relevant) attitudes towards the precise propositions and they have the same actions available to them, they will act in exactly the same way if they are perfectly rational. In what follows, I will offer some reasons to reject this strong version of the practical irrelevance thesis and thus, indirectly, some further reasons to reject accounts of vagueness that predict that attitudes towards vague matters have no important psychological role.

On the other hand, if you're an epistemicist who thinks that vagueness is just a matter of ignorance then you might expect PRACTICAL IRRELEVANCE to fail in a fairly drastic way. Our beliefs about things we are ignorant about are often extremely relevant to us in practical reasoning. If I'm uncertain whether it will rain then my degree of uncertainty will inform my decision about whether to bring a coat or not; being ignorant of some matter is no indicator of its practical relevance. In particular, ignorance about whether p doesn't necessarily prevent you from caring whether p: if vagueness were merely a matter of ignorance then there would be no reason why someone couldn't care intrinsically about the vague.

Thus, another pertinent precisification of the practical irrelevance thesis is the view that it is irrational to care intrinsically about vague matters, which appears to be violated by a purely epistemic account of vagueness. This principle bears directly on the problem we considered in the last chapter: to what extent do all classical accounts of vagueness collapse into some form of epistemicism? The answer I offer here is that it is not rational to care intrinsically about the vague. This, I think, is the precisification of PRACTICAL IRRELEVANCE that holds true, and it is explored in chapter 10.

I'll begin in sections 9.1, 9.2, and 9.3 by evaluating some different ways of understanding the practical irrelevance thesis in the context of a standard decision theory. After that, I'll reconsider the view that vagueness doesn't involve one being genuinely uncertain about anything, and argue that uncontroversial facts about preferences alone are enough to ground the judgement that vagueness involves uncertainty.

9.1 Vagueness and Decision Theory

Let me begin by outlining a standard formalism for representing an agent's beliefs, desires, and decisions—decision theory à la Jeffrey [73]. According to this framework each consistent proposition, A, is assigned a real number, $V(A)$, that represents the *expected value* (sometimes called the *news value*) of A. Roughly, $V(A)$ measures how positively you regard things to be on the supposition that A. At an initial parse, to evaluate $V(A)$, one first supposes that A is true, and then attempts to quantify how

---

[4] Of course, psychology, like any other special science, is couched in vague language. This is not to say that the subject matter of psychology must essentially involve vagueness in some way.

good things are for you. It is important to note that it is coherent to suppose things that one could never be in a position to learn: I can imagine what things would be like on the supposition that the total number of stars in our galaxy is even and that no one will ever know this, even though I couldn't ever come to know this proposition (since knowing the first conjunct entails the falsity of the second). Crucial for our purposes is the fact that, even though one could never be in a position to learn whether Harry is bald if you already know that it's borderline whether he's bald, it is still coherent to suppose that he is and to ask how that would affect how good things would be for you. (Formally, supposition will be represented by conditional probabilities: one can consider probabilities conditional on A even if one could never be in a position to conditionalize on A. An agent's conditional credence in B given A is determined by their credence in A and in $A \wedge B$—one must still assign these propositions credences, even if one cannot ever learn A.)

To get a better grip on this function we shall relate it to several other concepts. Firstly, we can state how it interacts with one's *credences*, another real-valued function on propositions which measures how confident you are in each proposition. A *partition* of A is a set of propositions, E_i, which are pairwise incompatible but whose disjunction is A. Intuitively a partition of A is an exhaustive list of more specific ways that A could be true. If Cr is your credence function and V your expected value function then the following ought to hold for any partition of A:

AVERAGING: $V(A) = \sum_i V(E_i)Cr(E_i \mid A)$.

That is to say, how positively you regard things to be on the supposition that A is a weighted average of how positively you regard things on the supposition of each of the ways A could be true.[5] You average the different ways A could be true by weighting them by your confidence in A being true in that way, under the assumption that A is true. When the partition consists of maximally strong consistent propositions (what we have been calling *indices*), i, we can think of $V(i)$ as determining what we value outright, before we take into account uncertainty. Informally, Jeffrey treats the maximally strong consistent propositions as possible worlds. However, nothing in his formalism requires this interpretation, and indeed, relaxing it allows us apply the theory to vague propositions. I will continue to refer to the maximally strong consistent propositions neutrally, as 'indices'.

(Technical aside: the following technical observation helps us get a more concrete mathematical model of V on the table. Note that we can introduce a random variable, u, taking value $V(i)$ at the index i: u here represents the utility of the actual index, unknown to us, and the expected value of A, $V(A)$, can simply be understood as the expectation, in the mathematical sense, of u given A, written $E(u \mid A)$. Thus, given a credence function, u and V are interdefinable.)

---

[5] When uncountable partitions are in play, we must use a generalized kind of integration; I shall set these technicalities aside.

We can secondly relate an agent's V function to what they are in a position to make true, and the normative notions of what they should and may make true. Each agent will be associated, at a time, with a set of propositions, $\mathcal{A}$, which represent the propositions that they are in a position to make true. Call this the agent's *action set*. The second role that V plays is thus:

MAXIMISE: If the agent is in a position to make A true, it is permissible to do so if and only if no member of the agent's action set has a greater value. It is obligatory if every other member has lower value.

In a slogan, one should maximize expected utility. Say that an agent *prefers A to B* iff $V(A) > V(B)$. Another way to state MAXIMIZE is that it is permissible to make a proposition true if there isn't a preferred action proposition, and obligatory if it is preferred over every other action proposition.

More likely than not, there will be many functions which satisfy the roles captured by MAXIMIZE and AVERAGING, so talk of a single function V is a bit of a fiction. For our purposes, however, this does not matter—the mere existence of a function satisfying these roles is enough to ensure important structural constraints regarding what it is permissible to make true.[6]

At any rate, the terms with which we shall be theorizing include the notion of an agent's credence, the agent's expected value, what an agent is in a position to make true, and what they should and may make true. I also used logical concepts when I talked of incompatible propositions and disjunctions of propositions. In order to make sense of this formalism—so that one can assume that rational credences are probability functions, that there are partitions, and so on—one must assume that the space of propositions has the logical structure of a complete Boolean algebra (see section 3.2).[7] As noted already, however, given these assumptions, nothing in this formalism requires that propositions be sets of worlds—they might be sets of world–precisification pairs, for instance, or something else.

It is crucial to bear in mind, in this regard, that the objects to which V assigns values are the objects of our desires, suppositions, and beliefs—V is not assigning values to physical objects, chunks of reality, or 'facts' in some inflationary sense. Moreover, if we are to avoid a revisionary story about the relation between vagueness and ignorance (see chapter 5), then we should not take the objects of thought to all be precise (some of these objects will have distinctive epistemic profiles, for example). Thus, since it is possible to desire, believe, and suppose that I'm rich, the value function will assign a value to the proposition that I'm rich in addition to the various precise propositions about my wealth. That there are vague propositions in the domain of V will be assumed in what follows.

---

[6] See the preference axioms in section 9.4.

[7] Or, at least, if propositions were not Boolean—if they were structured propositions or sentences for example—then the structure of the probability theory would allow us to partition these propositions into equivalence classes that did have a Boolean structure. Moreover, a strictly Boolean structure is needed to represent the value function in terms of a utility function.

Someone who endorses PRACTICAL IRRELEVANCE within this framework, then, should not take exception to the idea that we can assign values to vague hypotheses. The thesis is rather that these values are utterly otiose. While one can sensibly ask how good things seem to be for me on the supposition that I'm rich, this value is redundant for the purposes of decision making since one could as easily explain my actions in terms of my suppositions, beliefs, and desires in the precise. There are precise propositions about my wealth whose values are, at a more basic level, guiding my actions. Knowing an agent's values, and perhaps also their credence, in each precise hypothesis is all that we need to explain their behaviour, and all that they require to reason practically.

9.2 Vagueness and Action

Consider the following scenario. Emily decides to have a drink of water. The following things happen: Emily's elbow forms an angle of 116.3 degrees, her shoulder muscle contracts, and her hand moves a total of 19.7cm towards the glass of water waiting on the table in front of her. Her thumb and fingers contract by 1.3cm each and her elbow tightens until it makes an angle of 33.9 degrees, bringing the glass of water towards her mouth.

The above is a fairly precise description of the sequence of events that transpired after Emily decided to have a drink of water—it could, in principle, have been described in completely precise terms. It might be tempting, then, to think that actions are completely precise things.

An action certainly *consists* in a completely precise sequence of events—denying this sounds as though it would involve a commitment to vague objects or vague events of some kind. So there is certainly some sense in which actions are guaranteed to be precise. However, there is a way of formulating the question of whether actions are precise in the theoretical framework developed above where the question requires a little thought. The most straightforward way to reflect the idea that actions are precise is to maintain that the agent's *action set*—the set of propositions they are in a position to make true—consists only of precise propositions. Perhaps, for example, prior to deciding to have a drink of water, the precise proposition describing the sequence of events outlined in the first paragraph is one of the things Emily was in a position to make true.

The hypothesis that the agent's action set consists only of precise propositions entails a natural precisification of PRACTICAL IRRELEVANCE. If the action propositions are precise, then, according to MAXIMIZE, the action chosen will just be a function of the V-values we assign to precise propositions (assuming we choose rationally):

IRRELEVANCE TO ACTION: The proposition a rational agent makes true is determined solely by their preferences (or V-values) among precise propositions.

MAXIMIZE states that the action a rational agent makes true is determined solely by their preferences/V-values among action propositions; if the action propositions are always precise, then the above principle follows. In some sense this principle states that some of our attitudes toward the vague are epiphenomenal—you can have preferences among vague propositions, but they won't affect how you behave.

It should be noted that this principle is consistent with other attitudes toward the vague being relevant. For example, the value you assign to a precise proposition might be determined by attitudes, such as credences and desires, you have towards the vague. By AVERAGING, the value of a precise proposition can be written as a weighted sum of values of vague propositions.[8] It could turn out that what you really care intrinsically about is vague, and that the values you assign to the precise propositions only represent what you care instrumentally about—things that raise the expected value by making it more likely that the vague matters you care about are true. Even though you only need to know how the agent assigns expected values to the precise to know what she'll do, you still need to know her credences and desires about vague matters to know what those expected values are.

Be that as it may, I want to argue that even this modest form of PRACTICAL IRRELEVANCE is false. Vagueness seeps even into our actions, and consequently the action a person chooses cannot be determined purely by looking at their preferences among precise matters. It is important to be clear what this amounts to. It need not amount to denying that actions are constituted by precise events or anything like that, it is rather a claim about the kinds of things we are in a position to do. Indeed, while an aspect of an agent's behaviour consists of precise bodily movements, the central concept for decision theory is rather the intentional notion of doing, or making true:

VAGUE ACTIONS: The propositions we are in a position to make true (i.e. that are among our action propositions) at any given time are almost always vague propositions.

As we shall see shortly, we must distinguish this thesis sharply from the thesis that the action propositions are sometimes borderline (neither determinately true nor determinately false). Indeed, the action propositions for a determinately rational agent with determinate preferences will usually be determinately true or false, their vagueness notwithstanding.

For VAGUE ACTIONS to be plausible at all, we must distinguish sharply between what *happened* after Emily decided to have a drink of water, and what she *did*. Some of the things that happened after Emily decided to have a drink are relevant to the goodness of the outcome of her decision—if she had bent her elbow at a slightly different angle,

---

[8] The **Principle of Plenitude**, for example, guarantees that it's always possible to partition a precise proposition into vague proposition. For example, given a precise proposition p, it's possible to construct a proposition that takes up half of each cell p occupies, and overlaps no other cells. This proposition will be vague, since it cuts across cells, and it and its negation partition p.

for example, she would have spilt her drink. But lots of other things happened too: the crickets in her garden continued chirping, someone somewhere died in a car crash, and so on. This even applies to some of Emily's bodily movements: her heart continued beating, she continued to digest her last meal, and so forth—these are things that merely happened to Emily, not things she did. The difference is that most of these things are beyond her control, and are neither among nor are entailed by the things she was in a position to make true. As a rule of thumb, the things Emily made happen closely reflect the things we hold her responsible for—the kinds of things for which she can be subject to blame or praise.

Another issue in the vicinity is that the notion of making true that is relevant to decision theory is an intentional notion: for example, to have made it true that I ate the apple I must have intended to eat the apple. There is a purely causal way of understanding 'makes true' which therefore needs to be distinguished and ruled out. If I roll a die and it lands on a six then, in the purely causal sense, I made it land on a six; or similarly if I throw a dart at a dart board and it lands on a point, x, then I made it land on x in the purely causal sense. But since I didn't intend the dart to land on x—the most I intended was that it hit the board—I didn't intentionally make the dart land on x. The distinction applies even to my own bodily movements: if, after bending my elbow, it forms an angle of 116.3 degrees, this is rarely something I intentionally made true but merely something I caused.

So the first distinction we must be clear on is the distinction between the things that happened which Emily (intentionally) made happen, and those which she didn't (intentionally) make happen. Clearly only the former kind of things are the things Emily gets to choose from before she makes her decision, and are the kinds of things taken into account when we evaluate her actions after the decision.

The second thing we need to get clear on is how to think about the things she *did* do, and their role in principles like MAXIMIZE. The issue is that even if we exclude from our attention the things that happened that were beyond Emily's control, there are still a multitude of things that Emily did intentionally make happen in the above scenario. Some of the things Emily did in the above example include: picking up the glass, picking something up, moving her arm, and so on. The proposition that Emily picked up the glass and the proposition that Emily picked something up, for example, have different expected values: if the table had on it both a scalding hot plate and a glass of water, for example, then the value of picking something up is worse than the value of picking the glass up, provided she initially knows she's going to do one of the two things and assigns non-zero credence to picking up the plate. There is thus a puzzle about what one means by the actions Emily has available to her. Which of the things she made happen do we use to calculate the expected utility of her action? Or in other words, what are the things we are trying to maximize when we make decisions?

One awkward feature of the Jeffrey-style decision theory we've been using thus far is that the things that get assigned values are propositions, whereas phrases that denote actions, like 'picked up a glass', are not propositional. That said, actions clearly have some kind of logical structure—for example, picking up a glass entails picking up something but not vice versa, running to catch the bus entails running, and so on. It is therefore not entirely unnatural to regiment these actions as propositions: we can talk of the proposition that Emily is picking up a glass, that she is picking up something, that she is running to catch the bus, and so on. The presence of this logical structure gives us a straightforward answer to the question we just raised. In the above example, Emily made a number of things happen: it is surely the most specific thing that she did that we should evaluate her actions by, and the thing the value of which she should be attempting to maximize. It doesn't matter if the expected value of picking something up is low (because there's a scalding hot plate on the table) if Emily only picked up something in virtue of doing the more specific act of picking up the glass. In this framework, this can be represented by the conjunction of all of propositions she made true.

To summarize: in any given scenario, we can divide the propositions into those that Emily made true and those which she didn't.[9] To calculate the expected value of what would be made true in that scenario, we simply conjoin the propositions that Emily made true and calculate the conjunction's expected value. It is important to note, then, that an action proposition is not merely a proposition which it is possible (i.e. consistent with Emily's limitations) for Emily to have made true; it must be a proposition such that it is possible for it to be the *conjunction* of all the things that Emily made true.

With the action propositions thus delineated, MAXIMIZE tells us to make the action proposition with the highest expected value true.[10] With a clearer question in sight, we are now in position to return to our discussion of VAGUE ACTIONS.

Let us begin by noting that the precise proposition describing the events following Emily's decision, presented at the beginning of this section, does not primarily consist of things that Emily intentionally made true. While these are certainly things that happened, it is quite implausible to suppose that Emily made it true that her elbow contracted to an angle of exactly 116.3 degrees; this is simply too specific a thing to have been under her control, and she certainly didn't intend anything that particular.

On the other hand, she no doubt made some weaker propositions true. For example, she made it true that she contracted her elbow by some amount. But recall that an action proposition is not merely a proposition Emily could have made true: it's a proposition that could have been the *strongest* thing she made true. Presumably, then, the proposition that Emily contracted her elbow by some amount is not the strongest thing she made true—she has more control over her elbow than that. Nonetheless,

---

[9] There are many questions that we haven't addressed here. For example, there is a question of whether to model 'making true' as a factive normal modal operator. In which case, the propositions that Emily made true are closed under logical entailment and under conjunction introduction. Whether 'making true' behaves like this or not doesn't really affect our discussion.

[10] If 'made true' isn't closed under conjunctions (see footnote 9), then we must reword MAXIMIZE as telling us to make things true in such a way that the conjunction of things made true has maximum value.

might the strongest thing she made true still be a precise proposition? Perhaps the strongest thing she made true is the proposition that she contracted her elbow to an angle between 102.4 and 128.6 degrees? Again it seems as though this is just too precise a thing to have been the strongest thing she made true; what Emily made true is at least partly a matter of her intentions, and she certainly did not have intentions that precise. But more importantly, Emily's ability to control her movements isn't perfect. Someone who does not have perfect motor control cannot hope to bend their elbow to exactly 115 degrees, say, or to some precise interval around 115 degrees—the best they can do is aim to bend elbow at roughly 115 degrees.[11]

There is an extremely natural analogy to be drawn between the examples of imperfect motor control we have been considering here, and the example of imperfect perceptual faculties that we discussed in chapter 6. There, it was argued that when our perceptual faculties aren't completely precise, the conjunction of our evidence is typically vague. Similarly, for agents without perfect motor control, it is natural to think that the conjunction of the things we make true will also typically be a vague proposition. In the case of evidence, we saw this by observing that the evidential probabilities we'd expect to have after an inexact experience conform to a smooth curve that is different from the sharp curve you'd get from conditioning on a precise proposition.

In a similar way, when I decide to extend or flex my elbow, but I don't have perfect motor control, I am doing something that I think will result in my elbow making certain precise angles, although in a way that will leave me uncertain which precise angle I will end up making. If I'm trying to make a right angle, presumably I should be more confident that I will make an angle in the 85–95 degree range than the 95–105 degree range. Indeed, it is natural to think that the probability of each precise angle, conditional on my chosen course of action, will smoothly drop off either side of 90 degree.

However, if the strongest things we make true are always precise, then it seems as though we wouldn't get the smooth curve that seems to be predicted. The difference between the smooth curve and the sharp curve isn't a mere curiosity either—these probabilities are essential to applying decision theory. AVERAGING says that in order to calculate the value of this action we can look at the value of each precise angle and multiply it by the probability of it conditional on that action and sum them; yet these sums can change between the smooth and sharp curves even if our utilities remain constant.

---

[11] An argument can be given for the vagueness of the strongest thing Emily made true if we assume 'makes true' has a modal logic of S4. Namely, one can show, using p to denote the strongest thing Emily made true on a given occasion: if it's possibly borderline whether Emily made p true then it's possibly borderline whether p. (This follows because if p is the strongest proposition Emily made true, then p and the proposition that Emily made p true will be equivalent, assuming S4.) To show that it's possibly borderline whether p, it suffices to show that it's possibly borderline whether Emily made p true; I leave it as an exercise to construct a sorites starting with a scenario in which Emily clearly made p true and one in which she clearly didn't make p true.

In our discussion of vague evidence, we considered the idea that one's evidence isn't propositional at all, and that one should update by Jeffrey conditioning over a collection of precise propositions instead. There is an analogous move to be made here as well. Rather than treating actions as a choice between making a single, potentially vague, proposition true, we could imagine an action being a collection of precise propositions paired with probabilities representing the probability that we will make that precise thing true. Thus, perhaps, choosing to make certain elbow movements is a bit like choosing to enter a kind of lottery: the possible results are that I flex my elbow to an angle of exactly x degrees, and each of these results happen with probability q, where the result of graphing q against x conforms to the kind of smooth curve you would expect. Call such things 'mixed strategies'. Mixed strategies are thus the analogue of Jeffrey conditioning for the case of action: a choice, on this view, is a choice about which mixed strategy over precise eventualities to adopt rather than a choice about which proposition to make true.

Mixed strategies on this account play a fundamental role in thought and action. Indeed, much like the kinds of Jeffrey conditionings that arise through obtaining inexact evidence, there is a straightforward correspondence between mixed strategies and the things I called evidential roles—functions from maximally strong consistent precise propositions to $[0, 1]$. Since, on this picture, it is mixed strategies/evidential roles that play the proposition role—the things we know, believe, make true, desire, and so forth (see chapter 4)—there is little reason not to call these entities propositions. The result is a theory of vague propositions much like that outlined in chapter 6, or the expressivist account of vague propositions discussed in chapter 8.

Although I maintain that an agent's action propositions are often vague, it is important to distinguish this claim from the thought that the action propositions are often borderline (i.e. neither determinately true nor determinately false). Indeed, it is natural to think that determinately rational agents with determinate preferences always have determinately true or determinately false action propositions. The thought, to put it roughly, is that if A has the optimal value out of all of my action propositions (and this is a determinate truth) and A is borderline, then it's borderline whether I've satisfied MAXIMIZE, and, therefore, it's at best borderline whether I'm being rational.[12] We can turn this into an argument that the action set of any determinately rational agent with determinate preferences consists only of propositions which are determinately true or determinately false on the assumption that it's determinate that there are no ties among her preferences on actions. Let S be a determinately rational agent and V her value function.

1. For any pair of propositions A and B: either it's determinate that $V(A) < V(B)$ or it's determinate that $V(B) < V(A)$. (By the assumption that it's determinate what S's preferences are, and the fact that there are no ties.)

---

[12] Thanks to Cian Dorr for pointing this argument out to me.

2. For any proposition A, either it's determinate that A is an action proposition or it's determinate that A is not an action proposition. (It's determinate what S's action propositions are.)

3. For any action proposition A, determinately: A if and only if $V(B) < V(A)$ for every other action proposition B. (The agent determinately satisfies MAXIMIZE and the action propositions are pairwise incompatible.)

4. Therefore, for any action proposition A, either it's determinate that A (if A is optimal) or it's determinate that $\neg A$ (if A is not optimal).

The first and second premises are debatable, but are probably true in many relevant cases; for example, the first premise is entailed by the thought that it is a precise matter what credence and utility the agent assigns to a proposition. However, even though this is usually false—ascriptions of desire and belief are often slightly vague—the vagueness usually isn't enough to make a difference to the ordering of propositions. (3) follows from the claim that the agent is determinately rational and that it's determinate that there are no ties in A. If the agent is determinately rational and it's determinate that there are no ties, then she determinately makes the best proposition in her action set true: this makes the right to left direction true. If we assume that the action propositions are incompatible with one another, then the other action propositions must be false giving us the other direction.[13] Finally, from the precision of the right hand side of (3) we may infer that for every action proposition A, determinately: A iff [something determinate]. This ensures (4).

Summarizing, it follows that if A is an action proposition of a given agent, and it's borderline whether A, then either: (i) the agent is not determinately rational, (ii) the agent doesn't have determinate preferences, (iii) it's not determinate which propositions are the agent's action propositions, or (iv) it's not determinate that there are no ties. If none of the caveats (i)–(iv) obtain, then the action propositions will always consist of determinately true or false propositions.

Note, however, that this conclusion is completely compatible with VAGUE ACTIONS. As we have noted several times already, being vague is not the same as being neither determinately true nor false—the proposition that Bruce Willis is bald, for example, is a vague proposition despite the fact that it's determinately true. VAGUE ACTIONS just says that the action propositions are usually vague, and this is completely compatible with their usually being determinately true or false. Moreover, the mere fact that they consist of determinately true or false propositions does not mean that our attitudes toward the vague are irrelevant, because even if an action proposition turns out to be determinately true we won't typically know which it is prior to deliberation, and it's our ignorance, and in particular our ignorance about the vague, that changes how we act.

---

[13] The assumption that the action propositions are pairwise incompatible is not obvious, and does not follow from anything I've said so far. Although many people do make this assumption, it might be possible to avoid this argument by denying it. For the time being, however, I will grant the assumption.

Let us summarize the points we have made above and relate them to the practical irrelevance thesis. The thought we started off with was the idea that the behaviour of a rational person will always be a function of their rational desires and beliefs. Moreover, if the behaviour of an agent is always a precise matter, then the desires and beliefs which matter for the agent's behaviour will be ones with precise contents. Certainly, there are conceptions of behaviour, popular among early kinds of behaviourists, that identify behaviour fairly narrowly with precise bodily movements. However, it is quite clear that this conception of behaviour will neither play the role in decision theory we need it to play nor will it track the kinds of thing we care about when we make evaluative judgements about a person's behaviour. Here I have argued that behaviour, as it figures in decision theory, is best described in terms of a fundamentally intentional notion: that of making something true. Making true is a propositional attitude in the sense that it requires you to be in a certain state of mind, although like other externalist attitudes such as knowledge, it cannot be held unless the world around you complies.

On the alternative account of behaviour, our attitudes towards the vague are practically important. For example, it's possible that on a given occasion the only thing that you've made true is a vague proposition. In this case you simply cannot describe your behaviour purely in terms of the precise things you have made true. More importantly, for a rational person, which proposition is made true is determined by which proposition they assign the highest value to, so one's beliefs and desires in the vague determine which action they perform.

Thus, on this picture it is possible for two people to assign the exact same values to all precise propositions and be in a position to make the same propositions true, while rationally behaving in different ways.[14] If Alice and Bob, say, agree about the precise in the sense that they assign the same values to all the precise propositions, they might still end up behaving differently if they assign different values to the vague. The best proposition in Alice's action set might be different from the best proposition in Bob's action set—for Alice's best proposition might be one of the vague propositions Alice and Bob disagree about. This can happen even if Alice and Bob have the same action set, since all that agreement about the precise ensures is that they will both make the same proposition true in the special case where their action sets are the same and consist only of precise propositions.[15] Moreover, this will result in a concrete difference in behaviour: since Alice and Bob will choose the highest value proposition to make true, they will end up making different vague propositions true.

---

[14] Recall that the supervenience of vague beliefs on precise beliefs discussed in chapter 8 rules out the possibility of two people *with the same evidence* agreeing in their credences about the precise whilst disagreeing about the vague. However, it is still possible for two people with different evidence to agree about the precise whilst disagreeing about the vague.

[15] Talk of two agents having the very same action set is a bit of a fiction, because action propositions typically involve the things only one agent can control. The fundamental point could be restated by invoking a suitable notion of two action sets being isomorphic instead.

9.3 Vagueness and Preferences

MAXIMIZE and AVERAGING leave V fairly unconstrained. Indeed, for someone who is not in a position to make very much true, someone in a coma perhaps, MAXIMIZE imposes hardly any constraints at all.

It is natural to think that there is another concept we can relate V to that would further constrain an agent's value function. While it hardly seems plausible to think that there is any psychological reality to the claim that a person is matched with a particular real number, say 4.1954, that measures how good things seem for them on a supposition—after all, what about the agent's mental state could possibly ground a particular assignment of values and not some scalar transformation of it—one might still think that there is some reality to comparative judgements of value. That is to say, it seems perfectly sensible to ask whether things seem better for a person on the supposition of A than on the supposition of B. Here, A and B need not be propositions the agents are in a position to make true, since one can suppose many things to be true that are beyond your powers to bring about.

> SUPPOSITIONAL PREFERENCE: The agent considers things to be better for her on the supposition of A than on the supposition of B if and only if $V(A) > V(B)$. She considers things to be at least as good iff $V(A) \geq V(B)$.

Indeed, provided your suppositional preferences satisfy some natural structural constraints (to be discussed shortly), one can show that it is always possible to represent your suppositional preferences using a value function V, to some degree of uniqueness, relative to some probability function, also determined to some degree of uniqueness. The function V is, strictly speaking, a convenient fiction, but the fact that it's always possible to represent rational preferences in the above way entitles us to use it.

If we include suppositional preferences among the attitudes used to rationalize action, then one's attitudes towards vague propositions are not irrelevant. There are vague propositions whose supposition should cause you to rank precise propositions in a way that the supposition of no precise proposition would cause you to. Learning such a proposition puts you in a practical situation you could not be in without vague information.

Let us begin by noting that the supposition of a vague proposition can clearly change the order of your preferences. There are things that would be great to buy, on the supposition that you're rich, but, since they would bankrupt you otherwise, would not be a good idea to buy. Since the proposition that you're rich is vague, the truth of this vague proposition is relevant to your decision—if you were to learn it, you should behave differently.

It is far from obvious, however, that this shows that our attitudes towards the vague are not practically indispensable. There are closely related precise propositions that would produce this same reordering on their supposition, such as the proposition that you have more than a certain amount of money. What we want, then, is a vague proposition whose supposition induces a reordering of the precise propositions that could not be obtained by the supposition of any precise proposition.

Speaking purely formally, such examples are quite simple to construct. Imagine that there are two equiprobable maximally strong consistent precise propositions A and B whose values are $V(A) = 2$ and $V(B) = 1$. Conditional on any precise proposition the order of A and B will remain the same, with A beating B, provided they are ranked (if the precise proposition being supposed is inconsistent with a proposition, that proposition won't be ranked). Suppose furthermore that A is partitioned into two equiprobable vague propositions, A_1 and A_2 with $V(A_1) = 0$ and $V(A_2) = 4$. $B \vee A_1$ is a vague proposition, and on its supposition the ordering of A and B is reversed, since clearly $V(A \wedge (B \vee A_1)) = 0$ whilst $V(B \wedge (B \vee A_1)) = V(B) = 1$. Thus if I acquire the inexact evidence $B \vee A_1$, via some imprecise perceptual faculty, for example, then I will rank both A and B, and B will outrank A.

This is one type of formal counterexample to the thesis that attitudes towards the vague are practically irrelevant. Note, however, that in this type of example one had to care intrinsically about the vague. A_1 and A_2 had different values, even though they entailed the exact same precise propositions—they both entail A, and since A is a maximally strong consistent precise proposition, they have the same precise consequences. The difference in value has nothing to do with the value of some precise matter. Clearly there's nothing wrong with valuing vague things—one can care about being rich, or bald, or whatever—but these cases normally come with valuing some precise underlying parameter such as money in dollars, or hair number. The case being described here is one where, even once one has supposed all of the precise facts to be a certain way (i.e. one has supposed A), one still has preferences for the vague proposition A_2 over A_1.

Whether this kind of preference is rational is something we'll return to later. For now let me describe another class of examples that do not depend on this feature. In this case, consider three equiprobable maximally strong consistent precise propositions A, B, and C, with values 1, 2, and 5 respectively and suppose that these values remain the same for any proposition that entails A, B, or C (for example if X entails C then $V(X) = V(C) = 5$; this ensures that no one cares intrinsically about the vague). Now suppose that the propositions you are in a position to make true are B, $A \vee B$, and $A \vee C$.[16]

I claim that there is no precise proposition such that upon learning it you would rank $A \vee B$ below $A \vee C$, and $A \vee C$ below B. One can see this by noting that if a precise proposition is consistent with C, then $A \vee C$ is ranked above B conditional on it, if

---

[16] Note that there is nothing incoherent about the idea that both B and $A \vee B$ can be part of the action propositions. For example, I can choose to place a coin on the table heads up, or I could flip it ensuring that it would land either heads or tails.

they're ranked at all, and if it is inconsistent with C then, conditional on it, $A \vee C$ is ranked below or on a par with $A \vee B$.

On the other hand, however, there are vague propositions that could induce this ranking: Suppose that C is partitioned into two vague propositions C_1 and C_2 with C_1 taking up a quarter of the total probability of C, i.e. $Cr(C_1 \mid C) = 0.25$. Let V represent your values conditional on $A \vee B \vee C_1$—then $V(A \vee B) = 1.5$, $V(A \vee C) = 1.8$, and $V(B) = 2$, yielding the conditional ranking $A \vee B \prec A \vee C \prec B$.

This is a case in which the supposition of a vague proposition seems to alter the practical situation in a very distinctive way. These are not idle suppositions either—vague suppositions play an important kind of role in practical reasoning. If I'm deciding whether to wear my coat when I leave the house I consider the options of taking or leaving it on the supposition that it is cold, and on the supposition that it is not cold. In some cases it is better to wear the coat both on the supposition that it is cold and on the supposition that it isn't; in which case I can straightforwardly conclude that I should wear the coat. This kind of reasoning is an instance of Savage's 'Sure Thing Principle'.[17] The proposition that it is cold outside is vague, so the behaviour of preferences under vague suppositions cannot be ignored. Of course, similar reasoning can be carried out with precise suppositions, but there is nothing wrong with the reasoning described above, and it is moreover psychologically implausible to think that people suppose completely precise things when doing a quick piece of reasoning like that. It is more likely that the relevant supposition is whether it's sufficiently warm to go out without a coat, and it would be perverse to require a precise account of what being sufficiently warm amounts to.

Another way in which these observations can have real practical import is when we acquire vague information (see chapter 6). Suppose that you are taking part in a jelly bean contest. You get to look at a jar of jelly beans from a distance. Then, you must decide how full you think it is, and choose the option from a list of ranges that you think it's most likely to be in. The ranges are given in terms of the percentage of the jar fullness, and the ranges you can choose from are:

1. In the 70s,
2. Between 60% and 90%, but not in the 70s, or
3. Between 70% and 90%.

You stand to make $5 if the jar is in the 60s, $2 if it is in the 70s, and $1 if it is in the 80s; you get nothing if the percentage is not within the range that you chose.

Let's suppose that before you look at the jar you find being in the 60s, 70s, or 80s equally probable. Since your eyesight isn't perfect, and the jar is in the distance, you certainly do not learn, after looking at the jar, that the percentage to which the jar is full

---

[17] Note that the principle is only good when the relevant suppositions are probabilistically independent of the options. In this example, it seems plausible that whether I wear the coat is completely independent of the weather.

is in some particular precise range. Your evidence after looking won't be anything like, say, the proposition that the jar is between 65% and 75% full; it is more likely that your total evidence regarding the jar will be a vague proposition. Perhaps your evidence after looking is that the jar is pretty full. Being pretty full is perfectly compatible with the jar being in the larger of the ranges, although when it's in the high 60s, it's borderline whether the glass is pretty full. After learning that the glass is pretty full, I can't rule out the possibility that the percentage is in the 60s, but I become much less confident in it.

The case I have described above instantiates the formal example I gave earlier. My preference ranking after seeing that the jar is pretty full will plausibly be to rank option (2) below (1) and option (1) below (3) which seems to be distinctive to this piece of vague information.

9.4 Probability in the Absence of Uncertainty

Let us now turn to an application of the decision theoretic framework that we have set up here. What follows is something of a digression; readers who wish to move on may skip to chapter 10. In section 7.3.1, I noted that the Dutch book argument for probabilism is hard to get off the ground because it relies heavily on a fairly strong connection between belief and a specific form of physical behaviour (betting behaviour). However, decision theory gives us a much more general connection between belief and preferences (where the latter are understood as being attitudes that determine, but are not exhausted by, rational betting behaviour). In this setting a more general argument for probabilism is available.

Indeed, we will argue that even anti-probabilists, of the sort considered in chapter 7, can and ought to accept a sort of ersatz probabilism. The thesis that you should be uncertain about the vague, and that your uncertainties should be governed by the probability calculus, can be given a purely decision theoretic interpretation in which they come out true, even if one's official theory of credences is non-probabilistic.

Let's begin by considering an anti-probabilist objection to the classical account of vagueness-related belief: If borderline propositions do not have truth values (as philosophers have sometimes supposed), there is no relevant proposition whose truth we are uncertain about, so it seems to be incorrect to describe this as a case where we are uncertain about something. Similarly, if there is no fact about whether Harry is bald, and I know this, then there's nothing to be uncertain about so we shouldn't be asking how confident one should be about whether Harry is bald. If I know and am completely confident that it's borderline whether Harry is bald, one might object, then any assignment of confidence to the proposition that Harry is bald would be epiphenomenal. Unlike, say, a degree of confidence about whether it will rain tomorrow—which might inform my decision to bring an umbrella—any confidence about Harry's baldness would be practically inert. If I did have a credence in that proposition I could have pretty much any credence I like without negative consequences, and there would be no way to test what my credences were.

This objection rests on a couple of fallacies. First, this theorist is using the word 'true' in a way that prevents one from substituting p for 'p is true', whilst also equating uncertainty about whether or not p with uncertainty about whether p is true or false. By this theorist's lights, these are not the same thing, and it remains far from clear why the lack of a truth value should preclude uncertainty about whether or not p. More to the point, the objection appears to rest on a strong version of the practical irrelevance thesis. According to the considerations in the previous sections of this chapter, one's credences in the vague are not epiphenomenal: one can test a person's credences in the vague by observing their behaviour and indiscriminately changing your credences in vague propositions can result in actions with negative consequences, even if they leave your credences in the precise the same.

Now my opponent in chapter 7 could concede this point and still insist that there's no sense to be made of uncertainty when one considers there to be no fact of the matter. That is to say, the following concession is perfectly compatible with their view:

Although being genuinely uncertain in the face of vagueness is a mistake, vagueness-related attitudes can still rationalize distinctive practical behaviour.

In this section, I want to explore this kind of response. I will argue that, even if this theorist is right about there being no such thing as 'genuine' uncertainty due to vagueness, there will be some probabilistic notion in the vicinity of uncertainty that informs our practical reasoning. In particular, I shall argue that there is some kind of mental state, or mental feature that supervenes on your mental states, that *does* satisfy the axioms of the probability calculus, and moreover governs rational behaviour in the way we thought credences were supposed to. Whether we want to go one step further and call this thing a measure of 'genuine uncertainty' can be left up for debate. For most purposes, however, this conclusion is strong enough.

To illustrate the idea, let me draw an analogy with another debate in the literature. An extremely natural way to interpret the standard formalism of quantum mechanics (without augmenting it or taking it to be a partial description of the fundamental facts) is to treat the classical macroscopic reality around us—us and the things we observe— as constantly splitting into multiple 'branches' (see Everett [43]). So, for example, when I flip a coin which has a genuine chance of landing heads and of landing tails, we can expect that in reality there will be one branch in which the coin lands heads and another in which it lands tails. For the time being, let us set aside discussion of the merits of this idea, and assume that this is a correct description of the fundamental universe. According to this picture, there's a straightforward difficulty for making sense of probability: I know exactly what's going to happen in the coin flipping case— there's going to be (with certainty) a branch in which the coin lands heads and another branch in which it lands tails. Since all the potential outcomes actually occur on this picture, it's impossible to make sense of uncertainty about which outcome will occur.

This was a long-standing obstacle to this interpretation of quantum mechanics, and, for many, it was a decisive objection: the power of quantum theory comes from its ability to predict the chances of things happening, after all, so there is no place for an interpretation that cannot make sense of probability. However, David Deutsch [31] showed that, *even if* there's no room for uncertainty on this kind of hypothesis, one can still make sense of probability by looking at rational action.[18] More importantly for that project, he was able to show that a person who acts rationally (and knows some of the relevant quantum theory) will act *as though* she was uncertain about the outcomes and, moreover, uncertain to the degrees that quantum theory predicts (i.e. as given by the Born rule).

Note that if there's no such thing as genuine uncertainty in the many worlds interpretation of quantum mechanics, or indeed, in the case of vagueness, then there's a prima facie puzzle about how to interpret the decision theory, stated in terms of value functions, that I outlined. For example, AVERAGING explicitly appeals to credences, and if there is no real uncertainty in the cases of interest, then this appeal is problematic. This is solved in Deutsch's framework by dispensing with talk of values and credences, and just talking about an agent's *preferences* over various acts—things that have a direct behavioural interpretation. In our framework, we can take these to be given by our suppositional preferences over propositions, as defined in section 9.1. While our preferences over action propositions have the clearest behavioural interpretation, even our preferences over other propositions can manifest themselves as dispositional behavioural properties: counterfactuals like 'if I were in a position to make A or B true, I would make A true' can ground many of your suppositional preferences.

Just as Deutsch was able to do in the many worlds case, we can reconstruct talk of probability, without taking it for granted that people are uncertain in the face of vagueness. In order to do this one must make a few assumptions about the agent's preference relation: they must obey some natural constraints that seem natural for a preference relation to be rational. One's preferences ought to be irreflexive, asymmetric, and transitive, for instance. You cannot prefer A to B, and B to C without preferring A to C, you cannot prefer A over itself and if you prefer A to B you shouldn't prefer B to A. (Similarly obvious things can be said about the relations being 'preferable to or as preferable as', 'as preferable as'.)

Another constraint is that preferences be linear. Any two propositions are either as preferable as one another, or one is more preferable than the other. This principle is perhaps more controversial, although as in our discussion in section 7.3.2, its non-obvious appearance doesn't seem to be distinctive to vagueness. One can also state a version of AVERAGING purely in terms preference, without invoking credences. A straightforward consequence of AVERAGING is that if E_1 and E_2 partition A into two,

---

[18] These ideas are further developed by Greaves [67] and Wallace [149].

and $V(E_1) < V(E_2)$ then $V(E_1) < V(A) < V(E_2)$—this is obvious since $V(A)$ is just defined as a (weighted) average of $V(E_1)$ and $V(E_2)$. As a constraint on suppositional preferences, it seems intuitively correct even without appealing to the justification of it in terms of the probabilistic version of AVERAGING: if $E_1 \prec E_2$ then $E_1 \prec A \prec E_2$, provided E_2 and E_2 partition A.

The final two constraints are perhaps harder to understand, but no less intuitive once they are understood. According to the decision theory we've outlined, A and B can have the same value even if they are not equiprobable. However, in such cases, if you were to disjoin both A and B with a more preferable proposition C disjoint from both of them, it would pull their values apart. This is another consequence of AVERAGING, but this time we're making use of the fact that the average is weighted by probability: if A and B are not equiprobable, then their respective disjunctions with C will have different values because the A and B parts will be weighted differently. On the other hand, if A and B are equiprobable, then their disjunction with C will have the same value for every C of the kind described, because the weightings of the respective disjuncts will be the same. Thus we should have that if $A \approx B$, then either $A \vee C \approx B \vee C$ for no proposition C more preferable but disjoint from both A and B (in the case that A and B are not equiprobable), or $A \vee C \approx B \vee C$ for every proposition C more preferable but disjoint from both A and B (in the case that A and B are equiprobable).

The final axiom effectively states that the preference relation is continuous in the sense that whenever the supremum or infimum of a set of propositions lies in a certain interval (let us say, A is preferred to it and it is preferred to B) then some member of the set lies in that interval (i.e. some member is such that A is preferred to it and it is preferred to B).

The axioms listed above can be summarized as follows:

1. $\prec$ and $\preceq$ are transitive relations. $\prec$ is irreflexive and $\preceq$ is reflexive.
2. $\preceq$ is total: for any p and q either $p \preceq q$ or $q \preceq p$.
3. If $A \wedge B = \perp$ then
   (a) If $A \prec B$, then $A \prec (A \vee B) \prec B$.
   (b) If $A \approx B$, then $A \approx (A \vee B) \approx B$.
4. If $A \wedge B = \perp$ and $A \approx B$, then either $(A \vee C) \approx (B \vee C)$ for no C with $C \wedge A = C \wedge B = \perp$ and $A \not\approx C$, or $(A \vee C) \approx (B \vee C)$ for every such C.
5. If $A \prec C \prec B$ and $C = \bigvee_\alpha C_\alpha$, then for some β, $A \prec \bigvee_{\alpha > \beta} C_\alpha \prec B$. Similarly, if C is the infimum $\bigwedge_\alpha C_\alpha$, then for some β, $A \prec \bigwedge_{\alpha > \beta} C_\alpha \prec B$.

By only talking about preferences over propositions, then, we can formulate Jeffrey's decision theory without invoking the notion of a credence, and thus we can apply Jeffrey's decision theory without assuming that vagueness involves genuine uncertainty.

However, an important result due to Ethan Bolker [16] shows that if your preferences satisfy these constraints, then there is a unique (or uniquish) function V and probability function Cr which satisfy AVERAGING and represent your preferences in the sense that $V(A) < V(B)$ iff $A \prec B$ for every A and B. The above purely preference theoretic axiomatization of Jeffrey's decision theory shows that the appeal to the value function V and the credence function Cr are dispensable in favour of talk about preferences. However, Bolker's representation theorem shows that one can also reconstruct these functions from the preferences and use them as though they are part of the agent's psychology. In other words, even if uncertainty about the vague were problematic, we have decision theoretic substitutes for credences—the probability function that is generated by the representation theorem:

Theorem 9.4.1. *Bolker's Existence Theorem. Suppose that $\prec$ is a preference relation satisfying the above axioms. Then there exists a countably additive probability function Pr, and a real valued function u on the maximally strong consistent propositions such that:*

$$A \prec B \text{ iff } E(u \mid A) < E(u \mid B)$$

Where $E(u \mid A)$ represents the conditional expectation of u conditional on A, according to Pr. The function $V(A) = E(u \mid A)$ is a Jeffrey-style value function which satisfies averaging.

Theorem 9.4.2. *Semi-Uniqueness. Suppose that V represents $\prec$ and V satisfies averaging with respect to Pr, and similarly for V' and Pr'. Then there is a real number, λ with $\frac{-1}{\inf_X(V(X))} \leq \lambda \leq \frac{-1}{\sup_X(V(X))}$ such that:*

$$Pr(A) = Pr'(A)(1 + \lambda V'(A))$$
$$V(A) = V'(A)\frac{1 + \lambda}{1 + \lambda V'(A)}$$

Note that in the special case where your values are unbounded, then λ must be 0. So in that special case we get genuine uniqueness of V and Pr.

The representation theorem shows that we have decision theoretic substitutes for particular numerical credences. However, we can demonstrate the general strategy of substituting genuine doxastic attitudes with decision theoretic ones by considering a much simpler example: a decision theoretic substitute for the notion of being uncertain—i.e. having a credence strictly between 0 and 1.

Given that we can make sense of an agent's preferences between arbitrary propositions, it is also possible to make sense of what we might call *conditional* preferences. Just as you can ask whether someone prefers A to B, you can also ask whether they prefer A to B on the supposition of C. The former question amounts to asking whether things seem better to the agent on the supposition of A than on B, whereas the latter can be understood as asking whether things seem better to the agent on the supposition of both A and C than on both B and C. If we formalize a preference with the relation $A \prec B$, then conditional preference on C, written $A \prec_C B$, can be defined as simply $AC \prec BC$.

If the agent's preferences and preferences conditional on A are defined and identical, then we can say she is *practically certain* of A. Not only would her best action be her best action conditional on A, but her preferences over all propositions remain the same on the supposition of A. Thus for all possible practical reasoning, A is taken for granted.[19] The representing probability function assigns A an intermediate value iff A is practically uncertain (assuming the agent doesn't have completely flat preferences).

Let us now see how practical uncertainty can explicate uncertainty about the vague. The example will trade crucially on the fact that it is coherent to suppose things that are impossible to know. Recall that it's perfectly coherent to suppose that the number of particles in the universe at this moment is even and that nobody will ever know this fact. No one could ever know this conjunction, since it would involve someone knowing that the number of particles in the universe is even and also knowing that nobody knows this, which is jointly impossible. In the case of vagueness, it means that even if we know that it's borderline whether A, one can still ask what one's preferences are on the supposition that A, even if it is impossible to know A.

Here is an example in which uncertainty due to vagueness reveals itself in our preferences. Suppose that I have recently won a lottery, but I do not yet know exactly how much money I have won: I know that the total money I own, including the winnings, lies between $\$x$ and $\$y$. Suppose also that it's determinate that anyone that has between those two amounts of money is borderline rich. As it happens my credences are uniformly distributed over the possible amounts of money I could have between $\$x$ and $\$y$. Even though I'm in fact uniformly distributed over these values, on the supposition that I'm rich, I'm more confident that I have a larger amount of money than smaller, for this is exactly the kind of effect that conditioning on vague propositions has on my credences over the precise propositions. Thus, things are better for me on the supposition that I'm rich than on no suppositions, which means that my unconditional preferences rank the proposition that I'm rich above the tautology. However, trivially, I rank the proposition that I'm rich along with the tautology conditional on my being rich since, for any A, $V(A \wedge A) = V(\top \wedge A)$. It follows that I'm practically uncertain whether I'm rich. (Of course, I introduced the above preferences by talking about credences, although one doesn't need to be able to make sense of credence to ascribe that pattern of preferences.)

Note also that since it's in fact borderline whether I'm rich at t, I won't ever have evidence that I'm rich at t. However, as I have argued in section 9.1, this does not make these kinds of preferences epiphenomenal—they often will have real practical import.

Thus, it seems that one can make sense of probability talk even if you do not accept the thesis that vagueness involves any real uncertainty. One could, at this juncture call the resultant probabilities 'credences'—for after all, they play the right functional role by connecting with our desires and actions in the way decision theory prescribes. This would be my preference, although the conclusions I have drawn about the coherence of probabilistic talk does not depend on making this identification.[20]

---

[19] Related notions play an important role in some philosophical accounts of knowledge; see, for example, Fantl and McGrath [44].

[20] See Meacham and Weisberg [105] for more discussion of these issues.

10

Vagueness and Desire

In chapter 8, we noted that by treating vagueness-related ignorance as straightforwardly analogous to ignorance about the world, we concede much to the epistemicist way of thinking. If we are truly to think of vagueness as something more than mere ignorance, we need to say something to preclude a purely epistemic interpretation of borderlineness.

Given that the nature of ordinary 'worldly' uncertainty is clarified by its role in decision theory, it is natural to think that the decision theory for vague propositions developed in chapter 9 could shed some light on this issue. It is natural to think that a purely epistemic interpretation of vagueness would not distinguish in any meaningful way between the decision-theoretic role of a vague and a precise proposition. In this chapter, I shall argue that, although the formal theory of preferences might look the same, on a purely epistemic understanding of vagueness, PRACTICAL IRRELEVANCE fails in a quite spectacular way: in terms of practical reasoning, there is very little to distinguish ordinary ignorance about the things that affect what we desire and ignorance about these things that is due to vagueness. For if vagueness is just a special kind of ignorance, then there is nothing to prevent us from finding vague matters themselves to be intrinsically desirable, much like we find other outcomes that we are ignorant about to be desirable. On the other hand, by denying the decision theoretic analogy between vague and precise desires, I shall argue, we open up the possibility for a partially bouletic interpretation of vagueness that does not identify it as merely a species of ignorance.

10.1 The View that Vagueness is Merely a Kind of Ignorance

Why does decision theory look different on a purely epistemic understanding of vagueness? The thought, to put it in very broad terms, is that according to an interpretation in which vagueness is just a special kind of ignorance, there's really nothing to distinguish decisions made under uncertainty about factual things, and decisions made under uncertainty about borderline things: borderlineness just is a kind of uncertainty about factual things according to that view.

In this chapter I shall focus on a precisification of the practical irrelevance thesis that, I think, does capture an important truth about the difference between our beliefs and desires about vague outcomes and about precise outcomes: while we can care intrinsically about the latter, it seems as though there's something incoherent about caring intrinsically about the vague. Interestingly, this assumption does not look particularly well-motivated on the epistemic theory. If you think that vagueness is *merely* a matter of ignorance, and that, in general, nothing precludes you from caring about things you are ignorant about, then there is nothing to preclude you caring about the vague.

To demonstrate what this might amount to, let us consider an example. Suppose Bob has the telic desire to be bald, and, for simplicity, let us suppose this is the only thing he cares about. He could have this desire in a number of ways—for example, he might have the desire derivatively in virtue of his having the desire to have no hairs at all. Let us suppose that this desire is not had derivatively, and is the strongest proposition about his head that he desires: Bob wants to be bald and any other epistemic possibility about the state of his head that he finds desirable is already entailed by his being bald. There is, therefore, no specific number of hairs he wants; if he wants to have 98 hairs or less, this is only because he thinks this will make him bald and not because he cares about having a smaller number of hairs over a larger number of hairs. He does not care whether he has 98, 99, 100, . . . hairs unless it would make the difference between his being bald and his being not bald—that is, he cares about his numerical hair number only insofar as it contributes to making him bald.

Compare this with another case. Alice cares intrinsically about whether there is intelligent life elsewhere in the universe right at this second. Although Alice expects never to be able to know the answer to this question, I think that this kind of desire is still perfectly coherent. Perhaps Alice is a science fiction enthusiast, or assigns high value to worlds with more diversity and culture than worlds with less. Or perhaps she is a fundamentalist Christian, and disvalues possible worlds where humans do not occupy the role of being the sole bearers of intelligent life in the universe. Whatever the story, it seems like something it could be perfectly rational to care about (even if eccentric).

For simplicity, again, we may assume this is the only thing she cares about. Like Bob, she doesn't really care about other features of the universe. Let's also suppose that she's certain that if there is intelligent life, it's well outside the region spanned by her light-cone at any point in her life. Whether or not there is intelligent life out there is completely causally independent of her and is something she could never hope to verify. Nonetheless, she thinks that if there is intelligent life out there, that would be a pretty cool thing.

I think there are two kinds of reactions you could potentially have to these two examples.

Reaction 1. Bob's desire is just like Alice's desire. They both desire something which could never have a causal effect on either of them. Supposing that Bob is borderline bald, and that all intelligent non-human life is outside Alice's light-cone, the truth values of both the propositions they desire will never be known. The fact that Bob is bald (is not bald) is just like the fact that there is (is no) intelligent life in the universe. They are both perfectly coherent desires to have.

Reaction 2. Bob's desire is not at all like Alice's. While their desires are both unknowable and causally independent of their immediate surroundings Bob's desire is simply incoherent. Bob cares about whether he is bald or not when he already knows exactly how many hairs he has. Based on this desire Bob acts as if there is a fact of the matter out there, whether he's bald, just as there's a matter of fact about whether there's intelligent life in the galaxy; only Alice is entitled to act like this.

It is not unnatural to associate the first kind of reaction with an epistemicist who thinks that vagueness is *merely* a matter of ignorance, albeit a special kind of ignorance. For the epistemicist, facts about baldness in borderline cases are not especially different in kind from facts about life in the far reaches of the galaxy. Since it is perfectly fine to care about things you are ignorant about—even when that ignorance is impossible to remove—there is no distinctive reason why we could not care about whether a borderline case of baldness is bald or not.

The second kind of response is naturally associated with supervaluationists and other theorists who think that in borderline cases, there simply is no fact out there to be known, and no fact worth caring about. Propositions about life in the galaxy are different, since, while they are like borderline cases in the sense that they are unknowable, there is still a fact of the matter *out there* which one could reasonably care about. For this latter kind of theorist, Alice's desire is perhaps a little eccentric, perhaps even irrational in a loose sense, but it is not incoherent. Bob's preference, on the other hand, seems to be conceptually confused.

This observation is intended to help us with the problem we raised in chapter 8. If epistemicists and other 'factualists' think that vagueness is *just* a matter of uncertainty, then what is this extra thing, beyond uncertainty, that those who deny the factual nature of borderline cases think these cases involve? According to the theory I'm advancing, that difference is articulated by two further theses: (i) there cannot be rational differences of opinion that are solely differences of opinion about the vague and (ii) you shouldn't care intrinsically about the vague. The thought, in both cases, is that if vagueness amounts to nothing more than uncertainty, then there would be no reason to think there couldn't be rational disagreements about purely vague matters, or to think that one couldn't care intrinsically about vague matters. There is no general reason to think that one cannot disagree or care intrinsically about things we are uncertain about, so these two principles give us two ways in which vagueness amounts to more than the presence of uncertainty.

The first principle was formulated precisely in terms of the supervenience of vague on precise beliefs in chapter 8, and was defended there. We shall spend the rest of this section sharpening and defending the second claim.

Before I do that, it will be worth noting that the conjunction of (i) and (ii) have fairly drastic decision-theoretic consequences. For a view in which vagueness is merely the presence of uncertainty—i.e. an account in which neither constraint (i) nor (ii) are accepted—vague beliefs and desires, far from being epiphenomenal, can play a fairly concrete, even surprising, role in informing our actions. The cases that most starkly demonstrate this are cases which have the following structure. Suppose there are two rational agents, Alice and Bob, such that:

1. Alice and Bob have exactly the same evidence and assign the same credences to every precise proposition.
2. Alice and Bob have exactly the same desires (their utility functions are identical).[1]
3. Alice and Bob are in a position to make exactly the same propositions true.

If we accept both (i) and (ii), then it is pretty straightforward to show that Alice and Bob must assign the same V-values to *all* propositions, and thus make the same proposition true if they are rational (assuming, for simplicity, that there are no ties). However, a 'mere uncertainty' account of vagueness could allow both (i) and (ii) to fail simultaneously. If Alice and Bob intrinsically valued some vague matter and also had disagreements about vague matters which didn't appear as disagreements about precise matters then, I claim, it is possible to construct examples where Alice and Bob make completely different choices. For example, consider the following decision puzzle:

Alice and Bob are in an auction house and they are both bidding on a relatively inexpensive vase which is a shade of turquoise that is borderline between being blue and green.

Now suppose we know the following things about Alice and Bob. Firstly, they have exactly the same telic desires and have the same credences in the precise propositions. We also know the following facts. (1) As it happens they are both very peculiar people and they care intrinsically about owning green things. (2) Although they both know exactly what precise shade the vase is, Alice is more confident than Bob is that the vase is green.

The scenario described violates both (i) and (ii): it requires that Alice and Bob care intrinsically about the vague and that they disagree to some degree about whether the vase is green, even though they know the relevant precise facts such as its exact shade.

In the above scenario it is clear that Alice should bid more than Bob for the vase. Since they care about the same things, its value is the same for both Bob and Alice. However, since they know that it is borderline whether the vase is green, they are uncertain whether it is green and, therefore, the purchase is a gamble on its being green. However, according to Alice's subjective credences the odds that it is green are better than they would be according to Bob's. Therefore, Alice ought to outbid Bob.

---

[1] That is to say, the utility function mapping each maximal state to real numbers is the same in both cases. Note, however, that this doesn't mean they assign the same expected value to every maximally strong consistent precise proposition, because if Alice and Bob disagree about the vague, the supposition of some precise propositions could change their credences in different ways about how likely it is that vague things they care about will happen.

On the proposed view the kind of scenario described above requires Alice and Bob to have deeply irrational preferences, and at least one of Alice and Bob to have incorrect credences. On the other hand, the scenario would be permitted on the 'mere uncertainty' account of vagueness. According to that kind of factualism about vagueness, one can reasonably disagree about a proposition, even given all the relevant precise evidence, and, moreover, one can care intrinsically about the vague.

10.2 The Indifference Principle

It's obvious that you can care about matters which are vague: for example, you can rationally care about being rich, being happy, having lots of friends, and so on and so forth. The proposition that you're rich or that you're happy, and so on, are all vague propositions. While this is all compatible with the idea that one cannot care *intrinsically* about the vague, it is urgent that we explain the difference between caring about the vague and caring intrinsically about the vague.

Let us assume that the algebra of propositions is a complete atomic Boolean algebra. Every proposition can thus be thought of as a set of indices. As usual, we can also partition these indices into maximally strong consistent precise propositions: consistent propositions that are precise and which entail any other precise propositions consistent with them. (Here, as in earlier chapters, we assume that the precise propositions themselves form a complete atomic Boolean subalgebra of the algebra of propositions, as discussed in section 3.2.) For the supervaluationist, for example, the indices could be represented by ordered pairs consisting of a world and a precisification, and the maximally strong consistent precise propositions by a partition of pairs such that the members of any element of the partition all have the same world coordinate. But our characterization here is completely general, and our discussion applies to any theory that accepts the Booleanist assumptions outlined above.

With these notions in place, we can state the principle that correctly encodes the thought that one should not care intrinsically about the vague:[2]

Indifference. If, for every precise proposition, either both A and B entail it, or entail its negation, then one should be indifferent between A and B.

Note that this principle does not entail that you couldn't rationally care about being rich. For simplicity, let's assume you only care about your financial situation and let's also assume that whether you're rich supervenes on how much money you have. Given these assumptions, **Indifference** *does* entail that you should be indifferent between states that agree on how much money you have. If you furthermore know how much money you have, you should be indifferent between all the different maximally specific

---

[2] It should be noted that, since there is vagueness concerning which propositions are precise, there can be cases where it's borderline whether you may care about p.

ways things might, for all you know, be, even if they disagree about whether you're rich or not.

Recall that a utility function, u, maps each index to a value. In conjunction with a probability distribution representing your credences, the utility function determines the V-value of every proposition. This V-value will agree with the utility function when restricted to the maximally strong consistent propositions corresponding to the indices. There is an intuitive way to visualize the constraint that **Indifference** puts on your utility function, and, consequently, on your values. Again, we picture logical space as being partitioned into cells by the maximally strong consistent precise propositions. **Indifference** says that the utility function must assign the same value to indices in the same cell. Thus, the utility function can be completely represented by mapping each *cell* to a value: the value that that index in the cell would ordinarily by assigned. Now if A and B are such that, for every precise proposition, they either both entail it or entail its negation, then A and B are subsets of the same cell. So, in particular, by AVERAGING, $V(A) = \sum_{i \in A} V(i \mid A)V(i)$. Since $V(i) = u(i)$ is some constant value α, when i ranges over indices in a fixed cell, we have that $V(A) = \alpha \sum_{i \in A} V(i \mid A) = \alpha$ (since $\sum_{i \in A} V(i \mid A) = 1$). By completely parallel reasoning, $V(B) = \alpha$, so $V(A) = V(B)$, which is exactly what **Indifference** demands.[3] Conversely, **Indifference** says that when A and B are indices in the same cell, one must be indifferent between them: i.e. their values, and thus their utilities, must be the same.

10.3 Caring about the Vague

One might take exception to the **Indifference** principle on the grounds that there are phenomena that on the face of it seem closely related to vagueness, yet where it seems as though there's absolutely nothing wrong with caring in these cases. For example, according to some theories of time, most contingent propositions about the future are indeterminate. Yet it seems perfectly rational to care about what will happen to you in the future. Indeed, someone who didn't care about their future presumably would be indifferent between all options in a decision, and so people wouldn't ever feel the need to deviate from the status quo.

I think in this case it's tempting to turn the argument on its head. To my mind, this makes for a fairly compelling argument that truths about the future are not indeterminate, they are just epistemically inaccessible. Certainly, either there will be a sea battle tomorrow or there won't, and we do not know which, but that surely does not mean that there is no fact of the matter. The fact that it's perfectly rational to care about which outcome will obtain seems a reason to think that there is a fact of the matter for us to care about. Also, unlike in the case of vagueness, we do know some truths about the future—I know what I'm having for dinner tonight, and lots of

---

[3] Once we assume that each cell is uncountable, the preceding argument would need to be formulated more carefully, in terms of integration.

other mundane facts about the near future. If we combine this observation with the fact that we can care about the future, the analogy between the future and borderline cases quickly dissipates.

Other objections in a similar vein I think represent a misunderstanding of the **Indifference** principle. One might object that one ought to be able to care intrinsically about being happy, or sad, or about having pleasurable experiences and so on, yet each of these things are vague. For the sake of concreteness, let us focus on happiness. **Indifference** is certainly consistent with one caring about being happy— what **Indifference** rules out is that you care about being in states which count you as happy over states which count you as not happy, even if they agree about all the precise facts including the sorts of precise facts about your life that usually determine whether you are happy or not (the status of your personal relationships, how much money you have, etc.).

Let us consider a (highly idealized) sorites for happiness: suppose that you are going to give me a portion of a cake, measured in grams. If you give me a 1g crumb of cake, I will not be happy, whereas if you give me a large 200g slice, I will be happy. Keeping other variables relevant to my happiness fixed, there will be some amount of cake in the middle such that it is borderline whether I'd be happy if you gave me that much cake. Since it is borderline whether I'm happy in these scenarios, I don't know whether I'm happy when I get that much cake. Now consider two epistemic possibilities in which I receive that borderline amount of cake and which differ only in that in one I'm counted as happy and in the other I'm not. It seems to me that since I am receiving exactly the same amount of cake in both scenarios, and *ex hypothesi* the other factors relevant to my happiness are fixed, it would be bizarre to think that I should countenance a difference in value between the two cases. Happiness is not something you can directly pursue; it is a side effect of pursuing more specific things, and it is those things that you should care intrinsically about.

It is worth stressing that **Indifference** is for this reason a *substantive* postulate. It says that for every vague thing a rational person might find themselves caring about (such as being happy), there are some underlying precise matters which they care intrinsically about from which their vague cares arise. In the above case, I assumed for the sake of argument that it was the precise amount of cake I received in grams. But this assumption was made only for the sake of concreteness: in realistic cases it is reasonable to wonder what the precise things I care about that make me happy *really are*. Uncomplicated answers will most likely be phrased in vague vocabulary—after all, almost all language is vague. Nothing about the **Indifference** principle guarantees that it is possible to *articulate* the sorts of things you care about (I discuss this point further in section 10.4 below).

It is also worth emphasizing that the putative counterexamples to **Indifference** do not rest solely on judgements about what it is permissible to care about. They rest, rather, on the *conjunction* of those judgements with judgements about which propositions are precise. But we have seen that the naïve, pretheoretic notion of precision should not be the target of our theorizing (an issue we will return to, especially in section 12.2), and we have also suggested that precise propositions are rarely expressible by sentences of a natural language. One might wonder how we *can* properly theorize about precision without mooring it to those pretheoretic judgements, or to the sorts of things we can express in language. This is exactly where **Indifference** helps—it pins precision down by its role in thought. Once you have figured out which things it is permissible to care about, it is possible to figure out which propositions are the precise ones. We will give a precise recipe for doing this in chapter 13, but for the purposes of this discussion it suffices to note that theses concerning what it's permissible to care about may lead to interesting theses about which propositions are precise.

Some predictions of the **Indifference** principle apparently conflict with other aspects of the role of precision in thought, and these cases require further discussion. Let me take up an example discussed at length in Williams [154], drawing from the literature on personal identity: survival. I take it that survival is not always a completely precise matter. There are episodes that someone could undergo in which it would be clear that they would survive—perhaps if a few of their memories were erased, or some of their matter were replaced. And there are other episodes after which it would be clear that they do not—perhaps if all of their matter and all of their memories were replaced. It is a routine matter to construct a sorites sequence connecting the two kinds of cases, so it surely follows that there are episodes someone could undergo in which it would be borderline whether they survive: there is someone before the episode, call her Alpha, and someone after the episode, call her Omega, but it is borderline whether the person before and the person after is the same person or not.

Now, Williams argues convincingly that in such cases Alpha should care, at least to some degree, about what happens to Omega. For example, if Alpha had the chance to pay a small amount of money to prevent Omega undergoing a large amount of pain, she should probably pay that money. We could explain this, as Williams does, by supposing that the person before the episode cares intrinsically about what happens to *her*, and since she does not know whether she is the person after the episode or not, she should derivatively care about what happens to the person after the episode. This style of explanation would be disastrous for the **Indifference** principle, for it supposes that Alpha cares intrinsically about whether *Alpha* is in pain, and as we have seen, it is borderline whether Alpha undergoes the painful experience in the scenarios where Alpha doesn't pay the money. Thus Alpha cares intrinsically about the vague and seems rational in doing so.

However, I think there are a few complications with this argument that make it unclear whether it really conflicts with the **Indifference** principle. The complication is that the usual way to flesh out the scenario described above allows one to explain these preferences without invoking desires towards the vague, circumventing any violation of **Indifference**. The standard way to think about the vagueness that arises in this puzzle is that there are (at least) three salient temporally extended entities in the vicinity of Alpha and Omega: one relatively short one ending at the time of the episode which indeterminately disrupts Alpha's identity, call this A, another short one that begins at the time of that episode, call this thing B, and a long one coinciding with both A and B, call this C.

The 'standard' story locates the vagueness in the names 'Alpha' and 'Omega'—according to that story 'Alpha' is referentially indeterminate between referring to A and to C, and 'Omega' is referentially indeterminate between referring to B and to C; thus the sentence 'Alpha=Omega' is linguistically borderline. Of course, for a non-linguistic theorist a different story would be told about the *source* of the vagueness. However, they will nonetheless also accept the above thesis about the referential indeterminacy of 'Alpha' and 'Omega' and the linguistic borderlineness of 'Alpha=Omega' (recall that a sentence is linguistically borderline if it expresses a borderline proposition). Note also that what I am calling the 'standard story' hasn't taken sides on the mereological situation involved here: we could think of C as the fusion of A and B, or we could think of it as mereologically disjoint but coincident with both, or in some other way—it will not matter for our purposes.

Now, since, by hypothesis, Alpha is a self-interested person, and either Alpha is identical to A or Alpha is identical to C, it follows by logic that either A or C, or maybe even both, are self-interested people. On the assumption that Alpha is identical to A, A is a self-interested person and therefore cares only about what happens to A. However, since it is a completely precise matter what happens to A—it's determinate that A's life ends at the time of the disruption—A does not care intrinsically about the vague. Secondly, on the assumption that Alpha is identical to C, we get a parallel conclusion about C: C is a self-interested person, and so C does not care intrinsically about the vague since C cares only about what happens to herself, and it's determinate that she survives the disruptive event. Since, determinately, Alpha is identical to either A or C, and determinately A and C only care about the precise, it follows that, determinately, Alpha does not care intrinsically about the vague.

This is, I think, enough to deflate Williams' argument. Nonetheless, one might still be left with a feeling of puzzlement—why is it that it seems as though A ought to act as though she were uncertain about what will happen to her? Why does A seem to act as though she cares about what happens to the person after the disruption? Fortunately, we can explain this without invoking vagueness-related uncertainty or desire. At time t before the episode, A and C have exactly the same subjective experiences (or at least, things that are not determinately not subjective experiences), so it is natural to think that both A and C have *self-locating* uncertainty: for all A knows, she's C and is going to survive, and for all C knows, she is A and is not going to survive. So neither A nor C know whether *they're* the entity that survives, and thus will do things that benefit the entity that survives to the extent that they think that *they* are the entity that survives.

Thus, in the scenario described above, the only thing A cares intrinsically about is what happens to A. The problem is that A doesn't know whether she occupies the qualitative role A actually occupies or the qualitative role that C occupies, and, therefore, doesn't know whether she will survive. However, it is a perfectly precise matter what happens to A on either hypothesis so she does not care intrinsically about the vague; similarly for C.

10.4 Is it Always Possible to Articulate your Desires Using Precise Language?

So far, we have only considered specific examples in which it appears as though it is permissible to care about the vague. However, one might have a much more general worry about the **Indifference** principle. According to an extremely pervasive view, pretty much the only concepts that can be used to express precise propositions and properties are the concepts of logic, mathematics, and fundamental physics. On this assumption, however, the **Indifference** principle seems to entail two extremely implausible principles.

Firstly, it seems to entail that I can only care intrinsically about a very limited set of propositions. For, according to some philosophers, only the propositions of fundamental physics, mathematics, and logic are truly completely precise. Secondly, it follows that it was only until quite recently, with the discovery of modern physical concepts, that we were able to have thoughts about these precise propositions at all. That is to say, up until the birth of modern physics, pretty much everybody had irrational desires because the only desires they could have formed would have been formed using vague concepts.

This objection, if it has any force, rests on some presuppositions that are fairly specific to a linguistic conception of vagueness. The objector assumes, for example, that in order to have a vague belief or desire, one must first mentally do something with vague concepts, such as articulate a vague sentence in the language of thought. However, this picture is emphatically denied on the account of vagueness I have been sketching so far.

The idea that we can obtain a useful categorization of precise and vague propositions by looking at the kinds of sentences or concepts that express them is also a bad starting point. Indeed, on the view I have been endorsing, it is plausibly not true that the sentences of fundamental physics express precise propositions. (I will argue for this in more detail in section 12.2.)

The upshot of this is that the objector, in talking about our concepts, is assuming a radically different picture of vagueness. The lines between vague and precise propositions, on my view, simply do not correspond to the lines drawn by sentences (or concepts) that are used in a certain way; they are drawn by the role that those propositions have in thought. Indeed, according to my view pretty much the only sentences that pick out precise propositions are the sentences of mathematics and logic, and thus very few contingent precise propositions are picked out this way; arguably, then, most precise propositions are not expressible on this view.

Thus, to respond to the objection, while I agree that *sentences* of a public (or private) language rarely express precise propositions, it does not follow that we do not bear any interesting relations to precise propositions in the sense articulated by my preferred theory of propositional attitudes. The notion of a precise proposition can be implicitly introduced by the role it plays in a theory of rational propositional attitudes, a theory in which knowledge of physics is not a requirement of rationality.[4]

---

[4] It is also possible that the objection rests upon an equivocation with the word 'property'. There are two conceptions of properties: an abundant conception according to which properties are abstract objects, which serve as the semantic values of predicates and feature importantly as the objects of belief and other intentional attitudes, and a sparse conception according to which properties are physical things whose existence depends on fundamental features of the world and which are discovered by science. The property of being an electron, in the latter sense, may in fact *never* feature in the thoughts of an agent, even an agent who has all the concepts needed to state a final theory of fundamental physics.

PART III

Logical Matters

11

Vague Propositions

In the following chapters, we shall attend to a number of logical issues that arise concerning the treatment of vague propositions. The first order of business will be to outline a theory of vague propositions.

We will begin by observing that sets of possible worlds are not suitable for the modelling of vague propositions: the view that necessarily equivalent propositions are identical, along with the assumption that the vague supervenes on the precise, entails that all propositions are precise. Drawing from the supervaluationist literature, an extremely natural alternative to sets of possible worlds is sets of ordered pairs consisting of worlds and precisifications. I shall argue in chapters 12, 14, and 15 that there are problems with these sorts of theories as well. The purpose of this chapter, then, is to develop a theory of vague propositions that does not rest on either of these identifications. The approach adopted here can be traced back to the likes of Frank Ramsey and Arthur Prior, and might naturally be called the *propositions-first* approach. Instead of attempting to analyse propositions in other terms—as sets of worlds or world–precisification pairs—we simply take propositions as primitive. Theoretical entities, such as possible worlds or precisifications, are in good standing only if they can be understood as certain sorts of constructions out of propositions, and not the other way around.

In section 11.1, I begin by introducing a framework for discussing propositional fineness of grain: we introduce a binary individuation connective, $A \equiv B$, that can be read as saying that A is the same proposition as B. From this connective we can define a necessity operator that is extremely broad: one can show that it is broader than any other normal operator definable in the language, and that propositions that are necessarily equivalent according to that operator must be identical.

In section 11.2, I show that determinacy and metaphysical necessity are both independent sources of fineness of grain: propositions are not individuated by necessary equivalence or determinate equivalence. Moreover, due to higher-order vagueness, they cannot be individuated by any finite combinations of these operators either. I note that, as a result, the individuation connective and the broadest necessity cannot be straightforwardly defined from these operators.

In section 11.3, I outline my own theory of vague propositions, in which they are individuated by their role in thought. Finally, in section 11.4, we attempt to get clearer on the sorts of problems this theory is supposed to address.

11.1 Fineness of Grain

Taking propositions as primitive is consistent with a wide variety of hypotheses about how fine-grained propositions are: indeed, it is consistent with the coarse-grainedness assumptions embodied in the possible worlds theories, and with more fine-grained assumptions. A substantive theory of propositions would ideally provide some kind of independent criterion for individuating propositions. Here I investigate the general structure of questions of fineness of grain; I present my preferred criterion of individuation in section 11.3.

In one important respect, our discussion here will be less than fully general: I shall restrict our attention to views which accept *Booleanism*—the view that propositions form a complete Boolean algebra under the usual logical operations (see section 3.2). We are thus assuming that propositions are not so fine-grained as to distinguish a proposition from its double negation, for example (as a structured proposition theorist might maintain).[1]

In what follows, we shall be investigating various hypotheses within the Booleanist framework concerning the individuation conditions for propositions. The possible worlds theory, for example, corresponds to the condition that propositions are individuated by necessary equivalence, whereas the view that propositions are sets of world–precisification pairs can be approximated by the idea that propositions are individuated by some combination of determinate and necessary equivalence, involving all possible iterations of both determinacy and necessity together. (We will see shortly, however, that the true individuation conditions for this view cannot be straightforwardly stated in a language containing determinacy and necessity.) Such hypotheses about individuation conditions are not logically idle either. Identical propositions are intersubstitutable, so from the first hypothesis one can infer that if A and B are necessarily equivalent, they are intersubstitutable in all contexts, including contexts containing determinacy and necessity; a similar conclusion follows from the second hypothesis.

It makes sense, then, to begin our investigation by studying the logic of propositional individuation. To that end, let us thus introduce into a propositional language a primitive individuation connective, $A \equiv B$, to be read as saying that the proposition that A is identical to the proposition that B. The thesis that propositions are individuated by necessary equivalence, for example, can then be stated: $(A \equiv B) \leftrightarrow \Box(A \leftrightarrow B)$.

---

[1] It may be possible to add in extra fineness of grain later if we wanted to, but for now it is useful to abstract away from that. For the most part, this assumption is harmless, since the fine-grainedness due to vagueness is certainly not due to structure. Indeed I suspect the Boolean assumption is not central to the approach I am advocating for: if one had a more fine-grained theory of propositional content, one could attempt to recover entities with the Boolean structure that I am interested in by quotienting out the extra fineness of grain (i.e. by forming equivalence classes of the more fine-grained propositions under some suitable equivalence relation). With that said, however, I shall make no serious attempt in what follows to make this theory consistent with non-Boolean theories of propositions.

This connective can be given an intrinsic axiomatization. The most fundamental principle governing this connective is a variant of Leibniz's law—that identical propositions are substitutable in all contexts:

SUBSTITUTION: $A \equiv B \rightarrow (\phi \rightarrow \phi[A/B])$.

Two other principles governing individuation provide a more complete picture:

IDENTITY: $A \equiv A$.

RULE OF EQUIVALENCE: If $\vdash A \leftrightarrow B$ then $\vdash A \equiv B$.

If we were in a higher-order logic, which allowed quantification into the position that a sentence operator takes, then we could define $A \equiv B$ as $\forall O(OA \rightarrow OB)$. For given this definition all of the above principles would be provable from the rule of universal instantiation.[2] Note, in particular, that the RULE OF EQUIVALENCE encodes the assumption of Booleanism: each Boolean identity, such as $(A \wedge B) \equiv (B \wedge A)$, follows from the fact that the corresponding biconditional is provable from classical propositional logic—in this case the biconditional $(A \wedge B) \leftrightarrow (B \wedge A)$.

Given the notion of equivalence above, we can introduce extremely broad notions of necessity, consistency, and entailment (see Cresswell [29], Suszko [143]). The broad notion of necessity, which I will write LA, is defined by $A \equiv (A \equiv A)$.[3] Analogous notions of consistency and implication can also be introduced: consistency, written MA, can be defined as $\neg L \neg A$. We shall say that A entails B when A is identical to $A \wedge B$, or more formally: $A \equiv A \wedge B$. Given Booleanism, entailment can be expressed as a strict conditional: $L(A \rightarrow B)$ (see Lewitzka [95]).

Given the identity axiom, we can prove the T axiom, $LA \rightarrow A$, since by substitution we get $A \equiv (A \equiv A) \rightarrow (A \equiv A \rightarrow A)$, which, given $A \equiv A$, allows us to infer $A \equiv (A \equiv A) \rightarrow A$. Similarly, it is straightforward to prove the K axiom, $L(A \rightarrow B) \rightarrow (LA \rightarrow LB)$, and the necessitation rule which says that if you can prove A in this system, you can prove LA. Thus, L represents a reasonable notion of necessity.

What is perhaps most striking about the substitution axiom is that it guarantees that the notion of L-necessity is always the broadest kind of necessity in the language it is introduced in.[4] Let us say that an operator O is necessity-like in a theory if that

---

[2] In particular we can prove SUBSTITUTION as follows: $\forall O(OA \rightarrow OB)$ entails $(\lambda p.\phi)A \rightarrow (\lambda p.\phi)B$ by universal instantiation, instantiating O for $\lambda p.\phi$. By β-reduction we can then infer $\phi[A/p] \rightarrow \phi[B/p]$. IDENTITY follows straightforwardly. RULE OF EQUIVALENCE follows in any higher-order logic that admits the rule: if $\vdash A \leftrightarrow B$ then $\vdash \forall O(OA \rightarrow OB)$.

[3] You can also define it as $A \equiv \top$ for any tautology $\top$. However, my definition makes for simpler proofs in many places.

[4] Philosophers sometimes introduce the term 'logical necessity' to denote the broadest kind of necessity. This carries unfortunate connotations, since logical necessity is often associated with certain properties of sentences (such as being derivable, or being true in certain classes of models) and so cannot straightforwardly be compared with propositional notions such as metaphysical necessity. As far as I know, no one has noticed that L-necessity is provably the broadest kind of propositional necessity, giving precise meaning to this idea.

theory contains every instance of the schema $\mathcal{O}(A \equiv A)$. Being 'necessity-like' is a necessary condition on representing some kind of necessity operator, and is satisfied by all the candidates usually discussed when comparing the broadness of necessity operators: 'it's a conceptual truth that', 'it's metaphysically necessary that', 'it's known *a priori* that', 'it's determinate that', 'it's a logical truth that', and so on. The following theorem guarantees that the L operator is at least as broad as all of these operators:

PROPOSITION:　*L is the broadest kind of necessity*

Let T be any theory containing SUBSTITUTION in a language containing $\equiv$, and let $\mathcal{O}$ be any necessity-like operator in T. Then every instance of $LA \to \mathcal{O}A$ is a theorem of T.

That is to say, L is the broadest necessity operator. The proof here is quite straightforward: $A \equiv (A \equiv A) \to (\mathcal{O}(A \equiv A) \to \mathcal{O}A)$ is an instance of substitution. Given that $\mathcal{O}$ is necessity-like, it follows that we have $\mathcal{O}(A \equiv A)$, and so we can derive $A \equiv (A \equiv A) \to \mathcal{O}A$, which is $LA \to \mathcal{O}A$ modulo definitions.[5]

Some applications of this theorem are noteworthy. For example, we can immediately infer that if we add $\triangle$ and $\square$ to the language, with their usual necessitation axioms, then we can derive the following two theorems: $LA \to \square A$ and $LA \to \triangle A$. Indeed we can infer $LA \to \pi A$ where π is any sequence of $\triangle$s and $\square$s, since the sequence π also represents a complex operator that is also necessitatible. Thus we see that L is as broad as anything definable with a finite string of $\square$s and $\triangle$s.

Another upshot of this theorem is that L satisfies the characteristic S4 axioms, $LA \to LLA$. Note that since L is necessitatible—something we showed earlier—and since we can prove $A \equiv A$, we can also prove $LL(A \equiv A)$ by applying necessitation twice. Thus LL is a possible substitutend of $\mathcal{O}$ in the theorem, delivering $LA \to LLA$ as required. Curiously, one cannot prove within this system the characteristic S5 axiom, $\neg LA \to L\neg LA$.[6] This is perhaps a good thing: this principle would rule out the possibility of borderline L-necessities, which is something we might want to countenance.

As noted already, the RULE OF EQUIVALENCE encodes some of our Booleanist assumptions about fineness of grain, since every finitary Boolean identity is provable in the propositional calculus. It thus follows from this rule that propositions form a Boolean algebra under the relation of entailment. It is a further assumption that this Boolean algebra is *complete*—that arbitrary conjunctions and disjunctions of propositions exist—and that it is *atomic*—a condition that can be expressed in the

---

[5] The theorem proved shows that L is always the broadest necessity operator in whatever the language happens to be. A more general version of the theorem, that is not language relative, can be formulated in a higher-order logic: $\forall O(O\top \to \forall P(LP \to OP))$. Since writing this book I have explored this idea further in Bacon [3]; investigating this further here would take us too far afield.

[6] See Cresswell [29].

object language by saying that, of L-necessity, there's some truth that L-entails every truth (see Williamson [165] p. 201). I shall not challenge these further assumptions.

Given these assumptions, we can always talk about maximally strong consistent propositions: a proposition that is consistent and is identical to any proposition that entails it. In the theory of Boolean algebras, these are called atoms, but in the present context I shall refer to them as *indices*. It is a standard result about complete atomic Boolean algebras that each proposition can be represented isomorphically by the set of indices that entail it, and that the operations of conjunction, disjunction, negation, and entailment get cashed out in this representation as the set-theoretic operations of intersection, union, complementation, and subsethood.

In many ways, the resulting theory is much like the possible worlds theory, in that we can represent propositions as sets of entities. But unlike the possible worlds theory, we are not committed to the modal account of propositional individuation. Indeed, if desire and belief operators contribute to the individuation of propositions then the indices will not in general represent metaphysically possible ways the world could be.

A general point about the propositions-first methodology must be stressed at this juncture. Although it does not rely on possible worlds as ordinarily conceived, it is not supposed to be revisionary to ordinary semantic theorizing. Possible world semantics has enjoyed a great amount of success among linguists and is widely adopted in contemporary semantics. However, the success of this style of semantics has nothing to do with the fact that in many philosophical interpretations, the objects at which we evaluate sentences for truth are taken to be possible worlds; little would change if the worlds of the theory were interpreted differently. The proposed theory allows us to retain the thought that the meaning of a sentence is given by a non-linguistic entity, a 'proposition', whilst rejecting the claim that these things are individuated modally.

It's worth noting that even without bringing in considerations having to do with propositional attitudes, the style of semantics which employs indices and accessibility relations may not always allow one to interpret the indices as possible worlds. For example, a purely modal language can only allow the indices to be interpreted as representing possible worlds if we are assuming a modal logic including at least the S4 principle. If we are taking the index semantics at face value, then only the indices accessible to the index representing the actual world will be genuine 'possible' worlds and the other indices needed to represent the semantics cannot be understood as representing possible ways the world could be. Similar points hold for counterfactual logics which allow for non-trivial counterfactuals with impossible antecedents. Neither of these two examples relies on the features of propositional attitudes.

The possible worlds theorist identifies the things we have neutrally called 'indices' with maximally specific ways the world *could* (metaphysically) have been, and we have observed that this assumption is not essential to the success of this sort of semantics. But more importantly, the assumption that indices are maximally specific ways the world could have been seems to load the dice in favour of natural language operators like 'could' and 'possibly'. It is no surprise that on this way of construing indices, we run into trouble interpreting attitude operators like 'believes that' or 'desires that' in this framework. This assumption, however, is not forced on us; we could just as easily interpret the indices as being maximally specific ways the world could be believed, or desired to be.[7] One can still think of sets of such things as representing truth conditions: they are, after all, still *conditions* which can obtain or fail to obtain. The conditions under which a belief, say, is true could easily be *when Harry is bald* or *when Hesperus is Phosphorus*. For these kinds of things can be true or not as the case may be, and they can be believed to be the way things are, or the way things are not, and so on and so forth. There is no good reason to think that truth conditions must be individuated coarsely by their relation to adverbs like 'necessarily' and 'possibly'.

11.2 Individuation Conditions

Let us begin by studying the possible worlds theory of propositions. In the propositions-first setting, this view is characterized by the thesis that propositional individuation is given by necessary equivalence.[8]

$$(A \equiv B) \leftrightarrow \Box(A \leftrightarrow B)$$

Even for the modest purpose of modelling vagueness, the modal account of individuation is not satisfactory. Indeed, we can show that given the supervenience of the vague on the precise, the modal account of individuation entails that there are no vague propositions:

OBSERVATION: Suppose the following are true:

(i) Necessarily equivalent propositions are identical.
(ii) Supervenience: Necessarily, every truth is necessarily equivalent to some precise truth.

Then there are no vague propositions.

It should be noted that (ii) is an unnecessarily strong version of supervenience. A weaker way of articulating the supervenience idea, is the thesis that, necessarily, every truth is necessitated by some precise truth. However, given some plausible background assumptions—including the assumption that arbitrary disjunctions of precise propositions are precise—one can actually derive the stronger principle

---

[7] Where strength is measured by the notion of L-implication.

[8] Note that this claim is putatively weaker than another principle that has claim to formalizing the idea that propositions are individuated by necessary equivalence: $(A \equiv B) \equiv \Box(A \leftrightarrow B)$—i.e. A's being identical to B is just identical to A's being necessarily equivalent to B. I will show shortly that it can't be contingent whether $A \equiv B$. Moreover, given **S5** for $\Box$, it can't be contingent whether $\Box(A \leftrightarrow B)$, so the biconditional stated above implies the necessary equivalence of $(A \equiv B)$ and $\Box(A \leftrightarrow B)$, which in turn implies their identity. So, there is actually no logical distance between these two formulations.

(ii) we appeal to above.[9] However, given (ii), it straightforwardly follows that every proposition is necessarily equivalent to a precise proposition.[10] But by (i), that means every proposition *is* a precise proposition.

We can see the problem with the modal account manifest itself in many different ways. Necessarily equivalent things that are not determinately equivalent cannot always be substituted within the scope of a determinacy operator. Assuming that baldness facts supervene on facts about hair number, the proposition that the cutoff for baldness is 2,024 hairs is either necessarily true or necessarily false, although it's borderline which. Thus, this proposition is either necessarily equivalent to $0 = 0$ or $0 = 1$: but I cannot substitute the first claim for either $0 = 0$ or $0 = 1$ in 'it's determinate whether $0 = 0$' or 'it's determinate whether $0 = 1$' without changing their truth values.

It should be noted, by symmetrical reasoning, that determinate equivalence is not a good condition of individuation either, for there are determinately equivalent things that are only contingently equivalent. For example, it's determinate that Bruce Willis is bald if and only if Patrick Stewart is (since they are both determinately bald); however, this biconditional is contingent.

Less obvious is the fact that determinate necessary equivalence and necessary determinate equivalence will not do either. The reason is that second-order indeterminacy can prevent the intersubstitutability of necessary determinate equivalents and determinate necessary equivalents within determinacy contexts; similar problems arise for other finite combinations of 'necessary' and 'determinate' and vagueness at higher orders.

Let us now try to reformulate the supervaluationist theory in a propositions-first setting. Again, we shall achieve this by determining what individuation conditions are imposed by the stipulation (in the ordinary setting) that propositions are represented by sets of ordered pairs of worlds and precisifications. If A and B express the same set of world–precisification pairs, then, relative to any world–precisification pair, the sentence $\pi(A \leftrightarrow B)$ is true, where π is any finite sequence consisting of $\square$ and $\triangle$ symbols. (We will present the supervaluationist semantics in more detail in chapter 12.) This means that A and B are determinately equivalent and necessarily equivalent, so that they are substitutable in the context of one iteration of a $\square$ or a $\triangle$. But the above fact also means that A and B are determinately determinately equivalent, so we can also substitute A and B within two iterations of $\triangle$—indeed, because they are π equivalent for any string π of $\square$s and $\triangle$s it follows that we can substitute A for B in any context regardless of the number of iterations of $\square$ and $\triangle$.

If one had infinite conjunction in the language, one could define an operator $(\square\triangle)^* P := \bigwedge_\pi \pi P$ where π ranges over arbitrary finite sequences consisting of

---

[9] For any truth p, p is necessarily equivalent to the disjunction of precise truths that necessitate p.

[10] If p is true then (ii) entails that it's necessarily equivalent to a precise proposition. If p is false, then $\neg p$ is true and necessarily equivalent to a precise truth, q by (ii). So p is necessarily equivalent to $\neg q$, and $\neg q$ is precise since q is precise.

□ and Δ symbols. It is tempting to think that the supervaluationist individuation condition could then be articulated as:

$$(A \equiv B) \leftrightarrow (\square\Delta)^*(A \leftrightarrow B)$$

Indeed, with a suitably rich logic of infinite conjunction, determinacy, and necessity, it is possible to show that $\equiv$, when simply defined in terms of $(\square\Delta)^*$ by the above biconditional, satisfies the principles SUBSTITUTION, IDENTITY, and THE RULE OF EQUIVALENCE.

Although this might serve as a good approximation of the supervaluationist account of propositional individuation, it is not entirely accurate. There are supervaluational models in which A and B are $(\square\Delta)^*$-equivalent, but in which they do not correspond to the same set of world–precisification pairs. Such discrepancies arise in models in which there are at least two world–precisification pairs that cannot be reached from each other by following the □ and Δ accessibility relations any number of times: for if A and B are different but agree on the worlds that *can* be reached from (w, v) by following the accessibility relations, then they will count as $(\square\Delta)^*$-equivalent at (w, v). We discuss one such model in section 14.3; see also Figure 14.2. (Such models are not just curiosities either: models that do not validate the principle B for determinacy—a principle that plays a role in many of the paradoxes of higher-order vagueness—often have this feature.)

11.3 A Theory of Propositions

It follows that even for a supervaluationist there is no straightforward way to independently characterize the individuation conditions for propositions: in order to state them it seems one has to adopt the ideology of worlds and precisifications. It would be preferable to avoid putting so much theoretical weight on the coherence of these technical entities. In what follows, I shall propose instead that propositions be individuated by their role in thought. Roughly speaking, the role a proposition plays in thought is somewhat analogous to a conceptual role, although it is something which applies to propositions and not sentences or thoughts. At least in the context of the present theory of vagueness, it is natural to articulate this in terms of conceptually coherent priors and utilities.

The kernel of this idea can be traced to a broadly Stalnakerian way of making sense of possible world talk. According to Stalnaker [140], one begins with a set of indices understood as primitive objects playing a special role in a theory of rational agency. Whatever structure indices have is abstracted from this theory, and beyond this their nature is left open. From this sort of theory, one gets a handle on the individuation conditions for the indices, and thus a handle on the individuation conditions for propositions.

Stalnaker is not explicit about what this theory is exactly, but he does say the following:

What is essential to rational action is that the agent be confronted, or conceive of himself as confronted, with a range of alternative possible outcomes of some alternative possible actions. The agent has attitudes, pro and con, towards the different possible outcomes, and beliefs about the contribution which the alternative actions would make to determining the outcome. One explains why an agent tends to act in the way he does in terms of such beliefs and attitudes. And, according to this picture, our conception of belief and of attitudes pro and con are conceptions of states which explain why a rational agent does what he does.   (Stalnaker [140], p. 5)

On this picture of the metaphysics of indices, they are not things which are deeply tied to metaphysics, as a possible world in Lewis' sense would be: 'they obviously are not concrete objects or situations, but abstract objects whose existence is inferred or abstracted from the activities of rational agents' (Stalnaker [140], pp. 50–1).

From the above passage, you would be forgiven in thinking that Stalnaker individuates the 'possible outcomes' according to a rational agent's bouletic attitudes ('attitudes, pro and con') and doxastic attitudes. It is worth noting, however, that Stalnaker individuates propositions and possible outcomes modally, and indices, in the sense abstracted above, are identified with possible worlds. Stalnaker's reasons for doing this seem to have little to do with the picture outlined so far, and have more to do with his reductionist ambitions with respect to the problem of intentionality.[11]

For those who do not have any reductionist ambitions, there does not seem to be any barrier to individuating these objects epistemically. Indeed, we have compelling reasons for thinking that the entities we abstract from a theory of rational decision will be more fine-grained than possible worlds.[12] An astronomer who believes that Hesperus is a planet may display very different behaviour, both verbally and non-verbally, from someone who believes that Phosphorus is a planet. It furthermore seems perfectly possible that this astronomer may have very good evidence that Phosphorus is a planet, which is not also evidence that Hesperus is. In which case, it seems to me perfectly rational for this agent to believe that Hesperus is a planet without believing that Phosphorus is and rational for her to act accordingly.

Since I am arguing for a moderately fine-grained theory—I am claiming that propositions are more fine-grained than possible worlds but not as fine-grained as to distinguish logical equivalents—it's natural to ask at what point we stop individuating. How fine-grained are propositions? A broadly Stalnakerian theory provides a perfectly principled answer to this question: indices are as fine-grained as we need them

---

[11] He even considers the impossible worlds approach, writing, 'could we escape the problem of equivalence by individuating propositions, not by genuine possibilities, but by epistemic possibilities—what the agent takes to be possible?' but dismisses it on the grounds that although 'this would avoid imposing implausible identity conditions on propositions [...] it would also introduce intentional notions into the explanation, compromising the strategy for solving the problem of intentionality'. In the more recent Stalnaker [141], he disavows the project of reduction.

[12] Stanley [142], for example, makes the similar point that Stalnaker's approach to intentionality appears to be compatible with, and even motivates, a theory in which indices are epistemically possible worlds.

to be to specify the functional role of the belief that p as a function of the indices at which p is true. In other words, indices are whatever they need to be so that functional roles do not distinguish more finely than sets of indices.[13]

To formally represent the doxastic and bouletic attitudes of an agent, we shall make reference to a set Coh of pairs (Pr, V) consisting of conceptually coherent conditional probability functions and value functions (related by Jeffrey's equation). The sort of account of propositional individuation we endorse can then be stated:[14]

$A \equiv B$ if and only if $V(AC) = V(BC)$ and $Pr(A \mid C) = Pr(B \mid C)$ for every $(Pr, V) \in Coh$ and every proposition C.

Thus, two propositions are identical if they receive the same value and probability relative to any conceptually coherent agent. In what follows, however, we will be dropping the requirement that the propositions must agree in value, and will focus mainly on a simpler theory in which propositional individuation only references the role that propositions play with respect to conditional belief. Dropping the condition $V(AC) = V(BC)$ strengthens the individuation condition.

Might a proposition's role in desire not also be a source of fine-grainedness? This weaker theory might be natural for modelling moral propositions that play the same doxastic role, but differ in their motivational properties—to rationally believe them would require you to care about certain things. Such a theory might be useful for moral expressivists wishing to have some kind of lightweight theory of propositions. Although this is worth investigating, it goes well beyond my purposes here so I shall concentrate on the stronger but simpler version of the individuation axiom.

Let us now pin down some of the formalities. According to the Stalnakerian picture, indices are to be abstracted from a theory of rational action. The theory I shall adopt for this purpose contains two primitives. Firstly, a set of objects, P, to be understood informally as representing the set of propositions. However, although this will be the ultimate interpretation of P we do not assume that this set has any structure, Boolean or otherwise, from the outset. We shall not even assume that the standard logical operations are defined on P; definitions of conjunction, negation, and so on and their logical properties will arise out of the theory. Secondly, we posit a set, Coh, of binary functions taking two elements of P to a real number in $[0, 1]$. Coh informally represents the set of conceptually coherent conditional ur-priors (these terms will be

---

[13] Note that the functional role of a belief that p does not include the role that a belief that p plays in the cognitive workings of completely erratic and irrational agent (if we could ever determinately attribute a belief that p to such an agent). No two belief states have the same cognitive role across every possible agent, rational and irrational; the resulting notion of proposition would be uninteresting and useless for the purposes of giving a theory of objective information that can be communicated among different agents. For example, in order have a theory of communication, some degree of rationality must be assumed by both participants in a conversation if we are to be able to infer anything about what they believe from the sentences they are uttering.

[14] The presence of extra quantification over propositions, C, deals with cases where A and B only differ conditional on probability 0 events.

explained later) and each element of *Coh* can be informally thought of as taking two propositions and telling us how likely one of these propositions is conditional on the other according to that ur-prior. Lastly, one could also augment the theory with a set, U, of conceptually coherent utility functions—I suggested one reason why we might need this, relating to moral expressivism, earlier. However, for the basic theory that follows we won't need to appeal to U.

It is important to realize that the following theory axiomatizes these two notions together. The theory simultaneously entails the most important facts about the structure of P, the set of propositions, and also the behaviour of the elements of *Coh*, the rational priors—it is not possible to adequately characterize only the propositions without indirectly characterizing the conceptually coherent ur-priors, or vice versa. Effectively, the theory will entail that P has the structure of a complete Boolean algebra, and that *Coh* consists of 'normal Popper functions'—a slight generalization of the notion of a probability function that allows us to talk about probabilities conditional on zero-probability events. They are called 'normal' because they assign sensible probabilities conditional on all consistent propositions. The result is a theory that both describes a *role in thought*, whilst simultaneously postulating the existence of objects that play that role. The following theory is thus not unlike Popper's own theory (expanded on by Field in [48]), in which the logical properties of conjunction and the other Boolean operators are inferred from the account of probability, rather than simply assumed before we apply probabilistic notions to these entities.

With these primitives at hand, we can now introduce the concept of an *index* that has played a crucial role in our theorizing so far. Firstly, we must say what it means for a member of P to be consistent:

DEFINITION OF CONSISTENCY: A proposition p is *inconsistent* if and only if $Pr(q \mid p) = 1$ for every q and $Pr \in Coh$. p is *consistent* otherwise.

Intuitively, only the inconsistent proposition makes everything fully probable on its supposition.

DEFINITION OF INDEX: A proposition, i, is an *index* if and only if it is both consistent and $Pr(p \mid i) \in \{0, 1\}$ for every proposition $p \in P$ and every $Pr \in Coh$.

In other words, conditional on an index, every proposition is either certainly true or certainly false according to every ur-prior. Another important relation is the relation of a proposition being *true at* an index, which can be spelt out in this framework as follows:

DEFINITION OF TRUE-AT: p is *true at* an index i iff $Pr(p \mid i) = 1$ for every $Pr \in Coh$. p is false at i otherwise.

Note that there is a putative asymmetry between the definition of *truth at* and *false at*: a proposition is true at an index iff it has probability 1 on it relative to every coherent prior, whereas it is false at the index iff it has probability 0 for *some* coherent prior.

This leaves open the possibility that there are two coherent priors that both assign p probability 1 or 0 on the index i, but differ regarding which. This possibility will eventually be ruled out by the axioms: we can show that p has probability 1 on an index for some coherent prior iff it has probability 1 on that index for every prior.[15] From this result it also follows that p has probability 0 at an index for some prior iff it has probability 0 on that index for every prior.

This finally allows us to introduce the logical operations within this framework, as promised:

> DEFINITION OF CONJUNCTION: Say that p is a conjunction of a set $X \subseteq P$ iff p is true at exactly the indices at which every element of X is true.

> DEFINITION OF NEGATION: p is a negation of q iff p is true at exactly the indices q is false at.

Given conjunctions and negations we can also introduce the other logical operations in the usual way. Note that we cannot assume that every set of propositions has a conjunction and we cannot assume that every proposition has a negation, nor can we assume that, if it does exist, it is unique. These will all be derived facts. Finally, say that a set of propositions, X, entails another proposition p if and only if p is true at an index whenever every element of X is true at that index.

One of the key elements of this theory will be an axiom that tells us how to individuate propositions. Intuitively, this axiom tells us to individuate propositions no more finely than they need to be in order to satisfy their role in conditional belief.

> INDIVIDUATION AXIOM: If $Pr(A \mid C) = Pr(B \mid C)$ for every $Pr \in Coh$ and $C \in P$ then $A = B$.

An important consequence of this axiom is that it rules out interpretations of P in which it consists of structured propositions or sentences of some language. While the other axioms, to be listed shortly, are compatible with this interpretation (and indeed a linguistic interpretation was Popper's own interpretation of his theory), this axiom explicitly rules it out. For example, the axioms of my theory will require that for any two propositions A and B, there be another proposition, it's conjunction, whose probability is related to the probability of A and the probability of B in certain ways across all elements of Coh. However, these axioms allow that there be several propositions related to A and B in this way. Indeed, there is a model of the remaining axioms in which P is represented by a set of sentences in an infinitary language in which the sentences $A \wedge B$ and $B \wedge A$ and, indeed, any sentence logically equivalent

---

[15] The axioms entail the existence of a negation function, $\neg$, on propositions, such that (i) $\neg p$ is true at exactly the indices p is false at, (ii) $\neg\neg p = p$, and (iii) $Pr(\neg p \mid q) = 1 - Pr(p \mid q)$ for any q and Pr. Then the following things are equivalent: (1) the probability of p on i is 1 for every prior, (2) p is true at i, (3) $\neg\neg p$ is true at i, (4) $\neg p$ is false at i, (5) $\neg p$ has probability 0 on i for some prior, (6) p has probability $1 - 0 = 1$ on i for some prior.

to $A \wedge B$, are distinct conjunctions of A and B whose probabilities conditional on certain propositions are related to the probabilities of A and of B on those propositions in the ways described above. However, since any two conjunctions of A and B must have the same probability on any supposition to satisfy the relevant connections, the individuation axiom ensures that the conjunctions of A and B must be identical to one another.

The axioms of the theory can then be stated as follows:

BOTTOM: There is an inconsistent proposition.

CONJUNCTION: Every set of propositions has a conjunction.

NEGATION: Every proposition has a negation.

INDIVIDUATION: If $Pr(A \mid C) = Pr(B \mid C)$ for every $Pr \in Coh$ and $C \in P$ then $A = B$.

REFLEXIVITY: $Pr(A \mid A) = 1$ for every $Pr \in Coh$ and $A \in P$.

CONJUNCTION ELIMINATION: $Pr(X \mid B) \leq Pr(A \mid B)$ for every $Pr \in Coh$ for every conjunction of A and C, X.

MULTIPLICATION RULE: $Pr(X \mid B) = Pr(A \mid Y)Pr(C \mid B)$ whenever X is a conjunction of A and C, and Y a conjunction of C and B, for every $Pr \in Coh$.

ADDITIVITY: If C is consistent and B is a negation of A then $Pr(A \mid C) = 1 - Pr(B \mid C)$.

A notable absence from this theory is any form of countable additivity for Popper functions. Popper originally defined his functions without imposing countable additivity, although others have augmented the theory so as to include it (van Fraassen [148]). However, there is, I think, a decisive reason not to include countable additivity, namely that it is inconsistent with the existence of an infinity of probabilistically independent propositions with probabilities bounded by (a, b) with $0 < a < b < 1$. You couldn't have, for example, countably many independent coin flips.

Theorem 11.3.1. *No fully countably additive (i.e. countably additive on any condition) Popper function has countably many mutually independent propositions, A_n, with unconditional probability in (a, b) with $0 < a < b < 1$. (Unconditional probability here just means probability conditional on a tautology.)*

I put the proof in the footnote.[16]

---

[16] Without loss of (much) generality suppose each A_n has probability $\frac{1}{2}$. It will be useful to abbreviate the conjunction $A_n A_{n+1} A_{n+2} \ldots$ as X_n, and the disjunction $\bigvee_n X_n$ as X. Whatever one means by mutual probabilistic independence for infinite sets, it should entail that $Pr(A_n \mid X_{n+1}) = Pr(A_n) = \frac{1}{2}$. By the multiplication rule for Popper functions we have (1): $Pr(A_n X_{n+1} \mid X) = Pr(A_n \mid X_{n+1}X)Pr(X_{n+1} \mid X)$. Note, however, that $A_n X_{n+1} = X_n$ and $X_{n+1}X = X_{n+1}$. Making these substitutions in (1) $Pr(X_n \mid X) = Pr(A_n \mid X_{n+1}).Pr(X_{n+1} \mid X)$. Finally, we know that $Pr(A_n \mid X_{n+1}) = \frac{1}{2}$ by independence so making this substitution in (2) we have (3): $Pr(X_n \mid X) = \frac{1}{2}.Pr(X_{n+1} \mid X)$. This is only possible for all n if $Pr(X_n \mid X) = 0$ for all n. Yet, of course, $Pr(\bigvee_n X_n \mid X) = 1$.

While the pure mathematical theory puts important structural constraints on the space of propositions, there are many questions it does not settle. For example, it does not tell us whether it is conceptually coherent to assign necessarily equivalent propositions different conditional credences. Under the assumption that it is incoherent to do so, the individuation axiom entails, among other things, the view that propositions are individuated by necessary equivalence. More importantly, the proposition that Harry is bald would be identical to a precise proposition about Harry's hairline that it is necessarily equivalent to. This hypothesis about which priors are coherent therefore makes the theory unsuitable as a theory of vague propositions.

Conversely, one could also insist that it is conceptually coherent to have a different prior credence in the proposition that John is a bachelor than one has in the proposition that John is an unmarried man. So, by Leibniz's law, these would be different propositions. On my understanding of 'conceptually coherent prior', however, it is simply incoherent to assign these propositions different credences. So on my preferred interpretation of this theory, these two propositions would be identified.

The take-home message is that, although the formal axioms force us to make some choices—such as identifying logical equivalents—the informal notion of 'conceptual coherence' is also doing important work. Although I cannot hope to explicitly define the notion, it can be elucidated by examples which, I hope, should be enough to give the reader a reasonably good grasp on the notion (see also the discussion in section 8.3). For example, although I am sceptical of the idea that a sentence can be true purely in virtue of the meanings of its constituents, I suspect that many standard examples of analytic sentences express propositions that get probability 1 according to every conceptually coherent prior. A prior which assigns less than full credence to the proposition that vixens are foxes represents a conceptual confusion of some sort and according to our theory, that proposition's role-in-thought is the same as the tautologous proposition's role-in-thought.

Vagueness introduces more interesting examples. For instance, I take it that to be conceptually coherent, you should be certain that Harry is bald conditional on the proposition that he has no hairs at all. This seems like a fairly straightforward example of a proposition expressed by a conceptual truth, analogous to those mentioned above. However, conditional on the proposition that Harry has N hairs, where N is in the borderline region, I take it that it is conceptually incoherent to have anything other than some intermediate credence that Harry is bald. If you furthermore have a description of all the precise facts about Harry's head, then, as I argued in chapter 8, there is a particular credence which all conceptually coherent priors assign to the proposition that Harry is bald, conditional on Harry satisfying that description.

In the rest of the book, I will also apply the notion of conceptual coherence to utility functions, measuring how much people care about certain matters. I believe the notion of conceptual coherence, as it applies to desires, also has some pretheoretic appeal. Its relation and importance to the study of vagueness has already been discussed in literature. For example, in Field [55], Hartry Field contrasts two examples.

One involves a character, Roger, who thinks that if his bank account password has the same last digit as the number of nanoseconds Bertrand Russell was old for, then his life will go better. The other involves Sam, who thinks that his life will go better if the last digit of his bank account password is the seventeenth significant digit of the centigrade temperature at the currently hottest point in the interior of the sun. According to Field, while Sam's belief is thoroughly irrational, Roger's is intuitively even worse, as it is conceptually confused. The distinction between being merely irrational and conceptually confused will play an important role in the theory I am endorsing.

11.4 Moderately Fine-Grained Theories of Content

Let me end by making some general remarks about the scope of this project. Firstly, the account of fine-grained propositions I have proposed here does not solve, and is not intended to solve, all of the problematic issues surrounding propositional attitude verbs like 'knows', 'believes', 'desires', and so on. For example, the theory distinguishes between the proposition that Harry is bald and necessarily equivalent propositions stating the situation on Harry's head in more precise terms, but will identify the propositions expressed by any pair of tautologies, even if one tautology is significantly less obvious than another.[17] Call theories like this, that distinguish between necessary equivalents but not logical equivalents, *moderately fine-grained*. To see why a moderately fine-grained theory of propositions is motivated, I think it is important to distinguish two problems.

One of these is a problem in the philosophy of language about propositional attitude *reports*. This is a puzzle about how we succeed in reporting facts about a person's mental states in English using verbs like 'believes' and 'desires' and so on. In my view, this is fundamentally a puzzle about how we use words to ascribe attitudes—relations between people and propositions—and not a puzzle about the attitudes themselves. Indeed, many of the most promising solutions to the problem of attitude reports say something distinctive about the kinds of words we use to ascribe attitudes, whilst in principle remaining compatible with various theses concerning how fine-grained the arguments of these attitudes are.[18] A contextualist, for example, postulates the existence of a number of distinct but related propositional attitudes, which the word 'believes' can pick out in different contexts; when it appears as though we are making

---

[17] It also plausibly distinguishes other necessarily equivalent propositions whose distinctness has nothing to do with vagueness, such as the proposition that Hesperus is Hesperus and the proposition that Hesperus is Phosphorus.

[18] See Crimmins and Perry [30], Richard [120], Salmon [124], Soames [136]. Note, however, that although many of these authors also have views about how fine-grained propositions are, these views are theoretically separable from the solution to the attitude ascription. According to the view I favour, people typically stand in many different but related attitudes to many different but related propositions at any given time, and there's a certain amount of context sensitivity about which of these attitude words like 'believes' and 'desires' pick out which can be used to explain the linguistic data on attitude reports.

conflicting attitude reports about the same proposition, the contextualist says we are really attributing different attitudes to the same proposition.

This is how it should be—if we took all of our immediate judgements about differences between propositional attitude *reports* to show that there is a corresponding difference between the objects of the propositional attitudes, propositions would be as fine-grained as sentences, or maybe even more fine-grained.[19] According to the theories I just listed, one cannot infer much about the fine-grainedness of propositions from judgements about attitude reports. To make this vivid, note that these theories seem to have the explanatory power to reconcile our judgements about attitude reports with the view that propositions are sets of worlds, or even truth values![20]

The reasons I take to motivate a moderately fine-grained theory of propositions have nothing to do with accommodating attitude reports, or self-ascriptions of belief. The linguistic data on belief reports, despite the attention it receives, is only a small aspect of a full account of propositional attitudes. It is important not to forget that a person's propositional attitudes are also important for explaining and evaluating all kinds of behaviour, both verbal and non-verbal. An instructive example is the view that there are only two propositions: the true and the false. According to this view, it is impossible to explain a person's behaviour in terms of their beliefs and desires, for there are only sixteen types of people depending on the combination in which they believe or desire the two propositions, and this is certainly not enough to explain or evaluate all the possible kinds of behaviour they exhibit. Thus in order to explain or evaluate a person's behaviour in terms of their propositional attitudes, propositions must be more fine-grained than truth values. This line of reasoning plausibly generalizes to rule out sets of worlds as well—two people with necessarily equivalent beliefs and desires can rationally behave very differently. On the other hand, the fact that propositional attitudes determine rational behaviour gives us a defeasible reason to think that propositions cannot be so fine-grained as to distinguish logical equivalents. This is because our best theories of rational action—decision theory and probability theory—assume that propositions are moderately fine-grained. Decision and probability theory typically begin by assigning probabilities and utilities to indices of some sort and from this one assigns probabilities and expected values to arbitrary sets of these entities. To treat propositions as isomorphic to sets of entities is just to assume that propositions are structured like a Boolean algebra which, in turn, guarantees that equivalence in classical propositional logic suffices for identity.

---

[19] For example, we can judge that the sentence 'John said that it was hot out' is true without also judging 'John said that it was *hot* out' as true. In both cases we have the same embedded sentence with different emphasis. These judgements, most will agree, call out for a pragmatic explanation. However, someone who insisted on taking all of these judgements at face value would have to treat propositions as more fine-grained than sentences.

[20] For example, perhaps when we dissent from the attitude report 'Lois Lane believes that Clark Kent flies' we are denying that Lane stands in a relation to the true via a certain mode of presentation. Of course, there are plenty of problems with treating propositions as truth values, but I think that the problem of accounting for propositional attitude reports is not one of them.

(Note that it is possible to do decision theory and probability theory with a very fine-grained account of propositions by assigning values and probabilities to equivalence classes of logically equivalent propositions. Those who insist on such a fine-grained account of propositions can still accept my theory as an account of a kind of theoretical entity: entities that play the theoretical role equivalence classes play in the more fine-grained account.)

12

Vagueness and Precision

So far we have been theorizing in a fairly abstract and informal way about the distinction between vague and precise propositions. In chapter 4, I argued that there was an important propositional distinction to be studied, and, in chapters 6–10, I formulated some distinctive theses governing the propositional notion of precision and vagueness. However, I have yet to give anything like a formal and rigorous characterization of this distinction.

The formalism of choice for most linguistic theorists is the theory of supervaluations: a formalism with many applications, but which is most saliently associated with a certain semantical apparatus for dealing with vague languages.[1] Supervaluationist semantics is an extremely influential way to make precise the idea that vagueness consists in semantic indecision. A precisification of a vague language, according to this framework, is a way of making each word of that language completely precise. A precisification not only tells us what the extension of the predicate 'bald' is, and where the cutoff point is in a given sorites sequence, it also tells us what the extension would have been if those people had had different amounts of hair. They are things which, when given a possible world as input, tells us how to assign cutoff points (i.e. complete extensions) to each predicate (and words of other categories) at that world.

Since the rise of supervaluational semantics, however, it has become clear that it is not only theorists who identify vagueness with semantic indecision that can make use of the formalism of precisifications. Epistemicists will appeal to the notion of an interpretation of a language which is not knowable (for distinctive reasons) by the speakers to be incorrect (see Williamson [156]). Inconsistency theorists will talk of 'acceptable assignments of semantic values': precise interpretations of the language that come 'maximally close in satisfying the meaning-constitutive principles for the expressions involved' (Eklund [40]). The formalism, in one guise or another, is completely ubiquitous amongst classical approaches to vagueness. Given an appropriate interpretation of 'admissible' and 'precisification', these theorists can all accept the structural claim that a proposition or sentence is borderline

---

[1] The applications of supervaluationism have been very diverse: in the liar paradox (Kripke [85]), the philosophy of science (Mehlberg [106]), the semantics of conditionals (Stalnaker [139]), and empty names (van Fraassen [146]) to name a few.

iff it is true relative to some but not all admissible precisifications. Many of the things I say about supervaluationism generalize to other approaches that accept this formalism.

Although the theory of supervaluations is most commonly associated with linguistic theories of vagueness, it is natural to wonder if it might also be applied to a non-linguistic theory of vagueness. Barnes and Williams [9], for example, develop a metaphysical account of vagueness in which the supervaluationist's precisifications of a language are replaced by 'precisifications of the world'.

Supervaluational semantics can indeed be adapted to the kind of non-linguistic theory I have been developing. In addition to the comparatively fine-grained theory of propositions and properties we have been working with (which are not individuated by necessary equivalence), we could also accept the more coarse-grained ideology of possible worlds. A precisification may then be identified with a way of precisifying vague properties and propositions in much the same way that a linguistic theorist would think of one as precisifying predicates and sentences. A precisification in this setting would be a function telling us what the truth value of each vague proposition is at each possible world, telling us what the extension of each vague property is at each world, and so on.

Despite the formal availability of a supervaluationist semantics, I have come to realize that some substantive assumptions are hard-wired into this formalism that do not fit particularly well with the theory of vagueness developed here, and indeed, assumptions that may bring into question its suitability in the modelling of vagueness more generally. In the following chapters, I'll elaborate on why the supervaluational semantics is not suitable and explore the ways in which some standard assumptions about vagueness, often guided by the supervaluationist picture, need to be revised.

I'll start with the most basic problem for a supervaluational treatment of this theory: while supervaluational semantics is tailor-made to provide analyses of the 'determinacy' and 'borderlineness' operators, it does not directly give us a characterization of propositional vagueness and precision. In this chapter, I will argue that propositional vagueness and precision are not definable in terms of propositional borderlineness or determinacy. Thus, any theory that takes the latter as primitive (as the supervaluationist does) cannot provide a complete theory of vagueness. By contrast, at the end of the chapter, we will show that one can define propositional borderlineness and determinacy from propositional vagueness and precision.

This fact is particularly relevant to the present theory of vagueness. Recall that the core principles of this theory—**Plenitude**, **Rational Supervenience**, and **Indifference**—are all stated in terms of propositional vagueness and precision. Strikingly, the ubiquitous 'borderlineness' and 'determinacy' operators are completely absent from these principles; by all accounts they seem to be less theoretically central to the approach. This marks the first important departure from orthodoxy: according to my theory, the notions of precision and vagueness must be taken as primitive.

12.1 Borderlineness as Primitive

We are seeking a characterization of propositional vagueness and precision.[2] The distinction between a vague proposition or property and a precise one is importantly different from the distinction between a proposition that is borderline and a proposition that is definitely true or false, or the distinction between properties that have borderline cases and those that have no borderline cases. To illustrate the difference, consider the property of being a blue swan. This property is vague, not precise—one can imagine a sorites sequence of swans, beginning with a clearly green swan and ending with a clearly blue swan, without much trouble. However, the property of being a being a blue swan has no borderline cases because, as it turns out, it is a determinate fact that there are no blue swans. A completely analogous point can be made about propositions. The proposition that I am bald and the proposition that I have less than twenty hairs differ in a theoretically important way—one is a vague proposition, the other is precise. However, since I am, as of writing, definitely not bald, neither proposition is borderline.

It is clear that there are two types of properties that propositions can have—borderlineness and vagueness—and that they are related. Indeed, it is commonly assumed that one type can be reduced to the other. According to the standard line, vagueness can be defined in terms of the notion of being borderline. Alternatively one could maintain, as I do, that borderlineness can be reduced to the vague/precise distinction. But at any rate, given that we have this distinction between vague and precise propositions, it is perfectly acceptable to theorize in terms of it without yet taking sides on the direction of reduction.

The first thing to point out in this regard is that the distinction between precise and vague propositions is not itself a precise one. To see this, it is sufficient to note that one can construct a sorites sequence of propositions starting with propositions that are clearly precise and gradually changing until we have propositions that are clearly not precise. One such sorites consists of the sequence of conjunctive propositions that includes, for each natural number n, the proposition that Harry is bald and has n hairs. Note that the first proposition (when $n = 0$) is precise: it is plausibly a conceptual truth that someone with no hairs is bald, so the proposition that Harry has 0 hairs entails that Harry is bald, and, thus, the proposition that Harry has 0 hairs and is bald is equivalent to the proposition that Harry has 0 hairs (for a supervaluationist, for example, they correspond to the same set of world–precisification pairs). The latter proposition is precise and thus, given the sense of equivalence is sufficiently demanding, it follows that the former proposition is precise too. Yet as n increases we eventually arrive at propositions that are vague. This is, of course, intimately tied to the phenomenon of higher-order vagueness.

---

[2] Note that linguistic theorists attribute this distinction instead to sentences and predicates.

Given only what we have said so far, we can lay out a number of formal conditions that govern the behaviour of precise propositions. For example, it is obvious that the tautologous proposition is precise. Furthermore, it seems clear that if a proposition is precise, then so is its negation and, similarly, that conjunctions and disjunctions of precise propositions are also precise. For example, if the proposition that there are electrons is precise, then the proposition that there are no electrons is precise; if the proposition that there are three electrons is also precise, so is the proposition that either there are no electrons or three, and so on. Assuming that the space of all propositions forms a complete Boolean algebra (see section 3.2) under the Boolean operations of conjunction, disjunction, and negation, the preceding just means that the precise propositions themselves form a complete Boolean algebra of their own, albeit one that is smaller than the algebra of all propositions.[3] Our first constraint is thus:

Boolean Precision. The set of precise propositions forms a complete Boolean algebra, with the tautologous proposition being the weakest precise proposition.

Another important constraint we will include is that this is an atomic Boolean algebra—that every precise proposition is a disjunction of maximally strong consistent precise propositions—which is equivalent (given the axiom of choice) to stipulating that:

Atomicity. Given any jointly consistent set of precise propositions, X, there is always a consistent precise proposition that entails each member of X.

This guarantees, for example, the platitude that the conjunction of all the precise truths is consistent and seems like a reasonable general assumption to make.[4]

Next, we must say how the vague/precise distinction relates to the determinate/borderline distinction:

PRECISION TO DETERMINACY: Every precise truth is a determinate truth.

DETERMINACY TO PRECISION: Every determinate truth is entailed by a precise truth.

The first principle is fairly straightforward; an equivalent way to say it is that borderline propositions are vague. The converse, of course, is not true—we have already considered examples of determinately true propositions that are nonetheless vague. However, every determinate truth should be grounded in *some* precise fact: there couldn't be a determinate proposition stronger than every true precise proposition. (Note that there is a terminological confusion that should be avoided when reading

---

[3] We need the assumption that our background theory of propositions is Boolean. This could fail for a structured theory of propositions, for example.

[4] Atomicity is stronger than this, however: it additionally guarantees that it's not merely a truth, but an L-truth truth that the conjunction of all precise truths is consistent.

the second principle: some people reserve the word 'precise' for propositions that are precise at all orders in my sense, and there can be determinate truths that are not entailed by any truth that's precise at all orders.[5])

Finally, we should say how precision interacts with necessity. This will be the subject of chapter 15, but for now we may merely note that the following principle seems reasonable:

NECESSITY OF THE PRECISE: Precise propositions are necessarily precise.

Along with the above principles, this allows us to prove that precise propositions couldn't have been borderline, or, equivalently, are necessarily either determinately true or determinately false: $\square(\Delta A \vee \Delta \neg A)$.

12.1.1 The modal characterization of precision

It is somewhat striking that most philosophers—whether supervaluationist or not—tend to theorize with the notion of a proposition being borderline, or equivalently (modulo definitions), with the notion of a proposition being determinately true.[6] It should be clear, however, that we are also in need of an account of the distinction between vague and precise propositions.

There is a natural modal way to characterize this distinction which is widely adopted. For example, Fine [56] writes, attributing the distinction to Waismann [98], that 'a predicate F is extensionally vague if it has borderline cases, intensionally vague if it could have borderline cases. Thus "bald" is extensionally vague, I presume, and remains intensionally vague in a world of hairy or hairless men.' There is a corresponding characterization of propositional vagueness: the difference between the proposition that I am bald and the proposition that I have 0 hairs consists in the fact that, even though both are determinately false, the former proposition *could have been* borderline, whereas the latter could not. Call this the *modal characterization* of precision.

THE MODAL CHARACTERIZATION OF PRECISION

A proposition is precise if and only if it couldn't have been borderline.
A proposition is vague if and only if it could have been borderline.

It should be noted that the abstract principles we outlined earlier governing the concept of precision entail that no precise proposition could have been borderline. On the one hand, it's surely necessary that every borderline proposition is a vague

---

[5] Although every proposition that's determinate at all orders is entailed by some proposition that's precise at all orders. Related to this strong understanding of 'precise' is the notion of being a worldly proposition; worldly propositions are also not related to the determinacy operator in the way required by DETERMINACY TO PRECISION. Williamson [159], for example, uses 'precise' in a way that corresponds roughly to the notion of 'worldly' we introduced earlier.

[6] Given that Δ satisfies the modal logic K, we can prove that the definition of ∇p as $\neg \Delta p \wedge \neg \Delta \neg p$ works. Conversely, we can take ∇ as primitive (see Pelletier [110]).

proposition. It follows that if a proposition could have been borderline, then it could have been vague, and given THE NECESSITY OF PRECISION, it follows that it must in fact be vague. However, the converse claim—that every proposition that couldn't be borderline is precise—does not follow from those principles. It is consistent with our abstract principles governing precision above that there are vague propositions that couldn't have been borderline. The modal characterization, by contrast, rules this out: all vague propositions are possibly borderline.

Is this characterization adequate? The first order of business would be to show that our above formal constraints are satisfied. In other words, we must show (i) that the tautologous proposition is precise according to this definition, (ii) that negations and arbitrary disjunctions and conjunctions of precise propositions are precise, (iii) that every consistent set of precise propositions is entailed by a single consistent precise proposition, (iv) that precise truths are necessarily precise and (v) that precise truths are determinately true or false. I leave (i)–(iii) as exercises: the fact that negation and finitary disjunction and conjunction preserve precision is provable in a fairly minimal logic of vagueness and modality (the principles of the weakest normal modal logic K for both $\Box$ and $\triangle$), and the infinitary generalizations in a similar infinitary logic. Given the principle that what is necessary is necessarily necessary we can see that propositions that are necessarily determinately true or false are necessarily necessarily, determinately true or false, securing (iv). Given the factivity of 'necessarily' we can see that precise truths are always determinately true or false, securing (v). Thus the modal definition of precision at least meets the formal desiderata we listed above.

12.1.2 Supervaluationism

Supervaluationism provides us with a natural and relatively concrete way of simultaneously modelling the necessity and determinacy operators. It will thus be illuminating to see how the modal definition of precision plays out within a supervaluational model theory.

A supervaluational semantics invokes two basic sorts of theoretical entities: a set of possible worlds, W, and a set of precisifications, V. The notion of a possible world should be familiar already. As already noted, the precisifications belonging to V can be interpreted in a couple of ways. In the orthodox linguistic setting, a precisification is a function assigning precise extensions to every predicate of our language, relative to each world. A non-linguistic theorist can adopt a completely analogous interpretation of a precisification by thinking of it as assigning a precise extension to each vague proposition, property, and relation relative to each world instead. According to the latter interpretation, the truth of a vague proposition is completely determined once we have specified both a world and a precisification. Absent other forms of hyperintensionality, it will be for the most part harmless to simply identify each proposition with its representation as the set of world–precisification pairs at which that proposition is true.

The basic insight behind supervaluationism is the idea that borderline cases lack a certain semantic status: they lack a determinate truth value. Something is determinately true at a world if it is true relative to every admissible way of making each predicate of our language (or each vague property) precise, and is determinately false if it is false relative to every admissible way of making the vague predicates (or the vague properties) precise. Despite the existence of these apparent semantic gaps, the laws of classical logic are upheld because every way of making a language precise removes the gaps by assigning every sentence or proposition a classical truth value, even if this is in some cases achieved arbitrarily. It follows that a disjunction might be determinately true even if neither disjunct is—every precisification might make at least one disjunct true, but it might be a different one for different precisifications.

Thus, it is clear that supervaluationism gives us a straightforward way of capturing the distinction between a determinate and a borderline truth: respectively, a proposition that is true at all admissible precisifications, and a proposition that is true at some but not all admissible precisifications. Note that by varying the world coordinate we can also model modal notions. So, given the modal characterization of precision, we can also model the vague/precise distinction in a supervaluational model theory.

To demonstrate the idea, let us consider its application to a relatively simple language: the language of the propositional calculus with two unary connectives for expressing necessity, written $\Box A$, and a determinacy operator, written $\triangle A$. In this semantics, sentences will be evaluated relative to a pair of objects: a world (which I'll refer to with the letters w, x, y, z) and a precisification (which I'll refer to with the letters u, v). The worlds in our set, W, are related to one another by a reflexive accessibility relation, Rxy, meaning that everything necessary at x is true at y.[7] Many philosophers employing this formalism assume that R is simply the universal relation that holds between every pair of worlds. Thus, for most purposes, this relation can be ignored.

By contrast to W, V generally comes equipped with a non-trivial reflexive relation of relative admissibility, Suv. Relative admissibility requires a little more explanation. Generally, v is admissible relative to u only if everything determinate at u is true at v. On a linguistic interpretation, admissibility might amount to being compatible with the way English is used at the actual world.[8] Note that the concept of being admissible, as informally described above, is itself vague—there is vagueness, for example, concerning how English is used. Thus, one can precisify the notion in various different ways. Perhaps, according to one precisification of 'admissible',

---

[7] There is a problem with this informal gloss which I think is sometimes not appreciated by supervaluationists: it is not clear what it means for something to be true at a world *simpliciter*. The framework only seems to make sense of things being true relative to both a world and a precisification. One might similarly gloss the notion of relative admissibility (introduced below) as everything determinate at v is true at u—this suffers a similar problem.

[8] It is important to note that this guarantees that it is not contingent which precisifications are admissible. The results would be disastrous if, for example, we counted a precisification as admissible at a world of evaluation w if it was compatible *with the way English is used at w*.

v counts as admissible, and relative to another it doesn't. This is the relative notion of admissibility—of one precisification being admissible according to another—and it is a bit like an accessibility relation in modal logic. For short, we shall write Suv to mean that v is admissible relative to u.

An interpretation of our language, $[\![\cdot]\!]$, assigns each atomic sentence a set of world–precisification pairs. Let us write $x, v \Vdash \phi$ to mean that ϕ is true at the world–precisification pair x, v. Arbitrary sentences of our language are then evaluated at pairs of worlds and precisifications as follows:

(At) $x, v \Vdash p_i$ iff $\langle x, v \rangle \in [\![p_i]\!]$.

($\neg$) $x, v \Vdash \neg\phi$ iff $x, v \nVdash \phi$.

($\wedge$) $x, v \Vdash \phi \wedge \psi$ iff $x, v \Vdash \phi$ and $x, v \Vdash \psi$.

($\Box$) $x, v \Vdash \Box A$ if and only if $y, v \Vdash A$ for every y such that Rxy.

(Δ) $x, v \Vdash \Delta A$ if and only if $x, u \Vdash A$ for every u such that Svu.

Note that for necessity we keep the precisification fixed and check for variation with respect to accessible possible worlds. For determinacy we keep the world fixed and look for variation with respect to relative admissible precisifications.

Although we will typically ignore the modal accessibility relation, R, and quantify over all possible worlds, we cannot make a similar move with the S-accessibility relation. A decent logic of vagueness ought to make room for the possibility of second-order vagueness (propositions such that it is borderline whether they are borderline), or, more generally, higher-order vagueness. If every precisification was S-accessible to every other precisification, then it would always be a determinate matter whether a proposition is determinate or borderline, as embodied by validities like $\Delta A \rightarrow \Delta\Delta A$ and $\neg\Delta A \rightarrow \Delta\neg\Delta A$.

12.1.3 Degeneracy

What does the modal characterization of vagueness and precision amount to in this framework? To answer this question, our strategy will be to (i) take a set of world–precisification pairs (our proxy for a proposition) and use it to interpret the sentence letter A and then (ii) evaluate the formula $\Box(\Delta A \vee \Delta\neg A)$ at a given world–precisification pair $\langle w, v \rangle$. If the formula comes out true relative to $\langle w, v \rangle$, then that set of world–precisification pairs is precise relative to $\langle w, v \rangle$, otherwise it is vague relative to $\langle w, v \rangle$. Notice that our way of classifying propositions into vague and precise depends on the world–precisification pair we are evaluating the formula at—different propositions might be precise relative to different pairs. This is to be expected: as I noted earlier in section 12.1.2, due to higher-order vagueness about what is precise, the exact structure of the precise propositions will depend on the precisification.

To take a very simple example, suppose that there are only four possible worlds: $w_1, \ldots, w_4$. Then we can think of the space of propositions as divided into four

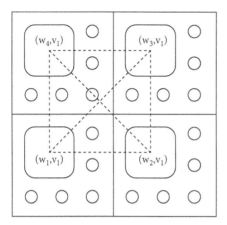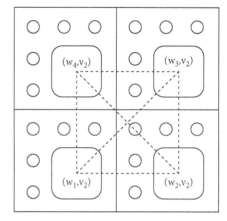

Figure 12.1. The space divided into four world propositions. The two diagrams represent two divisions into precise propositions depending on the precisifications v_1 and v_2.

quarters, as depicted by the four squares in either of the two diagrams in Figure 12.1, with each quarter representing the pairs that have a particular world as its first coordinate.

To evaluate a formula of the form $\Box A$ at $\langle w_1, v_1 \rangle$ we vary the world coordinate and see if A is true relative to each of the resulting pairs. Since there are only four worlds, there are only three world–precisification pairs accessible to $\langle w_1, v_1 \rangle$ other than itself: $\langle w_2, v_1 \rangle$, $\langle w_3, v_1 \rangle$, and $\langle w_4, v_1 \rangle$. Note here that even though our modal accessibility relation over *worlds* is universal, when we look at which *pairs* are accessible to one another, the modal accessibility relation forms a non-trivial equivalence relation. In our model, pictured in Figure 12.1, each equivalence class contains exactly four pairs, taken from each of the four quarters of our picture: we have represented the accessibility relations for two of these equivalence classes by the six dotted lines appearing in both the diagrams in the figure.

The rounded box surrounding (w_1, v_1) in the diagram on the left represents the world–precisification pairs that are S-accessible to (w_1, v_1). Note that all boxes are world–precisification pairs with w_1 as the first coordinate, which is represented by the fact that it is a subset of the lower left quadrant. The proposition corresponding to the set of points in the rounded box around (w_n, v_1) is determinately true at (w_n, v_1) and determinately false at each of the other accessible worlds–precisification pairs. Thus each of the rounded boxes represents a precise proposition (i.e. a proposition which is necessarily either determinately true or determinately false). Note that any world–precisification pair that falls outside of the rounded boxes is determinately false at all modally accessible worlds, and thus the singleton of any such pair will vacuously be a precise proposition. I have represented the singletons of these pairs by the smaller circles in Figure 12.1. Note that the existence of these degenerate precise propositions is inevitable given the presence of higher-order vagueness. If we were to maintain that the rounded boxes took up the entirety of their respective quadrants, we'd effectively be stipulating that each precisification was S-accessible to every other precisification.

We are now in a position to completely characterize the propositions that are precise relative to (w_1, v_1). Let's begin with the maximally strong consistent precise propositions.[9] As noted above, each of the rounded boxes is precise. It's also clear that any non-empty proper subset of the box around (w_n, v_1) will be borderline at (w_n, v_1), so no consistent proposition that is strictly contained in (i.e. stronger than) a box will be precise. As we noted earlier, the singletons of each world–precisification pair that lie outside the rounded boxes will be vacuously precise, and since they have no non-empty proper subsets, they automatically count as maximally strong consistent precise propositions. An arbitrary precise proposition is either the empty set or an arbitrary union of maximally strong consistent precise propositions.

I have illustrated the structure of the precise propositions according to the modal characterization using a supervaluationist framework. However, the conclusions are quite general: a similar structure will be shared by any theory which takes the notion of being borderline as primitive and defines precision using the modal definition.[10]

The precise propositions have a very strange structure according to this picture. Indeed, as we saw above, most of the maximally strong consistent precise propositions will be *degenerate*: a singleton of a world–precisification pair. This is problematic for both an intuitive reason and a more theoretical one.

The intuitive problem is that the degenerate maximally strong consistent precise propositions settle lots of seemingly precise questions: for example, they tell us the locations of the cutoff points for all the vague properties. Intuitively, no precise proposition should entail things like *the cutoff for baldness is exactly 2,049 hairs*.

The theoretical problem stems from the role that precision was supposed to play in the present theory. The **Principle of Plenitude** states that for any maximally strong consistent precise proposition and any proportion between 0 and 1, there's some vague proposition that takes up that proportion of the precise proposition: if p is a maximally strong consistent precise proposition, and $0 \leq \alpha \leq 1$, there is some vague proposition q such that $Pr(q \mid p) = \alpha$ for every rational ur-prior Pr. If there were any degenerate maximally strong consistent precise propositions, then **Plenitude** would fail: every proposition would either be entailed by it or be disjoint from it and, so, every proposition would have a conditional probability of 1 or 0 on it—no proposition will have a probability of $\frac{1}{2}$ conditional on a degenerate proposition.

Other core principles of this theory are affected too. **Indifference** states that one should be indifferent between any two propositions that entail the same maximally

---

[9] Here a set of ordered pairs is stronger than another if it is a subset, and is consistent if it is non-empty.

[10] For example, one will have degenerate maximally strong consistent precise propositions on any theory: take a maximally strong consistent proposition that is determinately false but not determinately determinately false. A natural supervenience thesis (see section 15.3) entails that any such proposition will be necessarily determinately false, and thus a degenerate maximally strong consistent precise proposition.

strong consistent precise proposition. **Rational Supervenience** states that all coherent ur-priors agree conditional on each maximally strong consistent precise proposition. The force of these principles is significantly lessened if there are degenerate propositions all over the place. For example, every coherent prior must agree conditional on a degenerate precise proposition, since it is the singleton of an index, and so **Rational Supervenience** is vacuously satisfied; these considerations also show that, for similar applications, **Indifference** becomes vacuous as well.

12.1.4 Doxastic features of vague propositions

More suspicion is cast on the modal definition by the observation that we cannot adequately capture the distinctive doxastic features that differentiate vague propositions from precise propositions if we adopt that account of precision. For example, I suggest that the propositions expressed by the following sentences are vague due to their distinctive doxastic features:

1. Harry is bald.
2. Either Jocasta is the mother of Oedipus or Harry is bald.
3. Patrick Stewart is bald.
4. Patrick Stewart is actually bald.

These propositions have the following feature that is quite distinctive to vagueness: in all cases, there is some precise hypothesis, p, such that one is rationally required to be uncertain in the vague proposition conditional on p (provided there is such a p consistent with your evidence). In the former case, for example, one should be uncertain whether Harry is bald conditional on the precise proposition that he has N hairs, whenever N belongs to a certain class of borderline cases.

Consider now Oedipus, who believes that Jocasta is not his mother. Suppose Oedipus also knows that Harry is in the borderline region for being bald: for Oedipus, both 1 and 2 are on an epistemic par. He should be uncertain about both propositions in that distinctive kind of way characteristic of uncertainty about borderline matters. What explains the presence of this distinctive kind of attitude in this case? It seems completely obvious that the explanation for the presence of this distinctive kind of uncertainty ought to be the same in both cases, and that the explanation has something to do with both propositions being vague.

Yet note that this conflicts with the modal characterization of the distinction between the vague and precise. The proposition that Jocasta is Oedipus' mother or Harry is bald could not have been borderline since it is an *a posteriori* necessary truth that Jocasta is determinately Oedipus' mother. In short, both 1 and 2 have features characteristic of vague propositions, yet only 1 counts as vague according to the modal characterization.

A similar argument can be run for 3 and 4. Since Patrick Stewart is determinately actually bald, it couldn't have been borderline whether he is actually bald. But surely all the reasons we have to think that 3 is vague extend to the claim that 4 is vague.

It might be tempting to insist, in response to these puzzles, that the world parameters should be replaced by epistemically possible worlds. But the resulting picture leaves no work for the precisification parameter to play. Since vagueness involves ignorance, there must be epistemically possible worlds where, say, Harry has the same amount of hair but which differ about whether he is bald. (Indeed, if we interpreted the modal operators with the (□) clause, we would encounter violations of the supervenience of the vague on the precise; we will return to these issues in section 15.3.)

This argument rested on a quite general principle, namely that if there's a (consistent) hypothesis, such that conditional on that hypothesis we should be certain that p is borderline, and thus have that distinctive uncertainty about p characteristic of borderline cases, then p is a vague proposition:

> CREDENCE TO VAGUENESS: If there's some consistent hypothesis, h, such $Pr(\nabla p \mid h) = 1$ for some conceptually coherent prior Pr, then p is vague.[11]

This provides us with another important connection between borderlineness and vagueness. For example, there are plausibly certain hair numbers such that it is a conceptual truth that people with that amount of hair are borderline bald in the sense that according to every coherent prior it's certain that Harry (say) is borderline bald conditional on his having that many hairs. Thus it follows that the proposition that Harry is bald is vague. The same goes for the proposition that either Jocasta is Oedipus' mother or Harry is bald: conditional on the hypothesis that Jocasta isn't Oedipus' mother and that Harry has N hairs, we should be certain that that proposition is borderline. The latter hypothesis in question is surely consistent according to any reasonably fine-grained account of propositions. There are, of course, very coarse-grained theories of propositions that predict that one cannot be uncertain about who Oedipus' mother is. However, I take it that most people accept enough fine-grainedness to make sense of this uncertainty, and these theorists will be in a position to conclude that 2 is vague from CREDENCE TO VAGUENESS.

12.2 Are the Propositions of Physics Precise?

It should be noted that CREDENCE TO VAGUENESS makes some surprising predictions about the classification of vagueness. For example, it is often taken to be a platitude among those theorizing about vagueness that the sentences of fundamental physics express precise propositions. I want to make the case, however, that these propositions are usually vague: there are epistemically consistent hypotheses conditional on which we should be certain that these propositions are borderline.

---

[11] I follow the convention of setting $Pr(p \mid q)$ to 1 when $Pr(q) = 0$. By a consistent hypothesis, h, I simply mean that $Pr(h) > 0$ for some rational credence function. Note, then, that CREDENCE TO VAGUENESS is equivalent to the claim that $Pr(\nabla p) > 0$ for some credence function.

We can illustrate this idea by entertaining a crude version of pre-Socratic physics. According to this picture, there are only four fundamental predicates: earth, wind, fire, and water. Given this account of the physics, the proposition that there is fire ought to be a precise proposition, for it passes our test of being a proposition expressed by a sentence of fundamental physics.

By modern standards the idea that the property of being fire is either fundamental or precise is not particularly plausible. Apply sparks to some kindling and at some point you'll have a fire, but the point at which there is fire is surely not a precise matter.

From this perspective, the proposition that there is fire is not a precise proposition. However, we can give an argument that even someone who was working under the false assumption that fire was a fundamental property should not consider this proposition to be precise. Since the hypothesis that there are borderline cases of fire is in fact true, it was certainly a hypothesis that was true for all the Aristotelian physicists knew. Thus, there is an epistemically possible hypothesis about fire conditional on which the Aristotelian physicist should have been certain that it was borderline whether there is fire. Given CREDENCE TO VAGUENESS, it follows that this is not a precise proposition.

If this conclusion does not strike you as particularly surprising, consider instead a proposition that you might find in a modern physics textbook such as the proposition that there are electrons. This might seem like the paradigm example of a precise proposition. Although we presently believe that electrons are basic and not composed of smaller particles, this is an empirical observation. There are consistent scientific hypotheses—hypotheses we might even wish to take seriously—in which electrons are composed of little clouds of smaller, more basic particles. One could also imagine that, just as with real clouds of water vapour, it can sometimes be borderline whether these smaller particles are close enough and dense enough to compose an electron.

It follows, then, that there are physical hypotheses about the world such that conditional on them the proposition that there are electrons has the epistemic profile distinctive to borderline cases. For example, suppose that α is a precise description of a possible arrangement of particles in which it would be borderline whether that arrangement of particles constitute an electron-cloud. Now consider the physical hypothesis that electrons are certain clouds of smaller particles and that there is only one cluster of such particles in the universe and that these particles are arranged in arrangement α. Conditional on this proposition, we should be uncertain in that distinctive way about whether there are electrons, and indeed we should be certain, conditional on this proposition, that it is borderline whether there are electrons. It follows, by CREDENCE TO VAGUENESS, that the proposition that there are electrons is a vague proposition.[12]

---

[12] Although the proposition that there are electrons is vague, one might be tempted to think that there are always closely related propositions that are precise. In this case, the conjunction: there are electrons and electrons are fundamental. Although this thought is indeed tempting, I will challenge it in chapter 14.

It should be noted that it is completely consistent with this conclusion that the proposition that there are electrons is necessarily determinately true or determinately false. This is not sufficient on our view for a proposition to be precise, however, and this leads to a radically different conception of the vague/precise divide—one that is not fixed by familiarity with particular examples but is instead fixed by the role the distinction plays in thought.

12.3 Vagueness as Primitive

The failure of the modal definition to capture precision, in my opinion, should make one doubtful that the notion can be defined from the determinacy operator at all. For example, some philosophers adopt a stronger definition of precision in which a completely precise proposition is one that not only couldn't have been borderline but couldn't have been borderline at any order. Given plausible assumptions, this characterization is also subject to the problem of degeneracy and the problem of vague propositions that are *a posteriori* necessarily determinate discussed in sections 12.1.3 and 12.1.4.[13] If this is right, then there is a serious lacuna in existing approaches to vagueness.

Following the treatment of determinacy operators in section 12.1.2, we might develop this idea by adding to a simple propositional language a primitive precision operator, written $\sharp A$. Formally, what we would like is some replacement for the supervaluational semantics: something that provides an illuminating analysis of the precision operator, as the supervaluational semantics does for the determinacy operator. We shall turn to this task in chapter 13 where we will introduce the notion of a 'symmetry', which plays a role somewhat analogous to the role a precisification plays in the analysis of determinacy operators.

For now, we'll just satisfy ourselves with getting a picture of the structure of precise propositions. We shall begin in the usual manner with a set of indices, to be thought of as maximally strong consistent propositions. Given the assumption that the set of propositions forms a complete atomic Boolean algebra, every proposition can be represented by a set of indices, namely, the indices that entail that proposition. It is convenient to talk about a proposition and its representation by a set of such indices interchangeably.

Given an index i, we wish to know what sets of indices will count as precise at i: which sets of indices have the property that $i \Vdash \sharp A$ when used to interpret A. Note that as before, which propositions count as precise depends on the index— without this feature there could be no higher-order vagueness. Given the assumptions **Boolean Precision** and **Atomicity** discussed in section 12.1, we know that the precise

---

[13] In the former case one gets degeneracy if there are at least two world–precisification pairs such that one cannot reach the other in any number of transitions along the S-accessibility relation. This can happen if the S-accessibility relation is asymmetric, or if one of the world quadrants can be split into two halves such that nothing in one half is S-accessible to anything in the other half, or vice versa.

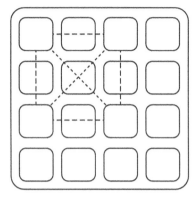

Figure 12.2. Logical space divided into cells of consistent propositions that are maximally strong among the propositions that are precise according to *i*.

propositions form a complete atomic Boolean algebra. As usual, the best way to specify a complete atomic Boolean algebra is by saying what its atoms are—i.e. by saying which the maximally strong consistent precise proposition are.

We can thus visualize the set of propositions that are precise relative to *i* by a partition of the set of all indices into lots of little cells, where each cell is a maximally strong consistent precise proposition. This structure is depicted in Figure 12.2. In general, the division of the whole space into cells will depend on the index: if it's borderline whether *p* is precise, there must be an index in which *p* is expressed as union of that index's cells, and an index according to which *p* cannot be expressed as a union of cells. Given the NECESSITY OF THE PRECISE, however, in the special case where *i* and *j* are modally accessible indices, they will agree about the cell structure for, otherwise, it could be contingent whether a proposition is precise or not, which is something we wish to rule out. In Figure 12.2, I have included the modal accessibility relation, represented by the six dotted lines. The cell structure depicted is thus the cell structure for any of the four modally accessible indices at the corners of the dotted lines.

Two differences between the present theory and the supervaluationist's are worth highlighting. The supervaluationist framework also gave us a similar partition of the space of world–precisification pairs into cells of maximally strong consistent precise propositions, but in that setting most of the cells were degenerate.[14] The kinds of models we have gestured at above are more general: they include models in which there are degenerate cells, but they also include models like the one depicted in Figure 12.2, where all of the cells are non-degenerate. The other point of contrast is that, in the supervaluationist setting, we also had a second partition of the space

---

[14] As mentioned earlier, strictly speaking, we can construct supervaluationist models where there are no degenerate cells, but these are very special models in which there is no higher-order vagueness.

into 'world propositions': sets of ordered pairs that shared a fixed world coordinate, represented by each of the four quadrants in Figure 12.1. This division doesn't seem to play much role in the supervaluationist account of determinacy and necessity and in, chapter 14, I'll consider some problems for taking the distinction seriously. In the models we have gestured at, this second way of dividing propositions is not available: the world propositions have been eliminated from this second picture.

12.3.1 Determinacy operators

Taking propositional precision as our basic notion does not mean doing away with determinacy operators altogether—these notions are still incredibly useful. Given propositional precision, we can define the determinacy and borderlineness operators as follows:

> DETERMINACY: It's determinate that p if and only if the strongest true precise proposition (i.e. the conjunction of all the precise truths) entails p.

> BORDERLINENESS: It's borderline whether p iff the strongest true precise proposition is consistent with p and with $\neg p$.

To find the strongest true precise proposition at an index, i, we do the following. Since i determines a division of logical space into cells, the strongest true precise proposition at i is just the cell in that partition that contains i. A proposition is determinate at i just in case that cell is entirely contained within that proposition. We can turn this into a more familiar Kripke semantics by stipulating that an index j is accessible to i iff j belongs to the same cell as i relative to the partition that i determines. Then the above just amounts to saying that a proposition is determinate at i iff it is true at every index accessible to i.

The definition of determinacy is also equivalent to the following slightly simpler formulation: a proposition is determinate if it is entailed by some precise truth. Thus, these definitions can be formalized in the object language, given enough expressive resources: Δp, for example, just becomes $\exists q(q \wedge \sharp q \wedge L(q \to p))$ where L is our broadest notion of necessity. Notice also that even someone who does not take precision as primitive can interpret these definitions as principles connecting precision and determinacy. A supervaluationist who endorses the modal characterization of precision will be able to interpret and derive the above principle.

In summary, by taking precision as our primitive instead of determinacy, we avoid the problem of degeneracy and are able to preserve the proper doxastic features of vague propositions. However, unlike the supervaluationist with their precisifications and the notion of relative admissibility, we do not yet have any analogous formal way of modelling precision operators. It is to this problem we now turn.

13

Symmetry Semantics

There is one important theoretical benefit that supervaluational views, and other views that take determinacy or borderlineness as basic, enjoy over the proposed alternative: determinacy operators can be given a very straightforward model theory in terms of the well-known Kripke semantics for operators governed by a normal modal logic. When combined with the intuitive notion of a precisification, this sort of semantics has a great amount of heuristic value; these kinds of theoretical virtues are at least part of the advantages of the supervaluationist package. By contrast, precision operators cannot be straightforwardly modelled by accessibility relations, and while the notion of a precisification seems tailor-made for analysing determinacy and borderlineness, it's unclear what the corresponding notion is for precision. To compete with the supervaluationist, it would be nice to have some alternative theoretical device. In this chapter I shall argue that there is such a device, the notion of a *symmetry*: a certain sort of permutation of propositions that preserves their logical role. Which permutations ultimately count as symmetries will depend on one's preferred analysis of precision. On this account, the notion of a precise proposition receives a simple characterization as a proposition that is fixed (mapped to itself) by every symmetry.

The abstract semantics, developed in section 13.4, is perfectly general and provides us with a semantic analysis of precision that is independent of any particular philosophical analysis of it. However, the present theory of vagueness can be shown to fall out as a special case when we additionally require that symmetries preserve the role each proposition plays in thought, as characterized by the conceptually coherent priors and utilities.

13.1 Where Things Stand so Far

Let me begin by bringing together the various theses about vagueness that I have defended throughout this book. That theory can effectively be summarized by the four principles listed below:

Boolean Precision. The precise propositions form a complete atomic Boolean algebra: conjunctions, disjunctions, and negations of precise propositions are precise, and every consistent set of precise propositions is entailed by a consistent precise proposition.

Plenitude. For any function, E, from the maximally strong consistent precise propositions to $[0, 1]$, there is a proposition p such that $Pr(p \mid w) = E(w)$ for every w and conceptually coherent ur-prior Pr.

Rational Supervenience. If p is any proposition and w any maximally strong consistent precise proposition, then $Pr(p \mid w) = Pr'(p \mid w)$ for every pair of conceptually coherent ur-priors Pr and Pr'.

Indifference. If p and q both entail a maximally strong consistent precise proposition, w, then you should be indifferent between p and q.

Let us try and put these ideas together into a coherent account of vagueness. **Boolean Precision** ensures that we can divide the space of propositions into a partition of maximally strong consistent precise propositions. Along with our assumptions from chapter 12, it also ensures that we can represent propositions using sets of indices. In this isomorphic representation, the partition of maximally strong consistent precise propositions will determine an equivalence relation that clumps the indices into non-overlapping cells, or equivalence classes, of indices that agree about all precise matters.

Although different coherent priors can disagree about the sizes of these cells—in the sense that they can disagree about how probable or improbable they are—**Rational Supervenience** guarantees that all coherent priors agree about what proportion of each cell is taken up by each proposition. If p takes up half of a cell according to one coherent prior, it takes up half the cell according to all coherent priors (see Figure 8.1). We may think of **Rational Supervenience** as telling us that every proposition has an *evidential role*. By those lights, then, **Plenitude** tells us that every evidential role is occupied by at least one vague proposition. Finally, **Indifference** ensures that indices in the same cell get assigned the same utility; thus each cell gets to be associated with a potentially different utility, although, within a cell, all the utilities are constant.

This, then, is the structure of rational degrees of belief and rational degrees of desire according to our account so far. The above principles provide us with the beginnings of a theory of vagueness: they relate the notion of precision (and thus vagueness) to other concepts such as the notion of rational belief and rational desire.

However, it is natural to ask if we can reverse the order of explanation to give a definition of precision in terms of these concepts. Is it possible to start with some theses purely about the structure of the coherent priors and utilities and arrive at an independent characterization of the precise propositions?

Our starting point will be the observation that the models of propositional belief and desire described above are closed under certain operations on the space of propositions which leave the structure of belief and desire, and the logical structure of the propositions, completely unchanged. This leads us to the notion of a *symmetry*, which we move to now.

13.2 Symmetries

The key idea that will be exploited in our account of vagueness is the notion of a *symmetry*. The notion of a symmetry of a theory is often appealed to in the philosophy of physics. To focus on a classic example, consider Leibniz's observation that uniform translation in a particular direction in Newtonian physics is a symmetry of that theory. The result of moving every object two metres in a particular direction will take you to a system that also obeys the laws of Newtonian physics, provided the original system obeys those laws. Moreover, this translation appears to preserve all physically significant facts. The distances between objects, their relative velocities, their shapes and sizes, indeed just about any observable property you can think of seems to be preserved by this symmetry of the theory.

The theory has other symmetries as well; spatial rotations and reflections, temporal shifts, time reversal (temporal reflection), and scaling the velocity of every object, as well as arbitrary combinations of these transformations all preserve physically significant properties. Formally, we say that these symmetries form a certain sort of algebraic object called a *group*.[1] All this means is that the operation of doing nothing is vacuously a symmetry, the result of performing one symmetry and then another is also a symmetry of the same type, and for every symmetry there is another that 'undoes' it, that is, takes you back to where you started. The symmetry group of a theory often gives us an important insight into the structure of the kind of objects that the theory characterizes; one can represent many things in this fashion—from operations on a Rubik's cube to the symmetries between the roots of a polynomial—and extract important structural insights.

The group in our above example divides the space of possible worlds into equivalence classes, known as *orbits*, where two worlds belong to the same orbit if they can be related to one another by some symmetry transformation. Thus, for example, two worlds containing only three equidistant colinear particles, stationary relative to one another, might belong to the same orbit because one can get from one world to the other by some combination of translation, rotation, and reflection. But a world with three particles in a triangular formation would not be in the same orbit, since symmetry operations preserve directly observable properties like colinearity.

The result is a picture according to which the space of worlds is divided into equivalence classes (orbits) of worlds which all agree with one another about the physically significant facts and differ only over the positions of those objects in absolute space. The relevant symmetries, therefore, do not preserve all facts—*de re* facts about which space–time points are occupied are not preserved. However, for

---

[1] Formally, a group is a triple $(G, \circ, e)$ such that $e \in G$ and $\circ : G \times G \to G$, and such that for all $a, b, c \in G$: $a \circ (b \circ c) = (a \circ b) \circ c, a \circ e = e \circ a = a$, and for every $a \in G$ there is a $b \in G$ such that $a \circ b = b \circ a = e$. In the present case G is the set of all transformations, $\circ$ the operation of composing two transformations, and e the trivial transformation that leaves everything alone.

those who find such facts suspicious or in some sense less basic, the structure that the group imposes gives us a precise way to distinguish between propositions that are straightforwardly factual in the relevant sense. Some propositions are preserved by the symmetries and are expressible as some union of orbits, and these propositions, because they are fixed by the symmetries, will not entail suspect claims about the *de re* locations of objects.

In a slightly more abstract setting, we can think of a symmetry as a permutation on a set of possible worlds or indices: a mapping from worlds to worlds such that no two worlds get mapped to the same thing and such that every world gets mapped to by some world. A translation of two metres in a certain direction, for example, determines such a mapping on the set of Newtonian worlds—translation by a given vector never takes distinct worlds to the same world, and every world is a translation of some other world by that vector.

This observation will be key to generalizing the notion of a symmetry beyond this limited example. We can also raise the notion of a symmetry to the level of propositions. The relevant notion for propositions is that of an *automorphism*: a permutation of propositions that preserves all the logical relations between propositions. A function, σ, from propositions to propositions is an automorphism iff the following things hold:

1. σ is a bijection (a one-to-one correlation).
2. $\sigma(\neg p) = \neg \sigma(p)$ for each proposition p.
3. $\sigma(\bigvee X) = \bigvee \{\sigma(p) \mid p \in X\}$ for any set of propositions X.

By the deMorgan laws, 2 and 3 ensure that automorphisms also preserve arbitrary conjunctions—indeed they preserve any logically definable operation. If we are thinking of propositions in terms of their representations as sets of indices, then a permutation of indices induces an automorphism of propositions by mapping a set of indices corresponding to p, to the set of indices each member of p is mapped to. Conversely, every automorphism determines a unique permutation by looking at its action on singleton sets. In what follows, I shall talk about permutations on indices and the corresponding automorphism on propositions interchangeably.

In our toy example, we saw that it was possible to characterize a particular class of facts in terms of permutations on worlds that preserve certain physical features. This is possible because the theory of Newtonian mechanics has symmetries. As we will see, the theory of rational belief and desire we have sketched above also enjoys symmetries in a somewhat analogous sense. A symmetry of our theory is an automorphism which preserves the preferences and credences of every (possible) rational agent. More precisely:

SYMMETRY: An automorphism of propositions, σ, is a rational symmetry if and only if $V(p) = V(\sigma(p))$ and $Pr(p) = Pr(\sigma(p))$ for every conceptually coherent ur-prior Pr and coherent value function V for Pr.

Given the averaging principle, from chapter 9, probabilities are representable by values; so, under that assumption, the clause stating that ur-priors must be preserved is redundant. Thus σ is a symmetry if and only if every possible rational ur-agent is indifferent between p and $\sigma(p)$.[2]

A symmetry not only preserves all the logical relations between propositions—it preserves the relevant rational attitudes. If p and q are related by a symmetry, then every coherent ur-prior must agree about p and q, and the conditional expected utility on p and on q respectively must be the same for any coherent utility function. The intuitive gloss is that two propositions are related by a symmetry iff one cannot coherently hold a certain kind of propositional attitude towards one proposition without holding it to the other. To employ a more familiar, albeit more contentious way of talking, p and q are related by a symmetry if a belief or desire that p has the same conceptual role as a belief or desire that q.[3]

As before, symmetries form a group: the identity automorphism is a symmetry, applying two symmetries in succession is a symmetry, and every symmetry has an inverse which is also a symmetry. It remains to show that our theory actually has symmetries. Note that the existence of a non-trivial symmetry implies that there will be pairs of propositions, related by the non-trivial symmetry, which have the same value and probability for all ur-priors. As a warm-up to the existence of symmetries, note that our present theory implies the existence of pairs of propositions like this. Since all priors will agree about what proportion of each cell a proposition takes up, it is possible to talk about *the* proportion of a cell that a proposition takes up, relative to every prior. Thus, take any two propositions, p and q, that take up the same proportion of each cell. Pairs like this can be generated by **Plenitude**—for example, take a proposition p that takes up half of every cell. Then its negation, q, will also take

---

[2] This fact is easy to show when the agent cares about whether p (i.e. $V(p) \neq V(\neg p)$). For supposing $V(p) = V(\sigma(p))$. Then, since it is a consequence of averaging that $Pr(p) = \frac{1}{1-V(p)/V(\neg p)}$ and of averaging and the properties of automorphisms that $\frac{1}{1-V(\sigma(p))/V(\neg \sigma(p))} = Pr(\sigma(p))$ we can conclude that $Pr(p) = Pr(\sigma(p))$. If the agent doesn't care about p, then consider the value function you get from the same probability with a utility function that does make p valuable and run the same argument.

[3] Although the term 'conceptual role' comes with a lot of baggage which I don't want to spend time on, it is certainly one a term one can make sense of without the baggage, and one that does a good job of describing what I have in mind here. I take it that the conceptual role of a belief or desire that p includes at least the role that belief or desire would play in a somewhat idealized psychological description of how a coherent rational agent's beliefs and desires would evolve over time in response to sensory inputs, and how these desires and beliefs in turn affect the agent's dispositions to act in certain ways. It seems very plausible, given our preference theoretic definition of symmetry, that a belief or desire that p has the same conceptual role as a belief or desire that $\sigma(p)$, for symmetries σ. Although not as direct, the notion of conceptual role presumably extends to other propositional attitudes such as hopes, fears, wonderings, and so on. Therefore, a natural further question of interest is whether symmetries preserve the conceptual role of these attitudes. For example, letting σ be a symmetry, whether the following holds:

A hope, fear, wondering that/whether p has the same conceptual role as a hope, fear, wondering that/whether $\sigma(p)$.

These additional theses bear further investigation; I shall not conduct them here.

up half of every cell. It follows that $Pr(p) = Pr(q)$ for any prior Pr, and since utility is always uniform within a cell it furthermore follows that $V(p) = V(q)$ for every value function based on Pr.

The existence of actual symmetries is really just a generalization of these ideas. Given the partition structure of maximally precise propositions, the idea is to find permutations that move indices around within each cell of the partition, and which only map a subset of a cell to another set of indices in that cell if both the sets take up the same proportion of the cell according to all ur-priors.

It is important to note, at this juncture, that a symmetry is not merely a permutation that preserves the precise facts. Not every permutation that moves indices around within cells satisfies the constraint that the permutation be measure preserving. (A mathematical example demonstrates this point: the x^2 function is a permutation of the unit interval, yet it maps the interval $[0, \frac{1}{2}]$ to the interval $[0, \frac{1}{4}]$ which is half its size.) The group of permutations that preserve precise facts has the same orbit structure as the group of symmetries, but it is larger. The group of symmetries therefore contains more information than the mere classification of propositions into precise and vague.

What is the relation between precision and symmetries? Given our earlier remarks, it is natural to say that every precise proposition should be fixed by every symmetry. Indeed, it is possible to prove this fact if we make the assumption that the set of coherent utilities and priors are sufficiently rich. If we assume, roughly, that assignments of utility and probability to maximally strong consistent precise propositions are relatively unconstrained, it is possible to show that precise propositions are fixed by all symmetries. Either of the following richness conditions would be sufficient:

RICHNESS OF PRIORS: For any probability function over the space of precise propositions, there is a conceptually coherent prior over the whole space that extends it. (Which, given **Rational Supervenience**, will in fact be unique.)

RICHNESS OF UTILITIES: Any function on indices which assigns constant values through each cell represents a conceptually coherent utility function.

Here is how richness helps. Suppose a symmetry maps an index, i, to another index j. i and j must be within the same cell: since they're related by a symmetry they must have the same value for every utility function, but unless they were in the same cell the second richness condition would allow us to make the values of i and j different. A similar argument can be given using the first richness condition. This time suppose there is a third index, k, whose value is lower than both i and j (a parallel argument would work with a k of greater value).[4] If i and j belonged to different cells you could choose a prior that made i more probable than j, and thus make the V-value of the disjunction of i with k greater than the value of the disjunction of j with k (since more

---

[4] That there is such a k follows from the minimal assumption that our preferences are not completely flat. For the value of i and j is the same (since they are related by a symmetry), and if every index had the same value as i and j, there would be no cases of preference.

weight is going to the higher valued index in the first case than in the second). So $V(i \vee k) > V(j \vee k)$. But we can also show that $V(j \vee k) = V(\sigma(i \vee k))$, by choosing the prior so that $Pr(k) = Pr(\sigma(k))$. For then $V(j \vee k) = V(j \vee \sigma(k))$ since $V(k) = V(\sigma(k))$ and $Pr(k) = Pr(\sigma(k))$, and $V(j \vee \sigma(k)) = V(\sigma(i) \vee \sigma(k)) = V(\sigma(i \vee k))$ since $\sigma(i) = j$. Thus we have that $V(i \vee k) > V(\sigma(i \vee k))$ violating the definition of a symmetry.

Either way, we have shown that every symmetry must take each index in a given cell to another index in the same cell. This is equivalent to saying that symmetries fix all precise propositions, as required.

What about the converse result? If a proposition is fixed by every symmetry, must it be precise? This is harder to prove, so I shall only sketch the idea. It's equivalent to showing that, for each vague proposition, there's a symmetry that doesn't fix it. We'll begin by showing the result for vague propositions confined within a single cell. By applying **Plenitude**,[5] we can construct another proposition, q, that is part of the same cell as p, is distinct from p, but has the same measure conditional on the cell. Indeed, given **Plenitude**, we can construct a measure preserving bijection, τ, between $p \wedge \neg q$ and $q \wedge \neg p$ (thinking of them as sets of indices), since these propositions also have the same measure.[6] Now consider the function that fixes the worlds in $p \wedge q$ and the worlds outside $p \vee q$, and behaves like τ on $p \wedge \neg q$ and like τ^{-1} on $q \wedge \neg p$. It is immediate that this function is a symmetry that does not fix p (it maps it to q). The result extends to arbitrary vague propositions, since if p is vague then there's some cell which it partially but not completely overlaps, and it is then possible to run the above construction.

In summary, then, a proposition is precise if and only if it is fixed by every symmetry. This result strongly suggests that it might be possible to reverse the order of explanation: to start off with a group of symmetries and from this *define* the notion of a precise proposition. By defining a precise proposition in this way, it may in fact be possible to *derive* **Boolean Precision**, **Rational Supervenience**, and **Indifference** from this definition. **Plenitude**, because it asserts the existence of certain propositions, cannot be derived from a definition. However, it would be interesting to see if our theory of vagueness can simply be reduced to **Plenitude** and a definition of precision in terms of a symmetries. I shall now turn to such a definition.

---

[5] The slightly stronger version we stated in chapter 6: for any subpartition P of the maximally strong consistent precise propositions, and function $E : P \to [0,1]$, there is a vague proposition q such that $Pr(q \mid x) = E(x)$ for each $x \in P$ for every conceptually coherent ur-prior Pr.

[6] Provided they have the same cardinality. More formally, this step can be substantiated by appealing to the thesis that every proposition with positive measure has the same cardinality—a substantive assumption, albeit one that holds of many measures of interest. The assumption is true, for example, of the Lebesgue measure on the real numbers—indeed, it is a consequence of the perfect set theorem. This argument breaks down in the special case when p has the same measure as the cell that contains it. In that case, for any q contained in the same cell, $p \wedge \neg q$ and $q \wedge \neg p$ have measure 0. In that case we look for a bijection of the indices in the cell not belonging to p with a measure 0 subset of p of the same cardinality, and extend that to a permutation of all indices by treating it as identity elsewhere.

13.3 Vagueness and Precision

Since indices are permuted within a cell in such a way that no index gets sent out of its cell by a symmetry, it follows that any proposition that is either a cell, or a union of cells will be left alone by each symmetry. In other words, precise propositions are fixed by symmetries. This forms the basis for a definition of precision:

PRECISION: A proposition p is precise if and only if $\sigma p = p$ for every symmetry σ.

A vague proposition is defined as a proposition that is not precise. Note that this definition meets the challenge, raised in chapter 12, of giving a direct account of precision that does not go via the modal characterization.

If we want a definition of determinacy and borderlineness we can get that from the notion of precision as in chapter 12; a proposition is determinate iff it is entailed by the strongest true precise proposition. There is also a direct definition in terms of symmetries:

DETERMINACY: It's determinate that p if and only if every proposition that p is mapped to under a symmetry is true.

A borderline proposition is defined as a proposition which is neither determinate nor has a determinate negation. This is equivalent to saying that borderline propositions are those that are both mapped to a truth and to a falsehood by some symmetry.

Our abstract analysis of precision in terms of being fixed by every symmetry entails some desirable structural features. For example, we can infer that the precise propositions form a complete Boolean algebra: negations and arbitrary conjunctions and disjunctions of precise propositions are also precise. For if a proposition is fixed by every symmetry automorphism, so is its negation by the properties of automorphisms. Similarly, if every member of X is fixed by every symmetry, so is the disjunction and conjunction of X.[7]

If we go beyond the abstract definition and invoke the particular notion of symmetry we have been using, the **Rational Supervenience** principle and **Indifference** principle (but not **Plenitude**), likewise fall out of our definition.[8] If i and j are indices that belong to the same maximally strong consistent precise proposition, then there must be some symmetry that maps i to j. Since symmetries preserve utilities this means $u(i) = u(j)$. This guarantees **Indifference**, which amounts to the claim that indices in the same cells have the same utilities. Showing the rational supervenience thesis is a bit more involved, but it is basically shown by proving that if the conditional probability of a proposition on a maximally strong consistent precise proposition is

---

[7] Atomicity, on the other hand, is not guaranteed. There is thus a sense in which the view can be developed without the atomicity assumption, although I think that atomicity is independently plausible and the assumption of atomicity is convenient from a technical perspective.

[8] This is hardly surprising since the **Principle of Plenitude** has existential import—no definition can guarantee its truth.

bounded by two rational numbers according to one prior, it is bounded by those numbers according to every prior.[9]

It thus follows that we can provide a fairly compact 'axiomatization' of the main principles defended in this book:

Symmetry. A proposition is precise if and only if it is fixed by every symmetry.

Plenitude. For any function from the maximally specific precise propositions to $[0, 1]$, E, there is a proposition, p, such that $Pr(p \mid w) = E(w)$ for every maximally strong consistent precise proposition w and conceptually coherent ur-prior Pr.

Let me now turn to some points of clarification in our analysis. The first point to observe is that while there are symmetries in the space of conceptually coherent priors and values, these may not represent symmetries in the credences and values of informed people. This fact is due to the possibility, defended in chapter 6, of one's total evidence being vague, for vague evidence can break symmetries that exist in the space of coherent priors. For example, consider a proposition that takes up half of every cell, so that there is a symmetry switching it for its negation. If this proposition was someone's total evidence then they would assign it full credence and its negation no credence, breaking a symmetry that existed among the priors.[10]

Despite the fact that symmetries can be broken amongst the credences of people whose evidence is vague, it is still natural to think that the space of rational values and credences of informed people are closed under a symmetry in an extended sense. If V is the value function of a possible rational person, so is the function $V(\sigma(p))$, which assigns p the value of its image under the symmetry σ. Clearly when V is a value for a prior credence function, these two value functions are identical. But even when V corresponds to the values of an informed person, symmetries do take you outside the set of possible rational values.

Another point that requires some clarification is the role that the preservation of rational desires is playing in our notion of symmetry. What would happen if we just worked with a notion of symmetry that preserves initial credences?[11] Evidently, the **Indifference** principle would no longer be a consequence of our definitions, so the principle is needed for my specific project. But would a definition purely in terms of credences give an extensionally adequate characterization of precision?

---

[9] For example, if $Pr(p \mid w) \leq \frac{1}{2}$ and some symmetry σ maps p to a disjoint subset of the cell w, then in order for every prior to assign the same conditional probability to p as $\sigma(p)$, every prior must agree that the conditional probability of p on the cell is not greater that $\frac{1}{2}$.

[10] This example demonstrates the general phenomenon. However, because this proposition is necessarily borderline, it is not particularly plausible that this proposition could be someone's total evidence. If we disjoin this proposition and its negation with some contingent precise fact we get a more realistic example.

[11] The notion of a symmetry preserving initial credences is naturally defined by the equation $Pr(p) = Pr(\sigma(p))$ holding for every coherent prior Pr. However, probability 0 propositions can have interesting features according to a Popper function that needn't get preserved according to this definition of symmetry. A better definition would be $Pr(p \mid p \vee \sigma(p)) = Pr(\sigma(p) \mid p \vee \sigma(p))$ for every coherent prior Pr.

In section 13.2, we effectively showed that, provided we accepted a richness condition on the space of coherent priors, one can characterize the set of precise propositions as those which are fixed by all symmetries that preserve all coherent initial credences. (A similar result applies for symmetries which preserve coherent utilities.) But what exactly is the status of the richness condition?

For a function to be a *rational* prior it is certainly necessary that it be probabilistically and conceptually coherent. But is this sufficient? Perhaps priors ought also to satisfy the principal principle, support sensible inductive hypotheses, respect the principle of indifference, and so on. Once these further constraints are taken into account, one might think that there are propositions that all initial priors must agree on. Indeed, at the extreme end of the spectrum there are some who think that there is exactly one prior which it is rational to adopt (see Carnap [25] and, more recently, Williamson [160]). On such views, the richness condition on rational priors fails quite dramatically.

This conception of a rational prior might be thought to pose a problem for my abstract analysis of precision. Let me focus on one potential example like this. Suppose that a principle of indifference regarding purely haecceitistic differences is in operation. Imagine, for example, a world exactly like our own except that two people have swapped qualitative roles: Bob has led a life qualitatively indiscernible from the life Bill actually has led, and conversely Bill has led a life qualitatively exactly like the one Bob has led. You might think that any rational prior should assign just as much confidence to the switched possibility as to actuality. This, in some sense, substantiates the idea that it is hard to distinguish between such possibilities without possessing any *de re* evidence. More generally, any permutation of individuals will naturally induce a permutation on qualitatively identical worlds—a general principle of haecceitistic indifference would require that these permutations preserve rational prior credence. Thus, although the proposition that Bob is exactly 175cm tall seems to be precise, assuming the haecceitistic indifference principle, there's a symmetry preserving rational prior credence that maps it to the distinct proposition that Bill is exactly 175cm tall. Thus, if we required that a symmetry preserve rational ur-priors, there is pressure on the characterization of precision as those things fixed by symmetries.

The account I have developed states things in terms of credences that are conceptually coherent, not in terms of a generic notion of rationality. Conceptual coherence is a fairly weak constraint: a prior which makes no purely conceptual confusions may still be irrational in the wider sense. There is nothing *conceptually* incoherent about a prior that supports strange inductive inferences, for example, but many would not count such a prior completely rational. Similarly, while it may, in some sense, be unreasonable to have priors that find it more likely that Bob has a certain qualitative role than that Bill does, there is nothing conceptually *incoherent* about this belief. It is not, for example, like having priors that assign the proposition that Harry is both poor and is a billionaire a high probability.

Provided we are clear about the distinction between rationality *in toto* and conceptual coherence, then it may indeed be possible to extensionally capture the notion of precision without invoking desires. However, I would imagine that there will be some who, despite my attempts at elucidation, find the distinction between a rational prior and a merely conceptually coherent one too obscure to be bearing the burden of explicating this important philosophical notion alone. For those wishing to theorize only with the notion of rationality *in toto*, it is more important that the constraint that symmetries preserve desires be included. It seems quite evident that I could coherently have haecceitist cares: that I could care about what happens to Bob but not about Bill, for example. This is even more striking when it comes to caring about oneself—surely Bob needn't be completely indifferent between what happens to Bob and what happens to Bill. The inclusion of bouletic notions means that even the neo-Carnapians, who hold that there is only one rational prior, can make sense of this account of precision, provided they accept a moderate kind of permissivism about rational desire.

Let me also point out that even if the definition of symmetry in terms of preserving a certain class of priors is extensionally adequate for characterizing vagueness, it might not do the job of a good explanatory theory. For example, if the principle that one should not care intrinsically about the vague is true, it calls out for an explanation. Presumably the explanation ought to have something to do with vagueness. An abstract analysis of vagueness purely in terms of coherent credences does not provide any such explanation, yet a theory that invokes bouletic notions could provide such an explanation. For example, according to my theory, in order for something to play the role of a particular vague proposition, your bouletic attitudes toward it ought to be aligned in certain ways.

Let me end this discussion by considering the question of whether our abstract analysis in terms of symmetries can act as a reduction of vagueness to more basic notions. In this section I showed how one could start off with nothing but a certain class of priors and utilities, and from these characterize the class of precise propositions. From the notion of a conceptually coherent prior and utility, we introduced a class of symmetries that preserved values according to every prior and utility, and from this I characterized a precise proposition as something that is fixed by every symmetry. Could this be considered a reduction of the notion of precision to the normative notion of being conceptually coherent?

There are reasons to resist this further reductive claim. In particular, it is hard to get a handle on the notion of a conceptually coherent prior without already having the concepts of vagueness and precision at our disposal. A crucial distinction that came up in our earlier discussion was the difference between being rational *in toto*, and merely being conceptually coherent.

To distinguish mere conceptual coherence from general rationality we noted that the former satisfied a richness condition: roughly, although one can have pretty much any opinion or preference about the precise matters without committing a conceptual confusion, not all such opinions are rational in the wider sense. Formally, for any probability function over precise matters there is a conceptually coherent prior over all matters that agrees with it; this may not hold when restricted to rational priors. Similarly, for any utility over the cells, there's a conceptually coherent utility that agrees with it and is constant within each cell. Explaining what a conceptually coherent prior or utility is to someone by appeal to these richness conditions would require them to already possess the concept of precision and vagueness.

If, like me, you think that completely reductive analyses are rare, the above conclusion is hardly surprising; the value of abstract analyses is of an entirely different nature altogether. What, then, have we gained from our analysis if not a reduction? At least one important result is to widen the circle of concepts that vagueness and precision are related to. Even if you don't have the concept of a coherent prior at your disposal, we have still succeeded in related vagueness and precision to overall rationality: we have discovered that it is simply irrational, for example, to care about the vague, even if we can't get the converse claim without invoking notions that presuppose the concept of vagueness. Note also that such analyses deliver important structural features of the target concepts. By analogy, analyses of counterfactuals in terms of similarity are not usually proposed as *reductive* analyses, reducing counterfactuals to an antecedently understood notion of similarity; rather similarity is usually taken to be a back-formation from our judgements about counterfactuals (see Lewis [90] and Stalnaker [140]). The value of these analyses is that they predict the validity and invalidity of inference patterns that have puzzled earlier philosophers thinking about counterfactuals, and subsume them under a more general theory.[12]

13.4 A Semantical Account of Precision in Terms of Symmetries

We shall now examine a semantics for language fragments of interest based on the above notion of a symmetry. To start with, we shall examine the language that results from adding a single unary operator, $\sharp$, to the propositional calculus. $\sharp A$ is to be

---

[12] In response to the charge that safety accounts of knowledge are circular and thus uninformative, Timothy Williamson writes: 'For comparison, think of David Lewis's similarity semantics for counterfactual conditionals. Its value is not to enable one to determine whether a counterfactual is true in a given case by applying one's general understanding of similarity to various possible worlds, without reference to counterfactuals themselves. If one tried to do that, one would almost certainly give the wrong comparative weights to the various relevant respects of similarity. Nevertheless, the semantics gives valuable structural information about counterfactuals, in particular about their logic. Likewise, the point of a safety conception of knowing is not to enable one to determine whether a knowledge attribution is true in a given case by applying ones general understanding of safety, without reference to knowing itself. If one tried to do that, one would very likely get it wrong. Nevertheless, the conception gives valuable structural information about knowing' (Williamson [164]). In the present case, our analysis delivers important structural features of precision and determinacy: that determinacy has a certain normal modal logic, for example, or the precise propositions form a complete Boolean algebra.

read as 'it is precise that A'. In more detail, our language consists of those sentences that can be built out of a set of atomic letters $\mathcal{A}$, with the connectives $\{\neg, \wedge, \sharp\}$. We shall call this language $\mathcal{L}(\sharp)$. Given a set I, we write $Aut(I)$ to denote the set of automorphisms of the Boolean algebra $\mathcal{P}(I)$. We shall abuse notation and often use $\sigma \in Aut(I)$ interchangeably with the permutation it determines on I, writing for example $\sigma(i) = j$ instead of $\sigma(\{i\}) = \{j\}$ for $i, j \in I$.

A symmetry frame for $\mathcal{L}(\sharp)$ is a pair $\mathcal{S} = (I, G)$ where

1. I is a set of indices.
2. $G : I \to \mathcal{P}(Aut(I))$.
3. $G(i)$ is a group for each $i \in I$. That is: $id_I \in G(i)$ and $\sigma, \tau \in G(i)$ implies $\sigma \circ \tau^{-1} \in G(i)$.

A model $\mathcal{M}$ based on a symmetry frame (I, G) is a triple $(I, G, \llbracket \cdot \rrbracket)$ where $\llbracket \cdot \rrbracket$: $\mathcal{A} \to \mathcal{P}(I)$. $\llbracket \cdot \rrbracket$ can be extended to arbitrary formulae as follows:

$$\llbracket A \wedge B \rrbracket = \llbracket A \rrbracket \cap \llbracket B \rrbracket$$
$$\llbracket \neg A \rrbracket = I \setminus \llbracket A \rrbracket$$
$$\llbracket \sharp A \rrbracket = \{i \in I \mid \sigma(\llbracket A \rrbracket) = \llbracket A \rrbracket \text{ for all } \sigma \in G(i)\}$$

We say that a sentence A is *true at an index* i iff $i \in \llbracket A \rrbracket$. We say that the sentence is *valid in a model* $\mathcal{M}$ if $\llbracket A \rrbracket = I$, *valid in a frame* if it is valid in every model based on that frame, and *valid* if it is valid in every symmetry frame.

Recall that a group of permutations of a set of indices I partitions the indices into cells where any index in a given cell can be mapped to any other index in that cell via some permutation in the group. These cells are called the *orbits* of the group. Any orbit of a group is fixed by every element of that group, so if the group in question corresponds to a set of symmetries, orbits of the group correspond to precise propositions. Indeed, the orbits of the symmetries correspond to the *maximally strong consistent* precise propositions, so that a proposition is precise iff it is a disjunction of such orbits. This suggests an equivalent way of describing the above semantics which will be sometimes helpful to bear in mind. Let $Orb(G)$ represent the set of orbits of the group G. (Note that $Orb(G)$ is always a partition of the set I that G is acting on.) Then, we can rewrite our semantics as follows by swapping the semantic clause for $\sharp A$ above for:

$$\llbracket \sharp A \rrbracket = \{i \in I \mid \llbracket A \rrbracket = \bigcup X, \text{for some } X \subseteq Orb(G(i))\}$$

That this clause is equivalent to the original is immediate from the above observations. The fact that they are equivalent means that any two symmetry models that agree about the sentence letters and have the same orbit structure agree about the value of every sentence.

PROPOSITION 1: Suppose that $\mathcal{M} = (I, G, \llbracket \cdot \rrbracket)$ and $\mathcal{N} = (I, H, \llbracket \cdot \rrbracket')$ are two symmetry models based on the same set of indices and the same valuation, and

$Orb(G(i)) = Orb(H(i))$ for every $i \in I$. Then $[\![A]\!] = [\![A]\!]'$ holds for every formula A if the identity holds for every sentence letter.

This fact is useful because it means that, for most practical purposes, all we have to care about is the orbit structure: if we know what the orbits of $G(i)$ are for each i, then we know what the semantic value of every sentence is, even if we do not know what the groups $G(i)$ are.

Note that for any possible orbit structure E—a partition of the set of indices—there is at least one group of automorphisms, G, that has that orbit structure: the group of *all* automorphisms that fix the elements of that partition. In fact, this is always the largest group of automorphisms that has that orbit structure: if H also has the orbit structure E, i.e. $Orb(H) = E$, then $H \subseteq G$. This leads us to the following additional proposition:

PROPOSITION 2: Let E be a function that maps each index in a set I to a partition of I. Then there is at least one symmetry frame (I, G) with the property that $Orb(G(i)) = E(i)$, given by $G(i) = \{\sigma \mid \sigma$ fixes every element of $E(i)\}$.

If we combine the above two propositions, we can see that in order to find a model of a formula, or a set of formulae, it suffices merely to say what the orbit structure of such a model has to look like. For, by the first proposition, once we know the orbit structure, then we know the value of every sentence, and by the second proposition, we know that there is at least one symmetry frame that has that orbit structure.

We can axiomatize the resulting logic as follows:

RE If $\vdash \phi \leftrightarrow \psi$ then $\vdash \sharp\phi \leftrightarrow \sharp\psi$

T $\sharp\top$

C $\sharp\phi \wedge \sharp\psi \rightarrow \sharp(\phi \wedge \psi)$

D $\sharp\phi \wedge \sharp\psi \rightarrow \sharp(\phi \vee \psi)$

I $\sharp\phi \leftrightarrow \sharp\neg\phi$

Intuitively, **RE** says that logical equivalence preserves precision, and the remaining axioms say that precision is preserved under the finitary Boolean operations (our language is not expressive enough to express the infinitary Boolean operations).[13]

13.4.1 Higher-order vagueness

I have argued that there is higher-order vagueness: there are some propositions that lie on the borderline between being vague and precise. We therefore do not wish to have the theorem $\sharp\sharp A$.[14]

---

[13] Soundness of the logic can be seen by inspection. I have not checked completeness, although that looks to be a fairly routine affair.

[14] Of course, we do not get the **S4** and **S5** axioms either: $\sharp A \rightarrow \sharp\sharp A$ or $\neg\sharp A \rightarrow \sharp\neg\sharp A$. However, these axioms are less natural for operators, like precision, that correspond to 'whether' clauses instead of 'that' clauses.

In the symmetry framework, this kind of phenomenon manifests itself in vagueness concerning which automorphisms of logical space are symmetries. The notion of a conceptually coherent ur-prior and utility, as we introduced them, are vague, so it stands to reason that it can be borderline whether an automorphism preserves all conceptually coherent ur-priors and utilities in the relevant way. To model this, you will have noticed, we allowed the group of automorphisms to depend on the index. $G(i)$ represents the set of all automorphisms that count as symmetries according to i: if it's vague which automorphisms the symmetries are, then there ought to be at least two indices that disagree about the relevant symmetries.

Recall that even if $G(i)$ and $G(j)$ are distinct, it doesn't necessarily follow that they generate different orbit structures. Since two indices do not disagree about which propositions are precise unless they disagree about the orbit structures, it doesn't quite follow that we have higher-order vagueness. However, since we have also placed very few constraints on the function G, we will in general be able to find models where the orbit structure differs between indices. Using this freedom, we can construct models that invalidate $\sharp\sharp A$.

By appealing to the two propositions proved in section 13.4, it suffices to say what the orbit structure of such a model would have to look like. For example, a simple finite model of higher-order vagueness is given by the following orbit structure: $I = \{i, j\}$ where $Orb(G(i)) = \{I\}$ and $Orb(G(j)) = \{\{i\}, \{j\}\}$. See Figure 13.1.

If the sentence letter A is true only at j, so that $\llbracket A \rrbracket = \{j\}$, then $\llbracket \sharp A \rrbracket = \{j\}$ and so $\llbracket \sharp\sharp A \rrbracket = \{j\}$, and so on. It follows that A is precise at all orders at j. It is also true that the following sequence of identities holds $\llbracket \neg\sharp A \rrbracket = \{i\}$, $\llbracket \neg\sharp\neg\sharp A \rrbracket = \{i\}$, and so on, so that A is vague at all orders relative to i: A is vague, the proposition that A is vague is vague, and so on. The preceding model thus not only demonstrates the existence of higher-order vagueness up to some level, it also demonstrates the existence of vagueness at all orders; an arguably desirable feature that many formal approaches to vagueness fail to secure.

PROPOSITION 3: There are symmetry models $(I, G, \llbracket \cdot \rrbracket)$ that accommodate vagueness at all orders: there are indices at which $\neg\sharp A$ is true, $\neg\sharp\neg\sharp A$ is true, $\neg\sharp\neg\sharp\neg\sharp A$ is true, and so on.

It is worth contrasting the picture we get with the supervaluational framework. One striking difference is that the supervaluational semantics drew two sorts of partitions of logical space. One corresponding to the maximally strong consistent precise

$Orb(G(i))$ $Orb(G(j))$

Figure 13.1. A simple model of vagueness at all orders.

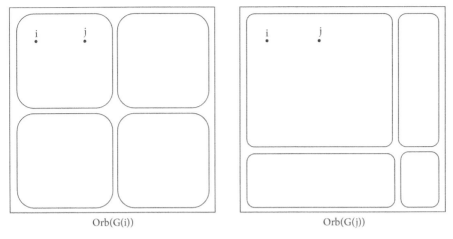

Orb(G(i))   Orb(G(j))

Figure 13.2. According to the indices i and j, logical space is partitioned into four maximally strong consistent precise propositions in two different ways.

propositions and another corresponding to the maximally strong consistent worldly propositions—propositions corresponding to the set of all world–precisification pairs that agree on the world coordinate. Unlike the partition of logical space into the precise propositions, the second partition is determinate and corresponds to the same partition relative to every world–precisification pair. In the present framework, by contrast, the latter division is absent. Like the supervaluationist, we can still make sense of maximally strong consistent precise propositions, but these seem to be a poor substitute for world propositions. It is vague which cells count as the maximally strong consistent precise propositions, and so to identify such propositions with worlds would be to postulate vagueness in the boundaries of the world. Perhaps this is an acceptable substitute for the supervaluationist's notion of a world proposition. However, it is worth emphasizing that it is at odds with the usual way of talking about 'the world' (we shall turn to this matter in more detail in chapter 14).

Pictorially, we can represent this sort of phenomenon as in Figure 13.2. In this example, the two indices i and j both partition logical space into four maximally strong consistent precise propositions, but do so in two different ways. In this case the indices agree about how many maximally strong consistent precise propositions there are, although this is not a feature of all models.

In our semantics, then, a proposition can be precise relative to an index, without the proposition that it's precise being precise relative to that index. Say that a proposition is precise^2 if it is both precise and the claim that it is precise is precise, formally: $\sharp^2 A := \sharp A \wedge \sharp\sharp A$. It is clear that this definition can be iterated, a proposition is precise^{n+1} if and only if it is precise^n and the proposition that it is precise^n is precise.

It is natural to wonder whether precision^2 (or the higher notions of precision) behaves like precision. In particular, does precision^2 satisfy axioms T, C, D, and I stated for precision: are arbitrary Boolean combinations of precise[2] propositions precise[2]? Surprisingly, the answer is no. We cannot prove these facts from the axiomatization above. There are simple finite countermodels.[15] One can see this more intuitively by considering other interpretations of ♯ that appear to satisfy our axiomatization of precision. Pretend that you are a logical mastermind and have the same opinion about logical equivalents. Then, it follows that 'you have an opinion about whether' satisfies our axioms. If you have an opinion about whether p and an opinion about whether q, then you have an opinion about whether $\neg p$, about whether $p \vee q$, and about whether $p \wedge q$. Notice, however, that if you have an opinion about whether p and about whether q, and also an opinion about whether you have an opinion about whether p, and an opinion about whether you have an opinion about whether q, it doesn't seem to follow that you have an opinion about whether you have an opinion about whether $p \vee q$. Your opinion about your opinion about p and about q may (incorrectly) be that you have no opinion about p and no opinion about q. So, it's logically open whether you have an opinion about $p \vee q$: if you have no opinion about whether the die will land on a 1 and no opinion about whether it will land on a 2, you could still happen to think it will land on a 1 or a 2, and you could also fail to think that it will land on a 1 or a 2. Since it's logically open whether you have an opinion about $p \vee q$, you could be agnostic about this, and thus have no opinion about whether you have an opinion about $p \vee q$.

It is worth contrasting our account of precision with the identification of precision with not being borderline as a matter of metaphysical necessity. On that account, we can define precision[2] as we introduced it above. Assuming metaphysical necessity obeys **S5** and determinacy obeys **KT**, one can prove that the precise[2] propositions are closed under finite conjunctions, disjunctions, and negations. The above observation about the behaviour of precision[2] therefore highlights one important difference between this account of precision and the view that takes it as a primitive notion.

Let us expand on one particular upshot of this result. For those who take the determinacy operator to be the primitive notion, and who takes it to be governed by a logic of **KT**, there is very little to distinguish determinacy from determinacy[2] (i.e. being determinately determinate). If the logic of determinacy is **KT** then so is the logic of determinacy[2] and, moreover, if we define precision in terms of necessity and determinacy, the logic of precision is the same as the logic of precision[2]. This problem is particularly acute for those theorists, considered in chapter 7, who additionally maintain that to assert or know a proposition it must be determinate at all orders. For these theorists, there is not even a difference between determinacy and determinacy[2] concerning the role they play with respect to knowledge and assertion. This is

---

[15] All one needs is to pick any function G which has the following orbit structure: $Orb(G(i)) = \{\{i\}, \{j\}, \{k, l\}\}$, $Orb(G(j)) = \{\{i\}, \{j\}, \{k, l\}\}$, $Orb(G(k)) = \{\{i\}, \{j, k, l\}\}$, $Orb(G(l)) = \{\{i, j, k\}, \{l\}\}$. Let $p = \{k, l\}$ and $q = \{j\}$. Then, at i the following is true: $♯p, ♯q, ♯♯p, ♯♯q, ♯(p \vee q), \neg♯♯(p \vee q)$. It is always possible to find a collection of groups with this orbit structure by proposition 1 and 2 from section 13.4.

puzzling, for it might lead one to wonder why philosophers have devoted so much time to the determinacy and borderlineness operators, when there are other operators that could explain the phenomena associated with vagueness equally well, such as second-order determinacy and the associated notion of borderlineness.

While it is true, on the present approach, that the logic of determinacy (as defined from precision) will be the same as the logic of determinacy^2, it is not the case that the logic of precision is the same as the logic of precision^2, as observed above. Indeed, the notion of a proposition p being both precise and being such that the proposition that it is precise is itself precise is not a theoretically central notion, since it is not formally well behaved—this property is not something that is transferred to disjunctions and conjunctions of propositions with that property. Thus, no parallel objection can be levelled at the present theory: precision is uniquely distinguished from its iterations by its logical role.

13.4.2 Determinacy and necessity

I noted that, given the ability to quantify into sentence position and to talk about propositional entailment, we can define the determinacy operator from the precision operator: a proposition is determinate if it is entailed by some precise truth. It is possible, however, to study this determinacy operator in isolation, without assuming that we have either precision or propositional entailment in the language. To achieve this, we shall expand our propositional language to include a determinacy operator Δ. Call the resulting language $\mathcal{L}(\sharp, \Delta)$ and the language with just Δ $\mathcal{L}(\Delta)$.

In the symmetry semantics, the above definition of precision is equivalent to saying that p is determinate iff every admissible automorphism maps p to a truth. It follows that we can interpret the determinacy operator in a symmetry frame by adding the following clause for Δ

$$[\![\Delta A]\!] = \{i \mid i \in \sigma([\![A]\!]) \text{ for every } \sigma \in G(i)\}$$

Roughly, the semantic value of ΔA is the set of indices which belong to the image A under the class of symmetries at that index.

Notice, however, that this clause can be reformulated so as to conform to a more familiar semantics for modal operators in terms of accessibility relations. Consider the following relation between indices:

$$Sij \text{ if and only if } \sigma(j) = i \text{ for some } \sigma \in G(i).$$

The pair (I, S) forms a Kripke frame in which, given an interpretation of the atomic letters a sentence, ΔA, is true at an index i iff it is true at every j S-accessible to i. Expanding our definition of S in terms of G, this amounts to saying that $\sigma^{-1}(i) \in [\![A]\!]$ for every $\sigma \in G(i)$, and this holds iff $i \in \sigma([\![A]\!])$ for every $\sigma \in G(i)$, which is exactly the clause for ΔA in the symmetry semantics.

Given standard results about Kripke frames, we can now easily investigate the logic of Δ in the symmetry semantics. First, S is reflexive: since $id \in G(i)$, it follows that there is always some element of $G(i)$ which maps i to itself. Furthermore, given any reflexive relation S let $G(i)$ be the set of permutations that permute members of $S(i)$, the set of indices accessible to i, and map indices not belonging to $S(i)$ to themselves: then Sij iff $\sigma(j) = i$ for some $\sigma \in G$, and G forms a group. It follows that every reflexive relation is generated by some indexed collection of groups and so sentences of $\mathcal{L}(\Delta)$ that are true in all symmetry frames are exactly those that are true in all reflexive Kripke frames; the latter we already know to be exactly the theorems of the normal modal system KT. (A full investigation of the logic of the language $\mathcal{L}(\sharp, \Delta)$ containing *both* precision and determinacy would take us too far afield; in addition to the rule RE and axioms TCDI for $\sharp$ and the axioms of KT for Δ, we'd need to consider interaction principles, such as $\sharp A \rightarrow (\Delta A \vee \Delta \neg A)$.)

It is natural also to consider adding a necessity operator, $\square$, to our language: resulting in the language $\mathcal{L}(\sharp, \square)$. We shall need this sort of framework in chapter 15, when we consider the interaction of vagueness and modality, but we have already come across one principle that requires metaphysical necessity: NECESSITY OF THE PRECISE, which formally amounts to the axiom:

$$\sharp A \rightarrow \square \sharp A.$$

To model necessity using symmetry semantics we need to augment it with an additional accessibility relation, R, representing the modal accessibility relation. A frame in this setting is now a triple (I, G, R), and the clause for $\square$ is given as below:

$$[\![\square A]\!] = \{i \mid j \in [\![A]\!] \text{ for every } j \text{ such that } Rij\}.$$

Since the indices belonging to I are in general more fine-grained than possible worlds, the modal accessibility relation is not usually the universal relation on I. If we are to ensure a modal logic of S5, we require that R be an equivalence relation.

Most importantly, to ensure the axiom $\sharp A \rightarrow \square \sharp A$ we impose the following constraint on a frame (I, G, R):

NECESSITY OF THE PRECISE: If Rij then $Orb(G(i)) = Orb(G(j))$.

This concludes our discussion of the logical aspects of the symmetry semantics. We will return to the interaction between precision, determinacy, and necessity in chapter 15.

What have we achieved? In section 13.2, I showed that given the present theory of vagueness, the set of propositions admits certain symmetries: mappings from propositions to propositions that preserve logical structure, and preserve all conceptually coherent preferences and probability assignments. I noted that every precise proposition is mapped to itself by every symmetry, and, moreover, that every proposition mapped to itself by every symmetry is precise. In section 13.3, I leveraged this fact to provide an abstract analysis of a precise proposition as a proposition fixed by every symmetry. I noted, moreover, that the thesis that a proposition is precise iff it's fixed by every symmetry is quite powerful: it entails **Boolean Precision**,

Indifference, and **Rational Supervenience**, allowing us to fully encode the theory of vagueness by the conjunction of this thesis with **Plenitude**. Finally, in section 13.4, I developed a semantics for a formal language involving precision operators, and briefly investigated the structure of higher-order vagueness and the interaction with metaphysical necessity in that setting. Next we shall investigate a couple of issues that arise in the context of supervaluational semantics, drawing out some differences between supervaluational and symmetry semantics.

14

Vagueness and the World

Despite the pervasiveness of vagueness, it is natural to find some version of the following schematic thought extremely compelling: that there is some definite collection of completely *factual* propositions upon which all truths supervene which are unaffected by vagueness. Supposing that one has the notion of a proposition being 'factual' in this sense, we can spell out what it means for it to consist of a 'definite collection of propositions unaffected by vagueness' a bit more precisely:

[F1]   Every factual proposition is either determinately true or determinately false.

[F2]   If a proposition is factual, then it is determinately factual.

[F3]   If a proposition is not factual it is determinately not factual.

Although I am sympathetic with those who are sceptical of this notion (see Dorr [33], Dreier [36], Field [50]), in this chapter I am interested in exploring views that theorize in terms of the notion of being factual and take it to be integral to the explication of vagueness.[1] The distinction employed by such philosophers abides by F1–F3: factual propositions are always determinately true or false and it is not a vague matter which propositions are the factual ones.

It should be noted that for F2 and F3 to be remotely plausible, one cannot simply take 'factual' to be a synonym for 'precise'. Due to the existence of higher-order vagueness, it is vague which propositions are the precise ones, and the analogues of F2 and F3 for precision are false. Factual propositions are supposed to be *rock bottom*—perhaps they are just the propositions about the values and properties of fundamental particles and fields at each space–time point (for example). Certainly every factual proposition is precise, but there can be some precise but higher-order vague propositions that are not rock bottom and thus not completely factual. Another candidate synonym is the notion of a *fundamental proposition*, for I think it is clear that only precise propositions are fundamental, and it is a precise matter which propositions are the fundamental.

What might some plausible candidates for these factual or fundamental propositions be? We have already ruled out the precise propositions due to considerations of

---

[1] See e.g. Fine [56], Fine [57], Russell [123].

higher-order vagueness. For similar reasons, the propositions which are precise at all orders will not do either, since, as we saw in chapter 13, it is also a vague matter which propositions are precise at all orders (see section 13.4.1).

What about the propositions of physics: propositions specifying the locations of fundamental particles, their charges, spin, and other fundamental properties? Let us suppose, for the sake of argument, that it is a completely precise matter which propositions are the physical propositions, and let us also assume physicalism. Even under these assumptions, the physical propositions will not serve our purpose once we have accepted the idea, defended in chapter 12, that the propositions of physics are not precise (although they are plausibly necessarily determinately true or determinately false).

Although the above considerations indicate that the idea of some collection of basic propositions of this sort is not a forgone conclusion, it is arguably implicit in a certain way of making use of the notion of a possible world. For example, on the face of it, the supervaluationist formalism, in which the truth of a formula is evaluated relative to a possible world and a precisification, appears to employ worlds in this way. The formalism suggests a clean separation between the kind of factors that cause variation in truth value that depend on facts to be found 'in the world' (things like height and hair number) and variation of truth value that is due to the arbitrary locations of cutoff points—differences in truth value that are merely a result of how we settle the cutoff points for things like tallness and baldness. In such a formalism, one can identify a vague proposition with a set of world–precisification pairs, and from among those one can distinguish a special class of 'worldly' propositions—sets which contain all ordered pairs involving a given world if they contain any ordered pairs involving that world. The semantics straightforwardly validates the claim that such propositions are precise and that it is a precise matter which propositions are worldly in this sense.

Despite the pervasiveness of possible worlds talk in philosophy, one might wonder what the alternative might look like. That is, one might wonder what it would be like for there to be vagueness all the way down: for there to be no definite collections of precise propositions on which all truths supervene or, perhaps, even, no definite collections of precise propositions whatsoever. According to this alternative, one can approximate the ordinary notion of a possible world by talking about propositions that settle all precise matters—that divide up logical space into cells of maximally strong consistent precise propositions—but unlike the usual way of thinking about worlds, it will be vague where these lines in logical space lie, for it is vague which propositions are precise.

In this chapter, I'll look at some reasons, relating especially to the paradoxes of higher-order vagueness, to reject the idea that vagueness bottoms out in this way. In doing so, I'll sketch an alternative to the supervaluationist picture, based on the symmetry analysis of precision, that eschews the usual notion of a possible world altogether.

14.1 Factual Propositions and Supervaluationism

The idea that some propositions are factual ('metaphysically first-rate', 'grounded in reality/the world', etc.) while others are not is an old one. In the context of vagueness, the view is most commonly attributed to some (but certainly not all) philosophers associated with the supervaluationist formalism.[2] According to that paradigm, vague propositions are not factual and when a proposition is borderline they say that there is no fact of the matter about whether it is true. (Indeed, some even see this to be the distinguishing feature of supervaluationism that prevents it from collapsing into a form of epistemicism.) The converse is not true, however: not all precise propositions are factual, because one can have precise propositions that are nonetheless higher-order vague.

The distinction is supposed to be a metaphysical one. By contrast, one might try to elucidate a difference between factual and non-factual propositions by looking at their role in thought—how the different kinds of propositions behave as the objects of belief, desire, and knowledge (see chapters 6–10, Field [53], and Schiffer [126]). A theory that elucidates the difference purely in terms of how we think with vague propositions is, I think, at odds with the dominant understanding of the distinction. For example, Barnett [11] writes that 'on the dominant view of vagueness [i.e. Fine-style supervaluationism], if it is vague whether Harry is bald, then it is unsettled, not merely epistemically, but metaphysically, whether Harry is bald'—it does not seem as though this metaphysical unsettledness could be explained purely in terms of how we think about things.

A natural place to start our investigation is to look at the way this notion gets formally cashed out in a supervaluationist framework. Recall that a vague proposition gets modelled within this formalism by a set of world–precisification pairs. Propositions are thus more fine-grained than sets of worlds. However, each set of possible worlds corresponds to a special kind of set of world–precisification pairs: the set of all possible world–precisification pairs whose world coordinate belongs to our original set of worlds. Intuitively these are propositions whose truth value can vary only through shifts in the world coordinate—no amount of shifting the precisification (i.e. shifting the boundaries of the vague predicates) will induce a change in truth value. Thus, in the supervaluationist setting a vague proposition is *factual* if and only if it corresponds to a set of possible worlds.

It is clear that in this framework the distinction is intimately tied to the notion of a possible world. In order to find out whether a factual proposition is true or not,

---

[2] This idea comes out especially clearly in the work of Kit Fine (Fine [56], Fine [57]). By contrast, supervaluationism is also employed by those who maintain that vagueness amounts to semantic indecision (see, for example, Keefe [78]). According to the latter view, all propositions are precise—a sentence is vague in virtue of the kind of relation it stands in to the candidate precise propositions it could potentially express. The existence of vagueness, according to this second kind of a view, lends no special weight to the idea that some propositions are metaphysically first-rate and others are not.

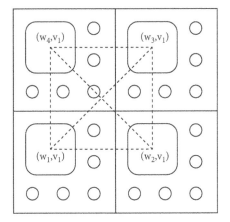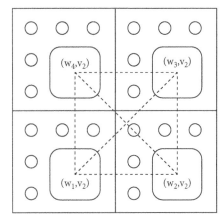

Figure 14.1. The space divided into four world propositions. The two diagrams represent two divisions into precise propositions depending on the precisifications v_1 and v_2.

one only needs to know which possible world is actualized—it doesn't matter which precisification accurately describes the distribution of cutoff points, since the truth of a worldly proposition won't depend on which precisification is accurate. The worldly facts are thus all and only those whose truth is determined by the world.

Recall also that due to higher-order vagueness, the factual/non-factual distinction is not the same as the distinction between precise and vague. This difference is made quite clear in the supervaluationist semantics: in Figure 14.1, the maximally strong consistent precise propositions are represented by the four rounded boxes and a collection of degenerate cells. By contrast, the worldly/non-worldly distinction defined in the previous paragraph corresponds to the four quadrants of the diagram in which the maximally strong consistent precise propositions reside. Thus, typically, the maximally strong consistent precise propositions are stronger than the maximally strong consistent worldly propositions. Moreover, notice that while the divisions between the precise and vague propositions can themselves be vague and may vary from precisification to precisification, it is built into the formalism that the boundaries between the worldly and non-worldly propositions remain constant.

It is only in very exceptional circumstances that the two distinctions align and, in the models in which they do align, there is no higher-order vagueness.[3] When explaining the role that possible worlds play in a supervaluationist semantics, it is tempting to fall back on something like the following characterization: 'Each possible world represents a (metaphysically possible) complete totality of precise facts and each metaphysically possible complete totality of precise facts is represented

---

[3] The absence of higher-order vagueness is not sufficient for the two distinctions to align: it could be that some world proposition is divided into two or more equivalence classes under the determinacy–accessibility relation. There would then be more precise propositions than world propositions even though there is no higher-order vagueness.

by some possible world. Precisifications settle the remaining facts that would have been borderline had those precise propositions obtained.' Given what we have said above, this is not correct—possible worlds correspond, rather, to complete totalities of *worldly* propositions.

We can introduce an operator $\mathcal{F}p$, stating that p is factual, into the language of chapter 12 as follows:

($\mathcal{F}$) $w, v \Vdash \mathcal{F}\phi$ iff $\{(x, u) \mid x, u \Vdash \phi\}$ corresponds to a set of possible worlds.

(As above, a set of world–precisifications pairs X *corresponds* to a set of worlds Y iff for all world–precisification pairs (w, u), $(w, u) \in X$ iff $w \in Y$.)

Alternatively, the right hand side amounts to saying that for any $x, y \in W$ and $u \in V$, $x, u \Vdash \phi$ iff $y, u \Vdash \phi$.

The supervaluational way of modelling the difference between factual and non-factual propositions has the distinctive feature that propositions can be modelled as sets of ordered pairs, and the notion of factuality corresponds to being invariant along one of the coordinates in the way described above.

14.2 The Problem of Higher-Order Vagueness

The notion of factuality captured by the supervaluationist semantics described above is subject to a straightforward paradox of higher-order vagueness. Consider the operator Sp that applies to p if it is entailed by some true factual proposition. We can introduce this operator by defining it from the notion of factuality and propositional identity, by employing propositional quantification: $S\phi := \exists p(p \land \mathcal{F}p \land L(p \to \phi))$.[4] Alternatively, we can introduce it directly in the semantics:

(S) $w, v \Vdash S\phi$ iff $w, u \Vdash \phi$ for every $u \in V$.

It is straightforward to check that if S is defined as above, it has the truth conditions described above in models where the propositional quantifiers range over all sets of world–precisification pairs. It is also straightforward to check that $Sp \to \Delta Sp$ and $\neg Sp \to \Delta \neg Sp$ will be true at any world–precisification pair. For, if, at a given world and precisification w, v, Sp is true, then p is true no matter what precisification we substitute for v and, thus, Sp is true relative w, u for any precisification u accessible to v. In other words, it is never borderline whether or not this operator applies or not.

Consider now the following scenario. Suppose that the world is divided into a large but finite number of spatio-temporally disconnected epochs, which we shall number by the order in which they occur. Suppose, moreover, that the nth epoch contains exactly n electrons, but is otherwise completely empty. Now, presumably, if

---

[4] Recalling that Lp is defined as $p = \top$.

any propositions count as factual at all, the proposition that there are no electrons in the 0th epoch is a factual proposition. Moreover, the proposition that there are no electrons in the 0th epoch surely entails that there is a small number of electrons in the 0th epoch. So we are committed to the following:

> Some true factual proposition (the proposition that there are no electrons in the 0th epoch) entails that there is a small number of electrons in the 0th epoch.

By contrast:

> No true factual proposition entails that there is a small number of electrons in the 1,000,000,000th epoch.

This follows because there simply isn't a small number of electrons in that epoch, and no truth can entail a falsehood.

But now, by completely familiar sorites reasoning, we can infer that there is a last number, N, such that some true factual proposition entails that there is a small number of electrons in the Nth epoch. On its own, this is not too surprising: the ordinary sorites has taught us to live with cutoffs like these. Although the sorites paradox forces one to accept the existence of a cutoff point, we ought to deny the existence of a *determinate* cutoff point. In symbols, this amounts to accepting $\exists x(Fx \wedge \neg Fx + 1)$ but denying $\exists x \triangle (Fx \wedge \neg Fx + 1)$—although there is a last epoch that contains a settled-to-be-small number of electrons, no epoch is determinately the last such epoch: it is vague where the cutoff point lies.

However, we noted above that according to the supervaluationist semantics it is always a determinate matter whether a proposition is entailed by a true factual proposition, and so the property of 'being an epoch in which it is settled that it contains a small number of electrons' cannot have any borderline cases. Thus, unlike the ordinary sorites paradox, it is a completely determinate matter which this final epoch is.

The existence of a determinate cutoff point is far worse than the existence of a cutoff point: if one accepts the existence of, say, a last bald person in a sorites, one can still maintain that it is borderline, and thus unknowable, which person that was. The conclusion that it is not borderline which epoch is settled-by-the-facts to be the last one containing a small number of electrons is, by contrast, paradoxical. For, without vagueness what barrier is there to finding out where that boundary is? The boundary between the epochs with a settled-to-be-small number of electrons is no more detectable than the boundary between the epochs with a small number of electrons: no amount of studying the epochs, or reflection about the nature of smallness, electrons, or factualness will allow us to figure out the locations of either boundary. Since both distinctions bear all the hallmarks of vagueness, it is extremely natural to want to associate both boundaries with vagueness—anyone who denies it in the former case owes us some explanation of the fact that we cannot discover that boundary.

14.2.1 Factual propositions

The puzzle, as I've presented it above, relies essentially on a particular way of cashing out the notion of a proposition being factual that used the supervaluationist formalism. However, we do not have any general argument that one cannot come up with an alternative, non-supervaluationist, theory of factuality that does the theoretical work that is needed of it.

In this section we show that it is possible to generalize the problem so that it applies to any theory of factuality that satisfies the constraints F1–F3:

F1   Every factual proposition is either determinately true or determinately false.

F2   If a proposition is factual, then it is determinately factual.

F3   If a proposition is not factual it is determinately not factual.

Recall that these principles capture the idea that factual propositions are not affected by vagueness: factual propositions are never borderline, and it is never borderline whether a proposition is factual.

The crucial feature of the supervaluationist's semantics was that we could show that if p was entailed by a factual truth, then it was determinately entailed by a factual truth, and if it wasn't entailed by a factual truth, it determinately wasn't. For, given those two features, we could derive the existence of a determinate cutoff point from the existence of a mere cutoff point.

But we can establish both of these features from F1–F3, given some very natural background assumptions. In what follows, the notion of entailment is defined in terms of propositional identity: P entails A iff $P \wedge A = P$.[5]

The argument has two parts. We must first show that if no factual truth entails A, it's definite that no factual truth entails A. Suppose, for contradiction:

1. No factual truth entails A.
2. It's not definitely not the case that some factual truth entails A.

Note that given the propositional Barcan formula, which captures the idea that the existence of propositions is a determinate matter, the second claim entails that there's some proposition, call it P, such that it's not definitely not a factual truth that entails A.[6]

3. If P is not factual, P is determinately not factual. (From F3.)
4. So P is factual. (By 2 and 3.)

---

[5] If one rejects Booleanism—the view that propositions form a Boolean algebra under entailment—this definition may not correspond to the intuitive notion of entailment.

[6] The Barcan formula for propositional quantification and determinacy states: $\forall p \Delta \phi \leftrightarrow \Delta \forall p \phi$. Here, we are contraposing and applying the duality of $\exists$ and $\forall$ to infer that if it's not determinately not the case that some P has a certain feature, then some proposition doesn't determinately lack that feature.

5. If not P, then determinately not P, contradicting 2. (Since by assumption P is factual and factual propositions are determinately true or false.)
6. So P.
7. $P \wedge A$ entails A, so $P \wedge A$ is either not true or isn't factual.
8. But if $P \wedge A$ is factual, then it's not true and, since it's factual, it's determinately not true (contradicting 2).
9. So $P \wedge A$ is not factual, and is thus determinately not factual (by F3).
10. Since P is determinately factual, $P \wedge A$ and P are determinately distinct.
11. So, determinately, P does not entail A, contradicting 2. (P entails A iff $P \wedge A = P$).

Note 10 is derived without assuming the determinacy of propositional distinctness.[7] We now turn to showing that if some factual truth entails A, it's definite that some factual truth entails A. Suppose for contradiction:

1. Some factual truth entails A.
2. It's not determinate that some factual truth entails A.

Letting P be the factual truth that entails A, we can reason as follows:

3. Since P entails A, it is determinate that P entails A. (As argued in section 11.1.)[8]
4. Since P is factual, it's determinate that P is factual.
5. Since P is true and factual, it is determinately true.
6. So determinately, some factual truth (namely P) entails A.

Thus, we have shown from F1–F3 that it is always determinate whether a proposition is entailed by the facts.

Let us finally add the premise that the facts entail that there are a small number of electrons in the 0th epoch:

F4   Some factual truth entails that there are a small number of electrons in the 0th epoch.

Then we can reason as in section 14.2. Since there isn't a small number of electrons in the 1,000,000,000th epoch, no truth—and a fortiori no factual truth—entails that there is. So there's a last epoch such that the facts entail that there is a small number of electrons contained in it. And since it is always determinate whether the facts settle something, there is a determinate fact about which epoch that is.

---

[7] Since propositions of the form LA can be borderline (see Bacon [3]), and since two propositions are identical iff they are L-equivalent, it follows that there can be borderline propositional identities.

[8] By a straightforward variant of Kripke's argument for the necessity of identity, it follows that all true identities are determinate (from $P = Q$ and the obvious truth $\Delta(P = P)$ it follows, by Leibniz's law, that $\Delta(P = Q)$). Since 'P entails A' is analysed as an identity, it follows that whenever something entails something else, it does so determinately.

A few remarks are in order about the notion of entailments employed in this argument. The proposition p entails q iff the conjunction of p with q is identical to p. First, note that for this definition to correspond to a plausible notion of entailment, the assumption of Booleanism, or something quite like it, must be in place. On a structured theory of propositions, for instance, identities like this would never be true and there would be no entailments by this account of entailment. Secondly, this relation of entailment is importantly stronger than a similar one defined in terms of metaphysical necessity. p necessitates q if the material conditional $p \rightarrow q$ is metaphysically necessary: $\Box(p \rightarrow q)$. Given Booleanism $(p \wedge q) = p$ can be written in a similar form, since it is equivalent to $(p \rightarrow q) = \top$, which, given our definition of L, amounts to $L(p \rightarrow q)$. As we noted in chapter 11, L is the broadest kind of necessity: for any normal operator, O, Lp implies Op. So, in particular, being L-necessary implies being determinately true, whereas being metaphysically necessary implies no such thing: whether a number is small or not is always a metaphysically necessary matter, but not always a determinate matter. Recall also that L-necessity implies determinate determinate truth, determinate determinate determinate truth, and so on, so that if a proposition entails another, the corresponding material conditional must be determinate at all orders.

It is important to bear this point in mind when evaluating F4. The claim that the proposition that there are a small number of electrons in the 0th epoch is necessitated by some factual truth follows from the plausible assumption that everything supervenes on the factual. The claim that this proposition is *entailed* by some factual truth is a more demanding condition. However, even once this distinction is recognized, it is hard to see why the stronger thing shouldn't be true either. We can reason this out in two steps. Firstly, the proposition that there are no electrons in the 0th epoch is surely a factual proposition: indeed, all of the words I used to express this proposition can be assumed to be expressing fundamental properties or properties that can easily be defined from fundamental properties ('there are no', '0th epoch', 'electron', etc.).[9] Secondly, the proposition that if there are no electrons in the 0th epoch, there are a small number of electrons in the 0th epoch seems to be a conceptual truth. They are on a par with things like 'scarlet things are red', 'having no money makes one poor', and so on: they are as clear examples of L-truths as we'll ever find. So it appears as though we can give a constructive argument that some factual proposition entails that there are a small number of electrons in the 0th epoch.

It seems clear that if there are any entailment facts between the factual and non-factual at all, then the entailment above is as good a candidate as any. One could, however, resist F4 in a principled way by denying that there are ever entailments

---

[9] There is tension here with my contention, made in section 12.2, that the propositions of physics are not even precise. But the assumption is dialectically effective here, since we are considering views that assume the existence of a layer of fundamental or factual propositions, which presumably include the propositions expressed by sentences in the language of fundamental physics.

between the factual and non-factual. This manoeuvre is not without its costs, however. For, given Booleanism and the plausible assumption that factual propositions are closed under Boolean operations, it follows that the disjunction of any factual proposition with an incompatible non-factual proposition will be non-factual.[10] But, given classical logic, it follows that this non-factual disjunction is entailed by the factual disjunct. So, to maintain this radical form of the independence of the factual from the non-factual, one has to either give up Booleanism or deny that factual propositions are closed under Boolean operations.

14.2.2 Fundamental propositions

Although I have targeted the ideology of a proposition being factual, or 'grounded in reality', the above argument generalizes to a great number of related distinctions. Many philosophers are quite happy to employ the closely related notion of a property or proposition being *fundamental*. This latter distinction is employed widely, even by those who eschew words like 'facts', 'grounded in reality', and so on.[11]

The above argument can also be deployed to make trouble for the notion of a fundamental proposition, since fundamental propositions plausibly are governed by analogues of F1–F3. It is clear that fundamental propositions are never vague, for they concern truths about the locations of electrons, the topology of space–time regions, and so on. On the usual understanding of the notion of a fundamental proposition, it is a completely precise matter which propositions are the fundamental propositions. Questions concerning whether a proposition describes the fundamental structure of the universe or not are not like questions that concern whether a person is bald: the former questions seem always to have determinate answers. Moreover, for the same reasons we gave in section 14.2.1, some fundamental proposition entails that there are a small number of electrons in the 0th epoch. Thus, we have secured F1–F4. We can reason that there is a last epoch which the fundamental truths settle to contain a small number of electrons, and that it is a completely determinate matter which this final epoch is.

Thus, I think, a similar conclusion can be drawn about the notion of 'fundamentality', at least as it applies to propositions: it is not in good standing.

14.3 Vagueness All the Way Down

A couple of points are worth bearing in mind when evaluating these arguments. Firstly, although they purport to show that the notion of a fundamental proposition (and cognate notions) are not in good standing, there is a great number of useful

---

[10] Suppose that p is factual, and $p \vee q$ is factual, and $p \wedge q = \bot$. Then $\neg p$ is factual and thus $\neg p \wedge (p \vee q)$ is factual, by the closure of factualness under Boolean operations. Given Booleanism and the fact that $p \wedge q = \bot$, $\neg p \wedge (p \vee q)$ is identical to q, so q is factual. So if p and q are incompatible and p is factual and q is non-factual then their disjunction is also non-factual.

[11] See, for example, Sider's notion of being *structural* in Sider [133].

related concepts that are not affected by the above considerations. For example, the ideology of 'fundamentality' is as often applied to language as it is to propositions (for example, when we talk about sentences formulated in the fundamental vocabulary of physics). Clearly, nothing I have said above suggests that there is anything problematic with the distinction between fundamental and non-fundamental language. Moreover, there is also a comparative notion of fundamentality that is not targeted by the above considerations. For all I've said, we can still talk about some propositions being more fundamental than others.

But as for the notion of a fundamental proposition (or, indeed, a factual one), it needs to be revised in light of the present theory of vagueness. One option is to reject the notion altogether. I have effectively argued that the very idea of a basic layer of facts completely unaffected by vagueness is not sustainable; yet this seems to be exactly what the idea of a fundamental proposition presupposes. Alternatively, one could attempt to revise the notion of fundamentality in such a way that it violates one of our premises, F1–F4. I take it that allowing some fundamental propositions to be vague would do too much violence to the notion. But, perhaps, one could employ a revised notion with the feature that it is vague which propositions count as the fundamental ones.

I am not entirely sure which of these two lines to take—rejection or revision—but I think that whichever line we take, it is clear the resulting picture is quite at odds with the dominant way of thinking about fundamental structure. The alternative might best be described as a view in which there is 'vagueness all the way down': a view in which truths don't bottom out in some determinate, basic layer of precise facts. We have seen this sort of idea in action already. The idea that there is such a basic layer of truths might seem inevitable given the fact that we seem to be able to readily come up with specific examples of such truths: the proposition that there are electrons, that gravity attracts, or what have you. But as we noted in chapter 12, these propositions are plausibly vague. It is epistemically open whether claims about electrons and gravity really are rock bottom, or whether they are just as emergent and subject to vagueness as propositions about tables and chairs. Once this conclusion is properly taken to heart, I think much of the motivation for postulating the notion of a fundamental proposition is undermined.

Let us now examine another notion that seems to be implicated in the paradoxes of higher-order vagueness: the notion of a possible world. We have already encountered several views that place special theoretical significance on the notion of a possible world. Both the supervaluationist described in section 12.1.2, and the expressivist we encountered in chapter 8 agree that, although in general propositions are more fine-grained than sets of possible worlds, there are special sorts of propositions—ones that are somehow 'metaphysically first-rate'—that correspond to a set of possible worlds. Similar ideas are frequently applied outside the literature on vagueness as well, for example in the context of moral expressivism (see e.g. Gibbard [64]), or epistemic modals (see e.g. Moss [109]). As we saw earlier, the idea that some propositions correspond to sets of possible worlds can be seen as one way of formally modelling the distinction between factual and non-factual propositions.

However, insofar as the notion of a possible world allows us to introduce a notion of propositional factuality, governed by principles F1–F4, the notion of a possible world is not in good standing either. Suppose that we have a primitive notion of a proposition *corresponding* to a set of possible worlds, allowing that some vague propositions may not correspond to any set of worlds. We can then say that a proposition is factual if it corresponds to some set of possible worlds. If it is always a precise matter whether a proposition corresponds to a set of worlds or not, and a precise matter which entities are the possible worlds, we can show F2 and F3 for our defined notion of factuality. If we also assume, quite plausibly, that only precise propositions can correspond to sets of worlds, we can also show F1 and F4 can be given a justification similar to the one we gave earlier.

The rejection of possible worlds is not something that should be taken lightly. After all, the use of possible worlds is ubiquitous in philosophy: how are we to reconcile our picture with the large body of valuable philosophical work that presupposes the notion of a possible world? In addressing this issue we must be careful to distinguish the characteristics of possible worlds that are essential to the theoretical work that involves them, from some of the superfluous metaphysical theses that are often associated with them.

Part of the issue is that the word 'world', as it appears in the philosophical literature, is actually ambiguous between at least two different notions: what I'll call the *logical notion of a world* and the *metaphysical notion of a world*. The first is the logician's concept of a world, which first made its appearance in an abstract kind of semantics for modelling a class of operators governed by principles of a normal modal logic— the Kripke semantics. The logical notion of a world does not carry commitment to anything like the notion of factuality discussed above. Indeed, for logical purposes, more neutral terminology is often employed in its stead such as *indices*, *states*, or *points*. In practice, the logical indices may not be metaphysically possible worlds, but world–time pairs, or world–precisification pairs, or perhaps things that can't be analysed in terms of pairing an (ordinary) world with some other entity (as on the present view). Note that if we are interested in languages containing operators representing metaphysical necessity, the logician's indices will come equipped with an equivalence relation that tells us whether a proposition (a set of indices) is possible or impossible relative to a given index. Thus, we can classify our indices into possible and impossible, although, if the equivalence relation is not the universal relation, it will be a distinction that depends on the index of evaluation. It is the notion of a logical index that is relevant to most technical work in the philosophy of language and linguistics.

According to the logical notion of a possible index, it is a vague matter which of the indices count as possible. For example, in a supervaluationist setting the indices are not possible worlds but world–precisification pairs. In that setting, a logical index

(x, u) is possible relative to (w, v) only if the two indices are modally accessible, which happens iff $u = v$. It follows that if u and v are Δ-accessible then it is borderline, at (w, v) whether (x, u) is metaphysically possible: if $v \neq u$ then (x, u) is not possible relative to (w, v) but is possible relative to (w, u) which is Δ-accessible to (w, v). If $v = u$ then (x, u) is possible relative to (w, v) but is impossible relative to any Δ-accessible pair (w, v') with $v' \neq v$. (This result is not specific to the supervaluationist either. It is a necessary matter whether a number is small, but it can still be a borderline matter. If N is borderline small, then there must be indices that disagree about whether N is small, and given that it is necessary whether N is small, at least one of these indices must represent a metaphysically impossible state of affairs. Although, it will be borderline which one.)

It is straightforward to introduce the notion of a proposition which corresponds to a set of possible indices, but since it is vague which indices are possible indices, we have no guarantee that the notion of corresponding to a set of possible indices so-introduced satisfies F2 or F3.

Although the notion of a logical index doesn't get us in to trouble, more problematic is the other use of possible worlds talk: what one might call, by contrast, the *metaphysical notion of a world*. Paradigm examples of the latter concept include Lewis' notion of a maximal spatio-temporally connected region of space–time or Plantinga's notion of a complete possible state of affairs. Such entities are typically *not* the same as the logician's indices. For those who theorize in these terms, it is quite common to model the logician's indices as ordered pairs of possible worlds and other things: times if we are doing tense logic, precisifications if we are modelling vagueness, and so on.

The assumption that indices can be decomposed somehow into a possible world and something else is not always innocent. In chapter 15, we will explore in more detail a context in which the logical notion of an index cannot be reduced to the notion of a possible world plus something else, for logical reasons relating to the combined logic of necessity and vagueness. But in the present context we can see that the assumption is already controversial, for if our semantics of vagueness is one in which the indices can be decomposed into pairs of metaphysically possible worlds and something else (e.g. supervaluationist precisifications), then we effectively have a variant of the supervaluationist semantics outlined in section 12.1.2, and we seem to be subject to exactly the same problems we encountered there: we reinstate a metaphysically significant division of propositions into factual and non-factual which is plausibly governed by the problematic principles F1–F4.

To be clear, then, it is the substantive metaphysical, and not the logical, notion of a possible world I am rejecting when I maintain that the ideology of worlds is problematic. It is natural to ask, however, whether it is possible to revise the ordinary metaphysical notion of a world so that it can do some of the work that it is sometimes put to without involving us in the paradoxes of higher-order vagueness? After all, the use of possible worlds is ubiquitous if only due to their considerable heuristic value. One might think, therefore, that it would be worth having some sort of story about what is going on when philosophers talk about possible worlds, even if it is not quite how these philosophers usually conceive of what they're doing.

I take it that it is part of the standard conception of a possible world, as they occur in the theories of Lewis, Plantinga, and so on, that it is a completely precise matter which things *are* the possible worlds:

For any object x, it is always determinate whether x is a possible world or not.

It is hard, for example, to imagine Lewis' conception of a world being subject to vagueness, since it appears to be definable from completely precise vocabulary (i.e. a 'maximal spatio-temporally connected concrete entity'). Possible states of affairs in Plantinga's sense are presumably also supposed to be taken as metaphysically basic and, thus, not subject to vagueness.

There is, however, an alternative way of thinking about possible worlds. We could, following Arthur Prior and others, attempt to recover talk of possible worlds by reconstructing them in terms of quantification over certain kinds of propositions. Prior was attempting to reconstruct worlds and times in terms of propositions, and although there are lots of ways one could attempt to do this, a quite natural approach would be to identify a world proposition with a maximally strong consistent eternal proposition.[12] An alternative approach, if we are modelling propositions as sets of world–time pairs, is to note that each possible world, w, corresponds to a proposition consisting of all pairs that have w as its first coordinate. These two definitions match up exactly—a world proposition (a maximally strong consistent eternal proposition) is a proposition that corresponds to a possible world in the sense just defined.

This idea extends naturally enough to the supervaluationist setting: each world w corresponds with the propositions consisting of all pairs with w in their first coordinate. Now recall that, according to the supervaluationist, a proposition is factual if it corresponds to a set of worlds, and a maximally strong consistent factual proposition is one that corresponds to a singleton of a possible world. Thus, we can alternatively define a world proposition in the supervaluationist setting as a maximally strong consistent factual proposition.

In the two cases above, we saw that we can either start off taking a metaphysically substantive notion of world as basic, treating propositions as certain sets of ordered pairs involving them, or we can treat worlds as certain kinds of propositions (encouraging a view in which the propositions are basic). Since we have expressed some scepticism about the metaphysical notion of a world, one might wonder whether we could introduce the notion of a world proposition in the latter way, without assuming a prior notion of possible world, or a conception of propositions as sets of pairs of worlds with something else. Of course, we cannot follow the supervaluationist in identifying them with maximally strong consistent factual propositions, because we have also

---

[12] An eternal proposition is a proposition that is always true or always false.

expressed scepticism about the notion of a factual proposition. However, we could instead talk about maximally strong consistent *precise* propositions:

A proposition p is a *world proposition* iff p is consistent, precise, and entails every consistent precise proposition that entails it.

The resulting notion of a world proposition bears some resemblance to the classical metaphysical conception of a world, but also has some notable differences. Firstly, note what maximally strong consistent precise propositions are: they are complete descriptions of the way the world could be in all precise respects. They settle all precise questions but leave open questions that would have been borderline given the precise way they describe things to be. As outlined in chapter 12, we can picture logical space as being divided into cells where precise propositions correspond to arbitrary unions of these cells, vague propositions can be viewed as propositions that cut across at least one cell, and the world propositions (the maximally strong consistent precise propositions) correspond to the cells themselves. See Figure 12.2.

In these respects, our conception of a world proposition is structurally exactly like the temporalist and supervaluationist conception of a world proposition, arising from a prior metaphysical notion of world, for both those conceptions impose a similar division of logical space into cells. However, unlike these notions, we have not motivated our definition by a prior conception of a proposition as a set of ordered pairs whose first coordinate consists in a metaphysical world. Indeed, in chapter 15, we shall look at some positive reasons to reject models of propositions that invoke ordered pairs in this way.

A more striking difference between our conception of a world proposition and the aforementioned ones, is that it is vague which propositions *are* the world propositions. That is, we can divide logical space up into cells in two different ways in such a way that it is vague which of the divisions correspond to the maximally strong consistent precise propositions. This is a result of higher-order vagueness: it is vague which propositions are precise, and, thus, vague which propositions are the maximally strong consistent precise propositions.

Thus, unlike Plantinga's states of affairs or Lewis' maximal spatio-temporally connected concreta, it is vague where the boundaries of our world propositions lie. One might be tempted to think of this as a form of metaphysical vagueness (see e.g. Barnes and Williams [9] or Wilson [166]). But this is, I think, to tie world propositions too closely to the concrete reality around us—our world propositions are 'worldly' in name only. It would be metaphysical vagueness if it were vague which entities were maximal spatio-temporally connected concreta, for then it would have to be borderline whether something was spatio-temporally connected, or borderline whether it was concrete, which are more familiar forms of metaphysical vagueness. Similarly, if it could be borderline whether an entity was a *fact* or a *state of affairs*, then one might similarly draw the connection to metaphysical vagueness. The vagueness I am describing amounts to nothing more than the vagueness one gets concerning which things are precise. Vagueness that we have already come to view as nothing out of the ordinary. I think a better moral to draw from the existence of this sort of vagueness, then, is that our notion of a world proposition is not a metaphysical one: it is not a distinction that carves at the metaphysical joints in the way that the distinction between fundamental or factual propositions was supposed to.

Let me end our discussion by briefly considering another contender for being the replacement for the notion of a world. Recall that a proposition p is precise* if (i) p is precise, (ii) the proposition that p is precise is precise, (iii) the proposition that the proposition that p is precise is precise is precise, and so on, at all finite orders. One might hope to limit the vagueness in our theoretical entities by theorizing in terms of maximally strong consistent precise* propositions instead. According to some theorists there's only one maximally strong consistent precise* proposition—the tautologous proposition—because on those views very little counts as precise at all orders.[13] So, this proposal is not open to those philosophers. However, even for theorists such as myself who accept a variety of precise* propositions there are no good reasons to think the boundaries between the possible maximally strong consistent precise* propositions are any more precise. One can see, in the four-index model depicted in Figure 14.2, that relative to indices i and k, the singleton of j is not a maximally strong consistent precise* proposition, but relative to j and l, it is (see also Figure 13.1).

(Note also that (a) the Δ-accessibility relation coheres with the partition into precise propositions at the four indices: an index x sees y if x and y both belong to the same orbit of $Orb(G(x))$, (b) that modally accessible indices impose the same orbit structure, and that (c) this model validates the product logic—indeed, it can be generated as a supervaluationist model with two worlds and two precisifications.)

What this model demonstrates is that there is no logical guarantee that the notion of a maximally strong consistent precise* proposition is any more precise than the notion of a maximally strong consistent precise proposition. Indeed, as I noted

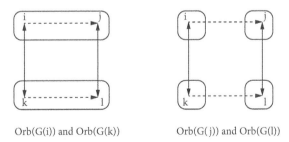

Orb(G(i)) and Orb(G(k))   Orb(G(j)) and Orb(G(l))

Figure 14.2. The left diagram represents the partition of logical space according to both i and k (the partition according to $Orb(G(i))$ and $Orb(G(k))$), and the right diagram according to j and l ($Orb(G(j))$ and $Orb(G(l))$).

---

[13] Dorr [35] defends this view in a linguistic setting, although Dorr's account is even more radical: even tautologies aren't precise*.

in section 13.4.1, I am doubtful that this notion plays a particularly interesting theoretical role, so I shall set it aside.

Let us summarize what we have done here. We have established that proponents of propositional vagueness ought to reject the notion that there is a basic layer of 'rock bottom' propositions that are completely unaffected by vagueness. This idea might be cashed out by the assumption that there is a set of fundamental or factual propositions, or a set of possible worlds that carve out a definite partition of logical space. We have proposed that we theorize instead with the notion of a precise proposition. The replacement carves logical space into a partition of precise propositions, albeit a partition whose boundaries are vague.

While these conclusions may sound radical and metaphysically revisionary, it is of interest to note that we were led to these conclusions primarily by epistemological considerations. In section 14.2, we argued that, since we can't know the cutoffs concerning what the fundamental propositions entail, these cutoffs bear all the hallmarks of vagueness. Similarly, in section 12.2, we argued that the propositions of physics are not precise due to the epistemic possibility that the propositions of physics correspond to less fundamental phenomena. Whether one wishes to classify these conclusions as deeply metaphysical is a matter of taste; I will not attempt to adjudicate that matter here.

15

Vagueness and Modality

In this chapter, we shall be investigating the interaction of metaphysical modalities with vagueness. As with the surrounding chapters, our main point of contrast will be the supervaluationist semantics outlined in section 12.1.2.

To address these issues, I shall adopt the framework of a modal propositional calculus with two primitive operators, $\Box$ and $\triangle$, representing metaphysical necessity and determinacy respectively. The standard way to model either of these operators in isolation is to specify a set of indices of some sort and an accessibility relation over those indices (see Kripke [84]). Each formula of the language is assigned a truth value relative to each index. A formula of the form $\Box A$ or $\triangle A$ is true at an index iff A is true at every index accessible to it. To generalize this sort of semantics to deal with both operators at once, one equips a single set of indices with two accessibility relations, one for each operator.

It would be nice to say something a bit more informative about what these indices are, and this is where matters become more complicated. In practice, we choose a set of indices that can model variation in the kinds of truths that the operator in question cares about. If we are studying the logic of a single modal necessity operator, the indices are usually worlds. If we are studying determinacy on its own, the indices are precisifications. If we are doing temporal logic, the indices are times. And so on. Propositions get identified with sets of worlds, precisifications, and times in each respective framework. Yet, I think it should be clear that none of these identifications are satisfactory. If propositions are sets of times, we run into trouble modelling contingency or borderlineness. If they're sets of worlds, there are difficulties modelling temporary or borderline propositions. And if they're sets of precisifications, we run into trouble with contingent or temporary propositions.

When we are considering logics that combine two or more of these operators, the indices cannot be either times, worlds, or precisifications, but something that is more fine-grained than any of these taken alone. A standard approach—the one adopted, for example, in the supervaluationist semantics we outlined in section 12.1.2—is to treat the indices for a multi-modal system as ordered tuples of the indices of each mono-modal system (see Marx and Venema [101], for example). In modal tense logic, for example, when we evaluate a tense operator at a world–time pair, we operate on the time coordinate and leave the world alone, and when we apply a modal operator, we operate on the world coordinate and leave the time alone. However, it is worth emphasizing that this approach encodes some fairly strong assumptions about the logical interaction of the two kinds of operators. In this framework, indices have two degrees of freedom, only one of which each operator can affect. This means that the two operators exhibit a kind of independence from one another that gives rise to a distinctive and controversial combined logic of vagueness and modality.

The strategy I have been pursuing in the last few chapters has been to take propositions as primitive, and treat the indices as maximally strong consistent propositions.[1] In this alternative setting, there is no guarantee that our indices can be decomposed into ordered pairs of entities in such a way that necessity corresponds to variance in one of the coordinates and determinacy to variance in the other. In this chapter, we shall investigate some of the logical freedom that this approach affords us.

15.1 The Interaction of Vagueness and Modality

Recall the supervaluationist truth clauses for $\Box$ and Δ relative to a pair consisting of a world x belonging to a set of possible worlds W and a precisification v belonging to a set V:

($\Box$) $x, v \Vdash \Box A$ if and only if $y, v \Vdash A$ for every y such that Rxy.

(Δ) $x, v \Vdash \Delta A$ if and only if $x, u \Vdash A$ for every u such that Svu.

Notice that R and S are relations over different domains, W and V respectively, and so, technically, the above semantics does not fit the usual definition of a Kripke model for a pair of operators. For that, one wants a single set of indices and a pair of accessibility relations defined over them. We can reformulate the above semantics so that it has this form as follows:

Let the indices be $W \times V$; the set of ordered pairs of worlds and precisifications.

Let S' be the relation that holds between (x, v) and (y, u) iff $x = y$ and Svu.

Let R' be the relation that holds between (x, v) and (y, u) iff $u = v$ and Rxy.

We say that $(x, v) \Vdash \Box A$ ($\Box A$ is true at the *ordered pair*, (x, v)) iff $(y, u) \Vdash A$ for every (y, u) such that $R'(x, v)(y, u)$. Similarly for ΔA. We can now observe that: (i) this clause is the same as the standard clause for a modality in a Kripke frame, and (ii) that A is true relative to an ordered pair (x, v) in the reformulated semantics if A is true relative to the world x and precisification v in the original semantics.

---

[1] The question of how to measure the strength of a proposition when there are multiple kinds of propositional operators in the language is a delicate question. One must distinguish between necessary implication, determinate implication, eternal implication, etc., and the various combinations of these notions, such as determinate necessary implication, necessary eternal implication, and so on. This question is solved if we have a maximally broad notion of logical necessity available to us. The L operator we defined from the identity connective in section 11.1 has exactly this feature.

According to most treatments of metaphysical modality, a proposition is necessary if it is true at *all* members of W, so that the $\Box$-accessibility relation R becomes the relation that relates every possible world to every other possible world. For this reason, it is omitted altogether from many formal treatments of modality. Note, however, that even if R is the universal accessibility relation over worlds, the corresponding R' relation is not a universal relation over the indices in $W \times V$. Rather, it partitions $W \times V$ into equivalence classes by the relation of sharing the same precisification coordinate. Our model thus contains indices that represent metaphysically impossible scenarios: scenarios represented by indices that are not $\Box$-accessible from the true index.

The first thing to notice about this framework is that we don't validate the principle that necessary truths are determinate:[2]

$$\Box A \rightarrow \Delta A$$

Although this principle is widely rejected, the principle has a degree of prima facie plausibility. It follows, for example, from the natural thought that metaphysical necessity is the broadest sort of necessity, or that borderlineness consists in a kind of contingency. But I think that on reflection neither of these ideas stand up to scrutiny. Consider that it is not a contingent matter how big a number is. It is thus surely a necessary matter whether a number is small or not—for whether a number is small or not surely depends only on how big it is, and this is not a contingent matter. On the other, hand there are numbers which are borderline small. Indeed, the last small number, N, is small, and so necessarily small, but not determinately small, contradicting the inference from $\Box A$ to ΔA.

More generally, assuming that there is at least one borderline proposition, the principle is inconsistent with the plausible claim that the vague supervenes on the precise. Suppose that it is borderline whether p, and suppose that either p or its negation is necessitated by some precise truth, q, in accordance with the supervenience principle. Thus $\Box(q \rightarrow p) \vee \Box(q \rightarrow \neg p)$. If we accepted the inference from necessity to determinacy, we could infer the disjunction $\Delta(q \rightarrow p) \vee \Delta(q \rightarrow \neg p)$. Since q is a precise truth we know that Δq, so we can infer, using the K principle for Δ, that $\Delta p \vee \Delta \neg p$.[3] This, of course, contradicts the idea that p was borderline. In the following, I shall follow the orthodox supervaluationist (and almost everyone who has written on this topic) in rejecting the inference from $\Box A$ to ΔA.[4] The failure

---

[2] To invalidate it consider the model $W = \{0\}$, $V = \{0, 1\}$ with $S = V \times V$, and interpret A as $\{(0, 0)\}$.

[3] The easiest way to make both of these inferences is to use reasoning by cases. Some supervaluationists reject this rule of proof, but they still accept the classical tautology $((A \vee B) \wedge (A \rightarrow C) \wedge (B \rightarrow D)) \rightarrow (C \vee D)$ which is sufficient to make both of the moves we made above.

[4] Juhani Yli-Vakkuri has explored (without endorsing) a view in which vagueness consists in a failure of supervenience on metaphysically groundfloor facts, so that a proposition A is borderline at a possible world w iff there is both an A and $\neg A$ world with same groundfloor facts that obtain in w (Yli-Vakkuri [168]). As far as I can see, this is the only sort of view that can keep the conditional $\Box A \rightarrow \Delta A$.

of this inference also explains the need for impossible indices: even if A is a necessary truth—true at all metaphysically possible indices—there must be indices where A is false if we also want $\triangle A$ to come out false.

According to the supervaluationist semantics, $\square$ and $\triangle$ behave like independent quantifiers—the former quantifying over $\square$-accessible worlds and the latter over $\triangle$-accessible precisifications. Thus intuitively we would expect determinacy and necessity to commute (much like the string of quantifiers $\forall x \forall y$ and $\forall y \forall x$), and that possibility commutes in one direction with determinacy (by analogy with the fact that $\exists x \forall y A$ implies $\forall y \exists x A$). These thoughts are summarized by the following interaction principles governing $\square$ and $\triangle$:

ND $\quad \square\triangle\phi \rightarrow \triangle\square\phi$

DN $\quad \triangle\square\phi \rightarrow \square\triangle\phi$

PD $\quad \lozenge\triangle\phi \rightarrow \triangle\lozenge\phi$

A little reflection on the supervaluational truth clauses for $\square$ and $\triangle$ reveals that these principles are indeed validated. Indeed, it follows by some standard results that the logic generated by adding these principles to S5 for necessity and KT for determinacy is complete with respect to the class of models described (assuming S is reflexive and R an equivalence relation).[5]

The last principle PD, sometimes called the 'Church–Rosser principle', states that possibility commutes with determinacy (in one direction only) and appears to exemplify an asymmetry between the two modalities. This asymmetry is illusory—it is easy to verify that its dual, $\neg\triangle\neg\square p \rightarrow \square\neg\triangle\neg p$, is in fact a consequence of PD by contraposition. What is *not* a consequence of PD, however, is its converse: $\triangle\lozenge\phi \rightarrow \lozenge\triangle\phi$. Indeed, the converse is demonstrably false. Suppose there is a bag of 1,000 balls of different sizes, ranging from balls that are clearly small to balls which are clearly not small, with the unclear cases in between as well. Since it is possible for me to pick any particular ball from the bag, it follows that, determinately, for any ball b it's possible that I have picked b from the bag. Since it's determinate that some ball is the largest small ball, it follows that, determinately, it's possible that I've picked the largest small ball. However, it is not possible that I've determinately picked the largest small ball. To do so I would have to be determinately holding the largest small ball, and since it is indeterminate which ball that is, it would have to be indeterminate which ball I am holding—which is not, I shall assume, a metaphysical possibility.[6] As noted already, the interaction between possibility and determinacy is formally analogous

---

[5] The use of S5 and KT for the logics of necessity and vagueness respectively are not required for this completeness argument. Analogous results can be formulated for any pair of the following logics: KT, KTB, S4, or S5. See Gabbay and Shehtman [59] and Kurucz [86]. In Gabbay and Shehtman, a class of logics, 'PTC logics', are defined for which analogous completeness results can be formulated for combinations of pairs of PTC logics, where the combination consists in adding the analogues of ND, DN, and PD.

[6] Note also that the dual of the converse of PD entails the converse of the final axiom (by contraposition) so the dual of the converse of the final axiom is also false for the same reason.

to the interaction between the existential quantifier and the universal quantifier: if someone is loved by everyone then everyone loves someone, yet the converse—that if everyone loves someone, someone is loved by everyone—is simply not true.

If we add these principles to a logic, L_1, governing $\square$ and another logic, L_2, governing Δ (in our case, S5 and KT) the result is called the product of L_1 and L_2. Whether the supervaluationist semantics is adequate will therefore depend on the truth of the theorems of the product logic. However, the validity of the product logic is far from obvious.

An important reason for scepticism is that modal operators like $\square$ and $\diamond$ themselves appear to be a source of vagueness; a thesis that I think has a great degree of plausibility independently of the plausibility of any putative example. Formulating this idea requires a little care, however. It would not do to simply say that necessity has borderline cases: that for some proposition, p, it's borderline whether p is necessary. For example, it's borderline whether it's necessary that 15 is small (for a number less than 100, say), but that is presumably due to vagueness in the proposition concerning whether 15 is small, and not due to vagueness concerning necessity. By analogy, suppose, for the sake of argument, that Mt Kilimanjaro is a vague object and that it is borderline whether it is exactly 19,341 feet tall. Then the property of being exactly 19,341 feet tall has a borderline case: Mt Kilimanjaro. But clearly this is not because the property of being exactly 19,341 feet is vague: it's because Mt Kilimanjaro is. A property is only vague if it has a *precise* object as a borderline case.

Similarly, a sufficient condition for an operator to be vague is if it has a precise proposition as a borderline case. If for some p the proposition that it's necessary that p is a borderline proposition, then that's either due to vagueness in the necessity operator *or* the proposition p. We must thus articulate the idea that necessity is vague as follows:

VAGUENESS OF MODALITY: For some precise proposition p, it's borderline whether p is necessary.

If it's borderline whether p is necessary, where p is precise, then the source of the vagueness must be the necessity operator. The idea that 'necessary' expresses a vague operator is plausible given both the sparsity of precise words in general, and the plethora of sorites puzzles that arise from 'Ship of Theseus'-type puzzles (I'll turn to these in a second).

It is worth noting, then, that VAGUENESS OF MODALITY cannot be maintained in conjunction with the product logic. Specifically, it is not consistent with the principles ND and PD.[7] For suppose that p is precise and that it is borderline whether p is necessary. If p is precise then p couldn't have been borderline, so we know both:

P1 $\square(p \rightarrow \Delta p)$

P2 $\square(\neg p \rightarrow \Delta \neg p)$

---

[7] DN, as far as I can see, is left intact by these considerations.

Furthermore, $\nabla\Box p$ entails $\neg\triangle\Box p$ and $\neg\triangle\neg\Box p$, and so, given the fact that $\Box p \vee \neg\Box p$, also entails:

[P3] $(\Box p \wedge \neg\triangle\Box p) \vee (\neg\Box p \wedge \neg\triangle\neg\Box p)$

But these three principles are inconsistent in the product logic. If the first disjunct of P3, $\Box p \wedge \neg\triangle\Box p$, is true then we get a contradiction from P1 and **ND**: P1 entails $\Box p \rightarrow \Box\triangle p$, which, given the first conjunct, allows us to infer $\Box\triangle p$. But, by **ND** this entails $\triangle\Box p$ which contradicts the second conjunct. If the second disjunct is true we get a contradiction from P2 and **PD**: $\neg\Box p$ entails $\Diamond\neg p$, which along with P2, entails $\Diamond\triangle\neg p$. Assuming **PD**, this entails $\triangle\Diamond\neg p$ which contradicts $\neg\triangle\neg\Box p$.[8]

Are there any plausible witnesses to VAGUENESS OF MODALITY—precise propositions whose necessity is borderline? There are in fact a class of puzzles that arise in the context of moderate forms of mereological essentialism that appear to be a good source of examples. Suppose, for example, that we have a chain necklace, called 'chainy', which was constructed by taking 100 links out of a box of links and putting them together. Now, according to a moderate form of essentialism, it seems as though chainy could have survived the loss of a single link.[9] For example, if chainy broke it would be possible to repair it by replacing one of the links. On the other hand, if we replaced all the links constituting chainy with new links, we would have another chain.

Let X be the set of links that chainy is in fact made out of. Note that the above observations provide the ingredients for a sorites sequence. We can formulate the claim that the chain could have survived 1 replacement but not 100 replacements as follows. Below, $\exists_n xFx$ means 'there are at least n F's'—this is a quantifier that can be defined in purely logical vocabulary.[10] $\leq$ represents mereological parthood and $\in$ set membership:

(i) $\Box\exists_1 x(x \in X \wedge x \leq c)$

(ii) $\neg\Box\exists_{100} x(x \in X \wedge x \leq c)$

(i) states that, necessarily, at least one member of X is a part of chainy and (ii) says that it is not necessary that every member of X be a part of chainy.

By classical logic, there is an n such that $\Box\exists_n x(x \in X \wedge x \leq c)$ and $\neg\Box\exists_{n+1} x(x \in X \wedge x \leq c)$. That is to say: it is necessary that at least n members of X are parts of chainy, although it is not necessary that at least $n + 1$ members of X be parts of

---

[8] Although the above argument might look on the surface as though it uses reasoning by cases, it is possible to formulate the argument so that the only rule appealed to is modus ponens applied to the theorems of K for determinacy and for necessity respectively.

[9] The sort of moderate essentialism I have in mind maintains that things have *some* properties essentially—I couldn't have been a boiled egg, for example—but also maintains that things don't have *all* their properties essentially—for example, I could have survived the loss of a fingernail. I take this to be the common-sense form of essentialism.

[10] $\exists_1 xFx$ may simply be defined as $\exists xFx$. $\exists_{n+1} xFx$ may be defined as $\exists x(Fx \wedge \exists_n y(y \neq x \wedge Fy))$.

chainy. Since this is a paradigm sorites argument, the truths at the boundary will not be determinate truths.

This much should, I hope, be uncontroversial provided one accepts the moderately essentialist metaphysics implicit in the set-up. The next step is to note that $\exists_n x(x \in X \wedge x \leq c)$ appears to be stated in purely precise vocabulary: it can be formulated only using identity, (unrestricted) existential quantification, set membership, and mereological parthood, all of which are plausibly precise, and two names: X introduced as a name for a particular set, and c, a name for a particular object. Since the proposition expressed by this sentence is precise, we know that it couldn't be borderline, and thus we have true instances of P1, P2, and P3. Thus, we have a precise proposition that is a borderline case of being necessary.

There are several plausible ways to resist the argument. One might, for example, deny that we succeeded in introducing a precise name for the chain in our example (a similar objection could be levelled at our name for the set of links). This response does not get to the heart of the issue: the premises P1–P3 are just as plausible if we existentially quantify into the position that the name c takes. The existential versions of these premises then rest on the following thought: that there's at least one chain such that it is borderline whether *it* could have had less n members of X as parts. The contradiction with ND proceeds just as before, except in the scope of an existential quantifier. One might similarly object to the idea that we could introduce a precise name for the set X, but again the premises are no less persuasive when X is replaced by an existential quantifier.

There are, of course, other ways of resisting the argument: one could deny the essentialist metaphysics that we need to get the sorites going, or one could deny that parthood is always precise. The principle that parthood is precise, for example, is sometimes rejected on the grounds that there are vague objects such as Mt Kilimanjaro: there are rocks and stones that are such that it is borderline whether they are a part of Mt Kilimanjaro. Thus there are pairs of objects such that it is borderline whether one is a part of the other. But to conclude from this that parthood is vague rests on exactly the sort of fallacy we exposed earlier and took care to avoid. We would not analogously conclude that being exactly 19,341 feet was a vague property on the basis that it has Mt Kilimanjaro as a borderline case of it. To show the vagueness of that property we must find a precise object such that it is borderline whether it is 19,341 feet, and Mt Kilimanjaro is not such an object. By analogy, if there is vagueness in the proposition that x is a part of y, then that is either due to vagueness in x or y or in parthood: one would need to find precise objects x and y that stand in indeterminate parthood relations to establish the vagueness of parthood.

Let me finish the discussion of this case by noting that these kinds of piecemeal responses fall far short of a general defence of the principle that modality is completely precise. The endless supply of metaphysical puzzles of this sort will require all kinds of ad hoc manoeuvres: to avoid postulating vagueness in the modal operators, we must postulate it in all kinds of other places. It is odd to imagine someone being willing to postulate vagueness in places one wouldn't normally expect, such as in the parthood relation, but be unwilling to acknowledge the vagueness of 'necessarily'. The vagueness of 'necessarily' is antecedently plausible, independently of our examples, since so few of our words express precise things anyway.

It is, of course, possible to modify the supervaluationist semantics in such a way as to invalidate these inferences. One could, for example, allow the Δ-accessibility relation to be sensitive to the world coordinate, as well as the precisification coordinate. We could similarly allow the $\Box$-accessibility relation to be sensitive to the precisification coordinate in addition to looking at the world coordinate. (Formally, this could be achieved by using ternary or quaternary accessibility relations over both worlds and precisifications so that the $\Box$-accessibility relation can be sensitive to the precisification coordinate, and the Δ-accessibility relation can be sensitive to the world coordinate.) I will defer a proper treatment of these proposals until section 15.4. However, let me note that many of these sorts of tweaks turn the supervaluational semantics into a notational variant of the general Kripke semantics of two operators: in these variants, the ordered pair structure plays no role in the semantics, and the ordered pairs could be replaced, without loss of generality, with indices that are not ordered pairs; this is exactly the sort of semantics we will investigate in section 15.2.

15.2 The Proper Logic of Vagueness and Modality

Let us now apply the framework of chapter 13 to the logic of vagueness and modality. Recall that a proposition p is determinate at an index i iff $\sigma(p)$ is true at i for every symmetry belonging to $G(i)$. Recall also that we can reformulate this semantics so that it takes the form of an ordinary Kripke model by introducing an accessibility relation as follows:

Sij if and only if $\sigma(j) = i$ for some $\sigma \in G(i)$.

Note that p is true at every j accessible to i iff $\sigma^{-1}(i) \in p$ for every $\sigma \in G(i)$, and this holds iff $i \in \sigma(p)$ for every $\sigma \in G(i)$. Thus, the clause for the truth of ΔA in the Kripke semantics is the same as the clause for the truth of ΔA in the symmetry semantics.

A slightly more intuitive picture of what is going on can be obtained as follows. Recall that each $G(i)$ partitions logical space into cells (orbits) corresponding to maximally strong consistent precise propositions and a proposition is determinately true at i iff the strongest true precise proposition at i entails it; i.e. iff every element of the cell containing i belongs to p. According to our definition of S, the indices accessible to i are just those that belong to the cell containing i, when the cells are determined by $G(i)$.

To add metaphysical modalities to our language, we must add another accessibility relation, R, over the indices. Putting this together, we get a Kripke frame $\langle I, R, S \rangle$. Here, I is our set of *indices* (maximally L-strong propositions), R the $\Box$-accessibility relation, and S the $\triangle$-accessibility relation as defined above in terms of symmetries. R and S range over a single domain of indices, and so are the counterpart of the two relations R' and S' that range over ordered pairs in the supervaluationist case.

In the supervaluational setting, however, the two accessibility relations would shift the world and precisification coordinates independently: R' completely ignores the precisification coordinate, and S' ignores the world coordinate. There is nothing corresponding to this constraint in the present setting.

Notice that much like the supervaluationist's $\square$-accessibility relation R', the $\square$-accessibility relation we have is not in general the universal accessibility relation that relates every index to every other index. In fact, if we are to invalidate the principle $\square A \rightarrow \triangle A$ which we argued against in section 15.1, R cannot in general be the universal relation, for otherwise $\square A$ would be true at an index only if A was true at all indices in I, thus only if $\triangle A$ is true at that index. In general, then, R will partition I into equivalence classes in which pairs of indices in the same equivalence class regard each other as metaphysically possible, and regard the inaccessible indices as metaphysically impossible (although, perhaps, not determinately false).[11]

As in supervaluational semantics, we can assign truth values to each atomic sentence relative to each world, and we can extend truth to arbitrary sentences relative to each index $i \in \mathcal{I}$ in the standard way. The crucial clauses for $\square$ and $\triangle$ are given as follows:

($\square$) $i \Vdash \square \phi$ if and only if $j \Vdash \phi$ whenever Rij

($\triangle$) $i \Vdash \triangle \phi$ if and only if $j \Vdash \phi$ whenever Sij

To ensure a reasonable logic of necessity and determinacy, we must place additional constraints on R and S. It's natural to want a modal logic of S5 and a logic of determinacy of at least KT. The distinctive feature of the former logic is that it rules out all kinds of higher-order contingency—if something's necessary it's necessarily necessary, and if it's not necessary it's necessarily not necessary. Due to the existence of higher-order vagueness, analogous principles for the logic of determinacy are not appropriate, and the logic KT reflects this; the only distinctive principle in this logic is the factivity of determinacy. To ensure that the theorems of these logics are true at every index, we must stipulate that R is an equivalence relation and S is a reflexive relation.

These constraints ensure certain features of the pure modal logic, and the pure logic of determinacy. However, neither of these constraints commit us to the principles ND, DN, and PD discussed in section 15.1. Indeed, to validate these principles one would have to impose the last three constraints listed in Table 15.1: none of these principles are required by the general semantics we have outlined here, based on the symmetry

---

[11] There are models in which R is the universal relation, but they are not particularly interesting. For example, in the supervaluational setting, the only models in which R' is universal are models where there is only one precisification. In other words, models where there is no vagueness.

Table 15.1. Interaction principles for vagueness and modality

$\Delta\phi \to \phi$	S reflexive	✓
$\Box\phi \to \phi$	R reflexive	✓
$\Box\phi \to \Box\Box\phi$	R transitive	✓
$\phi \to \Box\Diamond\phi$	R symmetric	✓
Supervenience^n	If Rij, Rik, $S^n ju$ and $S^n ku$ then $j = k$	✓
$\Box\Delta\phi \to \Delta\Box\phi$	$S \circ R \subseteq R \circ S$	✗
$\Delta\Box\phi \to \Box\Delta\phi$	$R \circ S \subseteq S \circ R$	✗
$\Diamond\Delta\phi \to \Delta\Diamond\phi$	If Rxy and Sxz then there's a w such that Syw and Rzw	✗

semantics from chapter 13.[12] In particular, the problematic combination ND + PD is not required in this sort of semantics.

It is natural to wonder whether there are there any further constraints that need to be imposed governing the interaction of R and S. I shall shortly show that if we want to expand the language to contain quantification into sentence position, and we want to make certain supervenience claims true, then we should additionally impose, for each n, the constraint labelled Supervenience^n in Table 15.1. If you compare the class of models in which R is an equivalence relation and S is reflexive with the subclass of those models in which the supervenience claim constraint is additionally imposed, no extra principles are validated in the propositional modal language containing $\Box$ and Δ (and even $\sharp$). However, it is presumably going to be a constraint that holds in the intended model, assuming the truth of the supervenience theses, and these constraints will also create new validities when the propositional quantifiers are added, so it is important not to omit these conditions from the model theory. In summary, constraints we are interested in are given by the first five principles in Table 15.1; the last three constraints are not imposed.

15.3 The Supervenience of the Vague on the Precise

Our investigation of the combined logic of vagueness and modality is not quite complete. According to a popular slogan, 'the vague supervenes on the precise'. For a linguistic account of vagueness, untangling this slogan is a tricky business. Supervenience is a modal notion relating one set of true propositions to another—the A facts supervene on the B facts iff, necessarily, every A proposition is necessitated by the truth of a B proposition. Yet according to linguistic theories of vagueness, the objects of vagueness and precision are not facts or propositions but representations.[13]

---

[12] The condition: $(Rxy \wedge Sxz) \to \exists w(Syw \wedge Rzw)$ is sometimes called the *Church–Rosser* property. $(R \circ S)xy$ means that x is an R of an S of y.

[13] See chapter 9 of Williamson [156] for a good discussion from the perspective of a linguistic theorist. There, things are set up in terms of the supervenience of the vaguely describable facts on the precisely describable facts. It is worth noting, however, that this sort of supervenience may be trivial if all vagueness

For the non-linguistic theorist, however, the supervenience idea gets a fairly straightforward treatment. In order to formulate it, however, we need a slightly richer language. Firstly, we must go beyond the determinacy and necessity operators and reintroduce the notion of precision; that is, we shall add the operator ♯ into our language. Secondly, to state supervenience we will need to quantify over propositions in the object language: this can be achieved by introducing devices for quantifying into sentence position, so that one can formulate generalizations like $\forall p(p \rightarrow p)$. Supervenience can be formulated as follows:

SUPERVENIENCE: Necessarily, for every truth, there is a precise truth that necessitates that truth.

$\Box\forall p(p \rightarrow \exists q(\sharp q \wedge q \wedge \Box(q \rightarrow p)))$.

SUPERVENIENCE states that all truths supervene on the precise; in particular, the truth values of the vague propositions supervene on the precise propositions (and the truth values of the precise propositions trivially supervene on the precise). A useful consequence of SUPERVENIENCE, which we will appeal to frequently, is that, given the claim that the precise propositions are closed under the logical operations, every proposition p is necessarily equivalent to a precise proposition, namely, the disjunction of precise propositions that necessitate p (in the formal language: $\forall p \exists q(\sharp q \wedge \Box(p \leftrightarrow q)))$.

Before we investigate the formal consequences of this principle, let's consider some informal applications of the supervenience idea. To deny that tallness supervenes on precise measurements such as height, for example, would be to countenance a metaphysical possibility in which someone has a certain precise height h and is tall, and another metaphysical possibility in which someone has exactly the same measurements but is not tall. Many philosophers share the intuition that these two scenarios are not both possible and are motivated to endorse a strong supervenience claim: that there couldn't be two metaphysically possible scenarios i and j that agree that objects x and y have the same precise height and measurements, but differ in that, according to i, x is tall and, according to j, y is not tall. A weak supervenience claim would state instead that any possible scenario i in which objects x and y have the same precise height and measurements is one in which x and y are both tall or both not tall. The weak supervenience claim is a straightforward consequence of the strong one.

It is important to distinguish supervenience, weak or strong, from a related epistemic claim. When Harry is borderline tall, I can know his exact physical measurements are m but still fail to know whether he is tall or not. In this case, it is *epistemically* possible for me that Harry is tall and has measurements m, and also epistemically possible for me that he is not tall and has measurements m. It would be a mistake to conclude that these epistemic possibilities are metaphysically possible. One of the two

is linguistic. For in that case, all propositions are precise, and in particular, all true propositions describable in vague language are precise, so it is trivial that the vaguely describable truths supervene on the precise.

possibilities is metaphysically impossible, although it is indeterminate which. Thus, strong supervenience is not undermined, even though a principle formally analogous to strong supervenience involving epistemic notions will presumably be false. (It is interesting to note, however, that it is extremely natural to think that the epistemic analogue of weak supervenience is true. In general, we can know that if x and y have exactly the same precise measurements m, they are either both tall or both not tall (even if we don't know which), due to the existence of a penumbral connection between the tallness of x and of y.)

The weak/strong distinction is only relevant when we are talking about one set of properties supervening on another. My main concern here, however, is the supervenience relations that hold between two sets of *propositions*—the vague and the precise—as captured by the principle SUPERVENIENCE.

Note that the supervenience of everything on the precise is entailed by theses such as physicalism, the thesis that everything supervenes on physical facts, in conjunction with the claim that physical propositions are precise. Indeed, one might think that physicalism is strictly stronger than we need: according to some philosophers, there is a basic layer of precise fundamental facts (whether physical or not) upon which all truths supervene. As I argued in chapter 12 and in chapter 14, I do not accept either of these motivations. The nearest thing to a basic layer of precise propositions are the precise propositions themselves. The supervenience of the vague on the precise is thus an important part of the view I have been advocating for, but it is a substantive hypothesis on its own: it is not entailed by some other more basic supervenience principle such as physicalism.

The constraint that ensures that SUPERVENIENCE holds at an index i, is that every maximally strong consistent precise proposition either contains exactly one metaphysically possible index, or none. Formally, this amounts to the condition that, for every $Z \in Orb(G(i))$, $|R(i) \cap Z| \leq 1$ (and we can say that supervenience is valid in a frame if this condition holds for every index). The rationale behind this constraint is quite intuitive: if some maximally consistent precise proposition, Z, was consistent with two distinct metaphysical possibilities, then there are two different ways things could have been with all the same precise facts obtaining, and this is tantamount to denying supervenience.

This constraint also has upshots for the interaction between the accessibility relations for the modal and determinacy operators. Firstly, recall that in chapter 12 we introduced the principle:

NECESSITY OF THE PRECISE: If p is precise, p is necessarily precise.

This corresponds to the following condition: if Rij then $Orb(G(i)) = Orb(G(j))$. The division of logical space into cells (orbits) of maximally strong consistent precise propositions is the same between modally accessible indices.

From these considerations we may derive a frame condition that applies to frames containing only the R and S accessibility relations, interpreting the language containing Δ and $\Box$ (but not containing precision or propositional quantifiers). Let us write $S(j)$ for the set $\{k \mid Sjk\}$. It follows that if Rij then since $S(j)$ is a member of the partition, $Orb(G(j))$ is also a member of the partition $Orb(G(i))$. By the supervenience condition, $|S(j) \cap R(i)| \leq 1$ whenever Rij. Since $j \in S(j) \cap R(i)$, this just becomes $S(j) \cap R(i) = \{j\}$. Furthermore, if we assume that R is an equivalence relation, we can infer that $R(i) = R(j)$ whenever Rij, so we can simplify this constraint to $S(i) \cap R(i) = \{i\}$ for all i. Putting this together with our earlier discussion, we can derive the following principle:

If Rij, Rik, Sju and Sku then $j = k$.

Suppose that Rij, Rik, Sju, and Sku. Since we required that it never be contingent what is precise, $Orb(G(i)) = Orb(G(j)) = Orb(G(k))$, so either $S(j) \cap S(k) = S(j) = S(k)$ or $= \emptyset$. Since $u \in S(j) \cap S(k)$, $S(k) = S(j)$. But since $R(i) \cap S(j) = \{j\}$ and $R(i) \cap S(k) = \{k\}$ it follows that $k = j$. Note that this condition rules out the possibility that some false but metaphysically possible index is not determinately false and other similar pathological scenarios, but it does not rule out the possibility that some false but metaphysically possible index is not determinately *determinately* false.

To rule out these apparently pathological situations, we might appeal to something stronger than the supervenience of everything on the precise. As we have noted already, the property of being a precise proposition has borderline cases; we cannot therefore infer from the fact that p is precise that the proposition that p is precise is itself a precise proposition. Recall that we called a proposition precise[2] if it is precise, and is such that the proposition that it is precise is itself precise, and that more generally that a proposition is precise^{n+1} if and only if it is both precise and the proposition that it is precise^n is also precise. This notion is definable in the object language by the iterative definition $\sharp^{n+1}p := \sharp p \wedge \sharp\sharp^n p$.[14] Finally we may say something is precise* if and only if it is precise^n for each natural number n (there is no analogue of this notion in the formal language we have been considering that permits only finite conjunctions). These definitions can be continued into the transfinite, although, for my discussion, this will not be necessary.

With these distinctions at hand, we can then introduce stronger forms of supervenience as follows:

SUPERVENIENCE^n: Necessarily, for every truth, there is a precise^n truth that necessitates that truth.

$$\Box\forall p(p \rightarrow \exists q(\sharp^n q \wedge q \wedge \Box(q \rightarrow p))).$$

Here 'n' can be substituted for any numeral (or for the '*' symbol to get the strongest form of the principle).

---

[14] Note that the first conjunct is required to ensure that every precise^n proposition is precise. The formula $\sharp\sharp p$ is consistent with $\neg\sharp p$ since it can be a completely precise fact that a proposition is not precise.

It is worth noting that some of the considerations that motivated the supervenience of everything on the precise extend straightforwardly to these strengthenings too. According to a natural thought (albeit, one I rejected in section 12.2), propositions about the physical—about the locations, velocities, and fundamental properties of particles and fields and so on—are not only precise, but precise^n for any number n. If this thought is correct, then physicalists, for example, should accept the strengthened supervenience claims. Analogous cases can be made that other putative supervenience bases are precise*.

Note, however, that some philosophers argue that very little is precise*. Williamson, for example, suggests that not even propositions stateable in the language of fundamental physics are perfectly precise (Williamson [156], chapter 6) and Dorr [35] argues that even the logical truths fail to be precise*. Such philosophers will not endorse the strengthened supervenience claims. Perhaps there are precise supervenience bases, and precise^2 supervenience bases, but as n increases the number of supervenience bases that are precise^n will decrease until, presumably, none are left. The truth of SUPERVENIENCE^n at an index i requires the following condition:

If Rij, Rik, $S^n ju$ and $S^n ku$ then $j = k$.

Where $S^n ij$ holds iff there is a sequence of n indices, $x_1, x_2 \ldots, x_n$, with $i = x_1, j = x_n$, and with $Sx_k x_{k+1}$.[15]

What are we to make of the supervaluationist who does not have a primitive distinction between the precise and imprecise at hand? For these philosophers, the relevant principle is that everything supervenes on the facts that couldn't have been borderline:

Necessarily, every truth is necessitated by a truth that couldn't have been borderline.

$$\Box \forall p(p \to \exists q(\Box(\Delta q \vee \Delta \neg q) \wedge q \wedge \Box(q \to p))).$$

This principle is a consequence of SUPERVENIENCE: since everything supervenes on the precise, and each precise fact couldn't have been borderline, everything supervenes on facts that couldn't have been borderline. Perhaps more surprisingly, if we defined Δ from $\sharp$ as in chapter 12, the above principle is true in a model only if SUPERVENIENCE is, removing any logical difference between my general statement and the above one.[16] Thus, SUPERVENIENCE is model-theoretically equivalent to a formula in the pure $\Box$, Δ language. The importance of this observation is that what seemed initially to be a principle only available to those who take precision as primitive can be stated equivalently by someone taking a determinacy operator as primitive instead.

---

[15] Unlike in the $n = 1$ case, it is unclear whether this condition is *sufficient* for the truth of SUPERVENIENCE^n at an index i when $n \geq 2$. It is possible that this question is more tractable if we assume that the precise^2 propositions form a complete Boolean algebra (a hypothesis that does not follow from the claim that the precise propositions form a complete Boolean algebra (see section 13.4.1)).

[16] Proof sketch: for each truth p, at i, there is a truth q, which couldn't have been borderline at i. To get a precise truth that necessitates p simply pick the proposition $q' = \bigcup \{S(j) \cap q \mid Rij\}$.

15.4 A Representation Theorem?

Given the assumption that the algebra of propositions forms a complete atomic Boolean algebra, we know that it is always possible to think of propositions as sets of indices and the modal and determinacy operators as certain kinds of mappings on these sets (see section 3.2). It is natural to wonder what these mappings and indices must look like for the indices to be representable by ordered pairs, and the modal and determinacy operators by mappings determined by supervaluational clauses involving variation over the two coordinates. It is also natural to ask when such a representation is unique. These questions are of clear importance to the supervaluationist, for if a representation of this sort is always possible under reasonable assumptions, then it is always possible to introduce worlds and precisifications as certain sorts of abstractions. On the other hand, if such a representation is not always possible, we get a clearer understanding of what substantive assumptions the supervaluational framework imposes on us. In this section we investigate these questions.

Recall that the supervaluational semantics is a special case of the general Kripke semantics. For given a set of worlds W, precisifications V, a relation R on W and S on V, we can turn this into an ordinary Kripke model by the following process. We begin by setting our indices, $\mathcal{I}$, to be the set of ordered pairs from W and V, $W \times V$. Then we introduce two relations, R' and S', on $\mathcal{I}$ defined so that $R'\langle w, v\rangle\langle x, u\rangle$ if and only if $v = u$ and Rwx, and $S'\langle w, v\rangle\langle x, u\rangle$ if and only if $w = x$ and Svu. The result is a Kripke frame that satisfies all the conditions in Table 15.1. Usually, R is the universal relation on the set of worlds, in which case the R' relation defined above will be an equivalence relation that partitions the indices (the world–precisification pairs) into a number of equivalence classes equal to the number of precisifications.

The converse to this result is certainly not true: there are plenty of Kripke models (I, R, S) that are not isomorphic to any supervaluational model. We already noted, for example, that a Kripke semantics does not in general validate the product logic. However, the unrepresentability of Kripke models in supervaluational ones remains true even if we insist that the Kripke models in question satisfy all the conditions in Table 15.1 corresponding to the product logic.

Let us examine two possible ways in which this can happen. The first has to do with the cardinality of the frame. In any finite supervaluational model based on ordered pairs of worlds and precisifications, the total number of ordered pairs is just the result of multiplying the number of possible worlds by the number of precisifications. So if the total number of pairs is a prime number then either there is only one world or one precisification, and it is thus a model in which either there is no indeterminacy anywhere in the model or no contingency anywhere in the model. On the other hand, for any Kripke model, there is an operation that increases the size of the model by any finite number but makes exactly the same modal formulae true.[17] Thus, take

---

[17] Given a Kripke model $\langle W, R, S, i, [\![\cdot]\!]\rangle$ such that $x \in W$ and $x' \notin W$, and such that R and S are reflexive, one can construct a new model whose cardinality is one bigger: $\langle W', R', S', i, [\![\cdot]\!]'\rangle$. $W' = W \cup \{x'\}$, $R'yz$ iff Ryz or $[y = x'$ and $Rxz]$ or $[z = x'$ and $Ryx]$ (and similarly for S'), and $[\![P]\!]'_y = [\![P]\!]_y$ if $y \in W$ and $= [\![P]\!]_x$ if $y = x'$. Repeat this operation as required to increase the cardinality by other finite numbers.

any finite model of your logic which makes the formula $\nabla A \wedge \neg \Box A \wedge \neg \Box \neg A$ true somewhere in the model (since this formula is clearly consistent—it just says that A is both borderline and contingent—it is natural to think that whatever constraints you put on your Kripke models there will be a model like this). Now, find a finite model which makes the same formulae true at each index but has a prime number of worlds: this can be achieved by repeatedly increasing the size of the model by one until you reach a prime number. This model cannot be represented by ordered pairs of worlds and precisifications since, as noted above, any such model with a prime number of indices is one in which there is either no contingency or no indeterminacy at any index, but by stipulation, the formula stating that A is contingent and borderline is true somewhere in this model.[18]

The second barrier to representing a Kripke frame by a supervaluationist frame arises when we allow failures of the product logic: that is, when we allow the last three constraints from Table 15.1 to fail. For example, consider the following model—the *one-way roundabout*. In this diagram the dotted lines represent the S relation, and the solid lines the R relation (both relations are implicitly assumed to be reflexive, although, to reduce clutter, I shall not represent this on the diagram).

This model satisfies all but the last three principles of Table 15.1. Notice that cardinality does not seem to be a barrier to our representing this Kripke frame. It looks as though it could be modelled using two worlds and two precisifications giving us a total of four pairs. But when we try to define the accessibility relations over these two precisifications we find that we cannot because the determinacy accessibility relation goes in opposite directions on the top pair than on the bottom pair.[19] Similar points apply to any Kripke frame that contains the one-way roundabout as a subframe.

Why should we be interested in these sorts of results? A supervaluationist who wanted to put the formalism of worlds and precisifications in good standing might want to have some kind of general guarantee of the following form: whatever the

---

[18] This argument does not depend on whether we model vagueness and modality with binary or ternary accessibility relations that we introduced earlier. This argument doesn't obviously generalize to infinite models, although one can establish the same point through different arguments there.

[19] Any supervaluational model that has four pairs must either have two worlds and two precisifications, four worlds and one precisification, or one world and four precisifications; only the first kind of model can represent the one-way roundabout. In fact, this argument works even if we try to represent it with a subframe of a supervaluational model: we know that the two leftmost indices have the same world coordinate, and analogously for the two rightmost indices, because they are related by R. Similarly, the top indices share the same precisification coordinate, and the bottom indices share the same precisification coordinate. So the indices of our model are of the following form: $\{w, x\} \times \{u, v\}$. Finally, note that we cannot define S over $\{u, v\}$ to produce the one-way roundabout model, due to the differing direction of accessibility between the topmost and bottommost indices.

intended model looks like, there is some canonical way of associating each index (e.g. each maximally strong consistent proposition) in the intended model with an ordered pair from a fixed pair of sets, W and V, and to reconstruct the accessibility relations purely from two independent relations on each of those two sets respectively. The guarantee that the supervaluationist is looking for is a *representation theorem*. For a supervaluationist, it would be extremely nice to have a representation theorem, because it would suggest that certain features of that semantics—such as the notions of a worldly proposition, possible world, and precisification—aren't just peculiarities, but are forced on us merely by the acceptance of certain logical principles governing determinacy and necessity.

Here is the idea in a little more detail. The intended model of the $\Box \triangle$ language can be described as follows:

1. The set of indices I is the set of maximally strong consistent propositions.[20]
2. A proposition p is true at an index i iff i entails p.
3. Rij hold iff every proposition that's necessary at i is true at j.
4. Sij hold iff every proposition that's determinate at i is true at j.

Here we assume a suitably broad notion of entailment,[21] that propositions form a complete Boolean algebra under that notion of entailment, and that the existence of propositions is neither contingent nor indeterminate. A routine argument establishes that there is an index $i \in I$ (the true index) such that a proposition is true *simpliciter* iff it is true at i in the Kripke model outlined above.

The intended model clearly has the form of a Kripke frame. Now, suppose that every Kripke frame was isomorphic to a unique supervaluationist frame. (That is, there are, up to isomorphism, unique sets W and V, a bijection $g : W \times V \to I$ and relations T on W and U on V such that: Txy iff $Rg(x, v)g(y, v)$ for every $v \in V$ and Uuv iff $Sg(x, u)g(x, v)$ for every $x \in W$.) Then, it follows that the indices of the intended model can be (uniquely) decomposed into ordered pairs of two kinds of entities: the first of which we might as well call worlds, and the second we might as well call precisifications.

What the above considerations seems to suggest is that no such representation theorem will be forthcoming if we are concerned only with the simplest version of the supervaluationist framework outlined in section 12.1.2. There can be no guarantee that we can associate the intended model isomorphically with a supervaluationist model. Reflection on the one-way roundabout model also seems to suggest that there can't even be a way to canonically transform (non-isomorphically) an arbitrary

---

[20] If the set of such propositions is too large to form a set, there are other techniques for describing the intended model using higher-order resources.

[21] Broad enough that a proposition being valid implies the result prefixing arbitrary strings of $\Box$s and $\triangle$s to that proposition. The supervaluationist we have been considering has such a notion available, for example, as does anyone who accepts a propositional identity connective (see chapter 11).

Kripke model into a supervaluationist model; in that case it is totally unclear which representable model the one-way roundabout model would get transformed to, or what the significance of the transformation would be if there was one.

There are, however, a couple of ways we could attempt to generalize the supervaluationist semantics to get around the two barriers to representation encountered earlier. The first problem was that every supervaluationist model with a prime number of indices either had no contingency or no indeterminacy: this arose because the number of indices is always the product of the number of worlds and precisifications. A natural way to avoid this is to allow for some ordered pairs of worlds and precisifications to be absent from the model. In describing a generalized supervaluationist frame we might relax the constraint that $I = W \times V$ to the constraint that:

> The set of indices I is a *subset* of $W \times V$ where W is the set of worlds and V the set of precisifications.

The second issue is that the supervaluationist semantics validates the product logic. But in section 15.1, we gave special reason to think that there are instances of **ND** and **PD** that are false and that, in particular, the intended model won't satisfy the conditions corresponding to these principles in Table 15.1. The product logic arises because the supervaluationist semantics imposes the constraint that the accessibility relations over ordered pairs had to be generated by two independently moving accessibility relations over W and V respectively: the modal accessibility relation, when given two pairs, only looks at the first coordinate of both pairs to decide whether it holds between them, and the determinacy accessibility relation only looks at the second coordinates. The most natural way to relax this constraint would be to have, instead, two quaternary relations over both domains both relating a world and a precisification to another world and another precisification where they both, as it were, can 'see' both coordinates of both pairs. Thus, for example, the modal and determinacy clauses become:

1. $x, u \Vdash \Box\phi$ if and only if $y, v \Vdash \phi$ for every y and v such that $Rxuyv$.
2. $x, u \Vdash \triangle\phi$ if and only if $y, v \Vdash \phi$ for every y and v such that $Sxuyv$.

R and S correspond to binary relations over ordered pairs in a straightforward manner.

I will not go into the resulting theory in too much detail. Suffice it to say that it does allow one to prove a representation theorem: given a Kripke frame F, the rough idea is to simply replace the indices in F with ordered pairs.[22]

However, the above argument just emphasizes that the relaxed supervaluationist semantics is no more than a notational variant of the Kripke semantics, in which we are simply stipulating that the indices are ordered pairs. It is not surprising that,

---

[22] More precisely, one finds sets $W \times V$ such that the cardinality of $W \times V$ is at least as big as I (which is always possible, even when I is finite), and a subset $I' \subset W \times V$ which has exactly the same cardinality as I. Let f be any bijection between I and I' and let $R'ij$ iff $Rf(i)f(j)$ and $S'ij$ iff $Sf(i)f(j)$.

given a Kripke frame, we can replace the indices with ordered pairs, if the ordered pair structure doesn't place any constraints on the accessibility relations. (In the abstract Kripke semantics, indices can in general be any old things, ordered pairs or otherwise.)

The representation theorem in the quaternary semantics is also radically non-unique. We could, for example, represent the intended model in a generalized supervaluationist frame that has only one precisification. Normally, this would imply that there was no indeterminacy, but the general supervaluationist semantics relaxes the connection between indeterminacy and precisifications: the Δ-accessibility relation doesn't have to correspond to variation in the precisification coordinate. More generally, every Kripke frame can be represented by a generalized supervaluationist frame with only one precisification. Similarly, we can also represent the intended model by a different generalized supervaluationist frame in which there is only one world. Usually, this would imply there was no contingency, but, as before, the generalized semantics relaxes the relation between worlds and contingency. This is all because the ordered pair structure isn't playing any role in this semantics whatsoever—constancy over the world coordinates needn't correspond to necessity, and constancy over the precisification coordinates needn't correspond to determinacy. We might as well have been modelling everything with unstructured entities and a pair of binary accessibility relations, as one does in the Kripke semantics.

Perhaps there is something like a happy medium in which some of the ordered pair structure plays a role in the semantics but not enough to prevent a representation theorem. If we replaced one or both of the binary accessibility relations in the supervaluationist model with a ternary relation, for example, we would have something intuitively in between the binary and quaternary proposal.

In what follows, I shall consider keeping the binary modal accessibility relation, but replacing the binary determinacy accessibility relation with a ternary relation $S \subseteq W \times V \times V$. The ternary accessibility relation, when given two world–precisification pairs, will look at both coordinates of the first pair, but only the second coordinate of the second pair, to work out whether it holds between them. The effect is to allow that S-accessibility depends on the possible world—one could think of this as allowing the notion of a precisification being 'admissible' to be contingent: u can be admissible relative to v according to some worlds but not others. The semantics would be the same as the supervaluationist semantics, except that the clause for the determinacy operator would be: $w, v \Vdash \Delta\phi$ if and only if $w, u \Vdash \phi$ for every u such that $Swvu$. As before, we also relax the requirement that once we have chosen W and V, every ordered pair in $W \times V$ must correspond to the index of a model: the indices are a (possibly proper) subset of $W \times V$.

Unfortunately, these modifications alone do not suffice for a representation theorem. However, if we restrict attention to Kripke models with certain structural features, then a representation theorem for the modified supervaluational semantics is possible:

Theorem 15.4.1. *Suppose that $\langle I, R, S \rangle$ is a Kripke model. Let S^+ and R^+ denote the transitive symmetric closures of R and S respectively.*

If $S^+ \cap R^+ \subseteq = (i.e.\ S^+(i) \cap R^+(i) \subseteq \{i\}$ for every $i \in \mathcal{I})$ then $\langle \mathcal{I}, S, R, i \rangle$ is isomorphic to a modified supervaluationist model $\langle I' \subseteq W \times V, R', S' \rangle$ where $R' \subseteq W \times W$ is a binary relation and $S' \subseteq W \times V \times V$ is a ternary relation and I', the set of indices for the model, is a subset of $W \times V$.

The proof of this representation theorem is not as straightforward.[23] It establishes the existence, for any Kripke frame, of an isomorphic supervaluationist frame of the required type, but not the uniqueness.

Does this representation theorem put the supervaluationist's distinction between worldly and non-worldly propositions, and the analogous distinction between precisifications, in good standing? I think the supervaluationist should find this representation theorem unsatisfactory for a few reasons. The most pressing is that, although not as bad as the quaternary proposal, we have no general guarantee that this representation will be unique. There could be two equally good ways of partitioning the sets of indices into world propositions and precisification propositions, such that the two binary relations of the Kripke model can be reduced to a binary and a ternary relation defined on the two respective partitions. In such cases, nothing bearing the title 'world' and 'precisification' can be uniquely recovered.[24]

Another problem, which I think is just as pressing, is that the assumptions needed to prove this theorem (i.e. that $S^+ \cap R^+ \subseteq =$) are incredibly strong. They are not guaranteed even by imposing all of the conditions listed in Table 15.1. The very simple frame below, for example, satisfies all those properties, but it is not possible to represent it even in the binary/ternary way described earlier:[25]

---

[23] The basic idea is to let the worlds be equivalence classes under S^+, and let precisifications be equivalence classes under R^+. We can then define (i) $\square$-accessibility and (ii) $\triangle$-accessibility relations R' and S' as follows. (i) $\langle [i]_{S^+}, [j]_{R^+} \rangle R' \langle [k]_{S^+}, [l]_{R^+} \rangle$ iff $[i]_{S^+} = [k]_{S^+}$ and Rik. (ii) $\langle [i]_{S^+}, [j]_{R^+} \rangle S' \langle [k]_{S^+}, [l]_{R^+} \rangle$ iff $[j]_{R^+} = [l]_{R^+}$ and Spq where $\{p\} = S^+(i) \cap R^+(j)$ and $\{q\} = S^+(k) \cap R^+(l)$. The resulting supervaluationist frame has the requisite form, and is also isomorphic to our original frame.

[24] Another theorem along these lines is proved in Kurucz and Zakharyaschev [87], Lemma 2.1: if $\langle I, R, S \rangle$ is a frame for **S5** and **KT** respectively, then it is the image via a bounded morphism of a subframe of a product frame. Unlike the representation theorem, we do not get a structure isomorphic to the original. Moreover, like the representation theorem, there will typically be several product frames which stand in this relation to the original, so there will be no unique way of assigning world propositions and precisification propositions to a frame like this.

[25] The top two points must have the same precisification coordinate, because they are modally related. Also the ternary relation guarantees that if a pair Ss another pair, then it Ss any other pair with the same precisification coordinate. Since the top left point Ss itself and has the same precisification coordinate as the top right, it must S that too which contradicts the diagram.

Lastly there is the issue that for this representation theorem to work, as we argued earlier, some ordered pairs must be left out of the model. But this seems to go against some of the original motivations for the supervaluationist semantics. Intuitively, a precisification is supposed to tell you not only where the cutoff points in fact are but where they would have been, had the worldly facts been different. For every way the world could have been, the precisification tells us where the cutoff points would have been had the world been that way, and so every way of combining a world with a precisification produces a coherent description of how things are. To make the representation theorem work, we have to drop some of these combinations from the model, which is to deny some of the heuristic value of the formalism that made it attractive in the first place.

Let us take stock. In section 15.1, we observed that the most straightforward version of supervaluationist semantics is committed to the product logic, and we saw how this conflicts with the plausible hypothesis that metaphysical necessity is vague. In the present section, we developed a more general supervaluational semantics which is capable of representing failures of the product logic, and proved a representation theorem. But even in this context we saw that the supervaluational semantics imposes fairly strong and non-obvious constraints on the structure of determinacy and modality. What's more, we observed that the structure of modality and determinacy alone was not enough to pin down the structure of worlds and precisifications. There are generally multiple ways of carving up logical space into worlds and precisifications that given rise to the same determinacy and necessity facts.

In section 15.2 and 15.3, we showed how the symmetry semantics introduced in chapter 13 does not impose these constraints, and we saw how various further theses, such as the product logic, and supervenience of the vague on the precise, could be imposed by adding certain constraints to the accessibility relations. In particular, we were able to see that attractive theses, like the supervenience of the vague on the precise, do not entail certain unattractive theses, like the precision of metaphysical necessity or the product logic.

16

Vague Objects

We observed in section 4.4 that an adequate linguistic account of vagueness must provide more than an account of sentential vagueness. One can also have vague predicates, like 'is bald', and further machinery beyond the notion of sentential vagueness is needed to analyse predicative vagueness. Indeed, vagueness can occur at almost every grammatical category: the quantifier 'some bald person', the operator 'it is known that', the determiner 'most', the name 'Mt Kilimanjaro' (and so on) are all vague—it is hard to think of a grammatical category that does not contain vague expressions. It is natural, then, to wonder what analogous kinds of vagueness arise on a non-linguistic theory of vagueness. Of course, in the non-linguistic setting we are not concerned with linguistic items belonging to certain grammatical categories, but with the sorts of things the linguistic items in those grammatical categories express. Reasoning about these sorts of things receives a more precise treatment in the setting of a typed higher-order logic: a theory that allows one to quantify into positions other than the position of a singular term (see, for example, Williamson [165]).

Central to our discussion of vagueness at other types will be the status of the non-linguistic analogue of vague names: *vague objects*. Indeed, we shall see that extending the theory of vagueness to objects, in addition to propositions, will be the key to understanding vagueness elsewhere in the type-theoretic hierarchy. The first question we must get clear on is therefore: *what is a vague object?* After outlining the general framework in section 16.1, I discuss some of the dominant approaches to this question in section 16.2 and 16.3, and in sections 16.4, I develop my own test for being a vague object.

This chapter offers less by way of substantive conclusions than the preceding chapters. One particular question—one might think the *central* question concerning vague objects—is left open: whether there *are* any vague objects. The non-linguistic theory of vagueness defended in this book provides a friendly environment for the discussion of vague objects whereas, by contrast, a linguistic theory of vagueness does not. However, this theory of vagueness does not straightforwardly commit us to the existence of vague objects either.

16.1 Vagueness Throughout the Type Hierarchy

Let us begin by investigating the nature of vagueness in other semantic categories. The natural background setting in which to frame such a discussion is type theory. In what follows we shall assume two basic types: e, representing the type of *individuals*—the semantic values of singular terms—and type t representing the type of propositions—the semantic values of sentences.[1] Complex types can be constructed from these in a recursive fashion: if σ and τ are types then so is $\sigma \to \tau$. In the usual set theoretic models for type theory, the domain of the latter sort of type consists of the set of all functions from the domain σ to the domain of τ, and I shall often refer to these things as functions for convenience. However, these models should not be taken too seriously: strictly speaking, sets, and thus functions, are individuals, and in the intended model they will all belong to the basic type e. Quantification at higher types cannot be straightforwardly thought about in terms of quantification over entities at all, set theoretic or otherwise. (Some of these issues are fraught; for a defence of the intelligibility of this sort of quantification, see Prior [112] and Williamson [161]).

In this setting, the semantic values of a variety of expressions can be represented quite easily. A predicate—'is red' for example—determines a mapping from individuals to propositions, in this case the function that maps x to the proposition that x is red. So, predicates belong to the type $e \to t$.[2] An operator expresses a function from propositions to propositions, $t \to t$, a quantifier takes a predicate to a sentence and so expresses a function in $(e \to t) \to t$, a determiner takes a predicate to a quantifier and so expresses things of type $(e \to t) \to ((e \to t) \to t)$, and so on. The logical expressions also fit into this hierarchy: conjunction (written $\wedge$) of type $t \to (t \to t)$, negation (written $\neg$) of type $t \to t$ and, for each type σ, a universal quantifier (written $\forall_\sigma$) over functions of that type, which has the type $(\sigma \to t) \to t$.

I began by noting that each of these grammatical types are occupied by vague expressions. On the sort of theory I am espousing, this vagueness results from vagueness in the semantic values of these expressions—of the functions belonging to the types $t \to t$, $(e \to t) \to t$, and so on. However, although I have given a treatment of vagueness in type t (i.e. an account of vague propositions) I have said very little to connect this with vagueness at other types.

The standard line is to define a predicate F as vague if it has a borderline case: $\exists x \nabla Fx$. Clearly, such a definition at best captures an extensional notion of vagueness for predicates.[3] But even setting that aside, it is subject to an objection, under

---

[1] Some authors take t to be the set of truth values and introduce intensionality elsewhere, such as in the construction of the function space (see e.g. Gallin [60], Montague [108]). In many cases, such approaches are just notational variants of the one I am presenting, although I believe the presentation chosen above to be conceptually clearer.

[2] In this framework n-ary relations are coded up in terms of unary functions. A binary relation would be a function of type $e \to (e \to t)$.

[3] One might hope to define the non-extensional notion by modalizing it—$\Diamond \exists x \Diamond \nabla Fx$—but such a strategy is subject to worries similar to those raised in chapter 12.

the assumption that there are vague objects. The property of being exactly 19,341 feet is intuitively precise, but assuming Mt Kilimanjaro is a vague object with an indeterminate height, it has a borderline case; i.e. there is some x (Mt Kilimanjaro) such that it is borderline whether x is exactly 19,341 feet tall.

A natural way to fix this is by taking the notion of precision for both of the basic types as primitive. That is, we will begin by taking for granted both the notion of a precise proposition and a precise object. With the notion of precision at type e and type t given, we can extend the definition up the type hierarchy recursively.[4] Suppose that we have already defined precision at types σ and τ:

> PRECISION AT HIGHER TYPES: An entity F of type $\sigma \to \tau$ is precise if and only if F maps every precise thing of type σ to a precise thing of type τ.[5]

The informal statement hides the fact that, to state, it we need to employ higher-order quantification. To spell out the idea in a bit more precision, suppose that we have introduced higher-order predicates $\natural_\sigma$ and $\natural_\tau$ of type $\sigma \to t$ and $\tau \to t$ respectively. Then $\natural_{\sigma \to \tau} := \lambda X (\forall y (\natural_\sigma y \to \natural_\tau Xy)$. Let me emphasize that here we are not theorizing with the extensional notion of precision. This has an important conceptual advantage: although the extensional notion of precision for propositions—determinacy—seems clear enough (and can be defined in terms of precision), it's not entirely clear what the extensional analogue of being a precise object is. Indeed, the way we defined determinacy from precision in the propositional case relied on features that were distinctive of propositions: that they had a logical ordering given by entailment (recall that we defined a proposition to be determinate if it is entailed by a precise truth). There is no straightforward analogue of entailment in the objectual case, and thus no obvious analogue to our notion of propositional extensional determinacy for objects.

A compelling principle governing linguistic expressions states that a complex expression is vague only if one of its constituents is vague or, conversely, that if the constituents are precise, so is the composite. Our definition of precision at higher types entails a similar principle in the non-linguistic setting: combining precise things can only deliver precise things. However, this definition makes sense even given a relatively coarse-grained theory of content. The notion of a 'constituent' is absent

---

[4] The following way of extending the notion of vagueness to higher types is essentially due to Bertil Rolf [121]. However, it should be noted that, in contrast with this project, Rolf is characterizing the extensional notions of vagueness and precision. Moreover, Rolf attempts to fix the problem with the naïve definition not by restricting the quantifiers to precise objects (understood as a new primitive notion), but by restricting quantification to objects with precise spatial boundaries (thus, in principle, allowing us to define vagueness at type e in terms of the type t notion of borderlineness and an occupation relation O: x is vague iff $\exists y \nabla Oxy$). As I will argue shortly, however, it seems unlikely that the notion of a vague object can be understood solely in terms of its standing in vague location relations.

[5] A slight modification to this definition is needed if you believe some individuals, propositions, or higher-order entities exist only contingently: F is precise iff it's necessary that, for every precise x of type σ necessarily, Fx is precise at type τ. On the assumption of higher-order necessitism—that things at all types exist necessarily if at all—this definition should be equivalent given the plausible assumption that things at type e and t are necessarily precise if precise at all, and necessarily vague if vague.

from the definition—it is phrased in terms of functional application—and so we do not need to assume that propositions have constituents in the same way that sentences do.

In summary, to generalize the propositional notion of vagueness up the functional type hierarchy we need to take the notion of a vague object and vague proposition as primitive. Although we have paid considerable attention to the latter notion, we have said relatively little about the former. We will now turn to some popular accounts of the distinction.

16.2 Vague Identity

In 'Can there be vague objects?' [42] Gareth Evans interprets the question of whether there could be vague objects as a question concerning whether there could be borderline identity statements: true statements of the form $\nabla a = b$. As Evans formulates it, the claim at hand involves applying a propositional borderlineness operator to a statement; it is therefore a claim one can perfectly well understand on the present theory of vagueness, although it is not the sort of thing a linguistic theorist can straightforwardly make sense of. Note that if Evans' characterization of vague objects is correct, then it may be possible to define a vague object in terms of the propositional notion of vagueness and the identity relation, eliminating the need for a primitive notion of a vague object.

Care must be taken here to distinguish Evans' thesis from the following linguistic claim: there can be sentences of the form '$a = b$' such that it is borderline whether they are true (i.e. it is borderline whether they express a truth). The latter is possible even if there are no vague objects: it could, for example, be vague which precise object the name 'Princeton' refers but not (as) vague which precise object 'Princeton Borough' refers to, so that it is borderline which proposition 'Princeton = Princeton Borough' expresses. In particular, it is borderline whether this sentence expresses a false proposition (stating the identity of two distinct geographical regions) or a true proposition (stating the identity of two identical geographical regions). So, it is borderline whether this identity sentence expresses a truth, even on the assumption that all the relevant objects are precise.

Of course, it is manifest that there are vague names and thus, presumably, identity sentences such that it is borderline whether they are true. But in such cases we have no reason to think there are truths of the form $\nabla a = b$ (only truths of the form: ∇ '$a = b$' is true) and so, by Evans' criterion, no reason to think that there are vague objects. Indeed Evans famously gave an argument that there can be no truths of the form $\nabla a = b$.

1. $\nabla a = b$.
2. $\neg \nabla a = a$.
3. $a \neq b$ (from 1 and 2 by Leibniz's law).

What this argument uncontroversially establishes, granting Leibniz's law and the claim that it's not borderline whether a is identical to itself, is that if it's borderline whether a and b are the very same object, then a and b are, in fact, not the same object. But this falls short of an argument that it's not borderline whether a and b are the same. For all we've said it could be that a and b are two distinct objects that are borderline identical (or, in other words, a and b are distinct but not determinately so).[6] Anyone who accepts excluded middle must make peace with the fact that there are truths of the form 'p but it's borderline whether p', so the mere fact that all vague identity claims have this form is no argument against them.[7]

A simple model demonstrates the consistency of this vague identity with Leibniz's law and the determinacy of self-identity ($\neg \nabla x = x$). A model will consist of a set I of indices, a reflexive accessibility relation $S \subset I \times I$, a domain D, and, for each index i, an equivalence relation $\sim_i$ on D representing the identity relation. Each n-ary atomic predicate P gets an interpretation $[\![P]\!]_i \subseteq D^n$ consisting of a set of ordered n-tuples relative to each index. The atomic letters are required to abide by the restriction that if the ordered tuple $(\ldots, x, \ldots) \in [\![P]\!]_i$ and $x \sim_i y$, then $(\ldots, y, \ldots) \in [\![P]\!]_i$. In other words atomic predicates cannot distinguish between entities that are considered identical at an index. (Note that since the interpretation of $=$ at i, $[\![=]\!]_i$, is $\sim_i$, it automatically satisfies this constraint in virtue of being an equivalence relation.) Finally, we require that if Sij then $\sim_i \subseteq \sim_j$—that is, if i sees j then every identity that holds at i holds at j.

Notice that, although we have stipulated it in the case of the atomic predicates, there is no *a priori* guarantee that *non-atomic* predicates will be unable to distinguish between identical entities. This is what the last condition is there to ensure. Given the last condition, one can prove, by a straightforward induction, non-atomic predicates do not distinguish identical entities.[8] Once this is established, Leibniz's law $x = y \rightarrow (\phi \rightarrow \phi[x/y])$ is guaranteed to hold for any ϕ.

Now consider the model consisting of two indices, i and j, in which each index sees itself and i sees j, and a set of individuals $\{x, y\}$ which bear $\sim_j$ to one another but not $\sim_i$. Since this model has the requisite features, it follows that every formula has the feature that it can't distinguish between 'identical' entities at an index and, consequently, that Leibniz's law is satisfied. Moreover, $x = y$ is false at i and true at j; since i sees an index where $x = y$ is true and false $\nabla x = y$ is true at i.

Notice that the accessibility relation in our model is not symmetric, and so the principle B—$A \rightarrow \Delta\neg\Delta\neg A$—is invalidated. In fact, it turns out that given B one

---

[6] See McGee [103].

[7] I am here using 'truth' in the deflationary sense. The claim follows since given excluded middle, ∇p entails $(p \wedge \nabla p) \vee (\neg p \wedge \nabla p)$, as we noted in chapter 2.

[8] Suppose ϕ and ψ are formulae with free variables $x_1, \ldots, x_n$ that have the property that, for any i, whenever the sequence $\ldots x \ldots$ satisfies ϕ at i and $x \sim_i y$ then $\ldots y \ldots$ satisfies ϕ. It follows quite straightforwardly that $\neg\phi$, $\phi \wedge \psi$, and $\forall x \phi$ also have this feature. Moreover, suppose that $\ldots x \ldots$ satisfies $\Delta\phi$ and $x \sim_i y$ and consider any j such that Sij. It follows by the clause for Δ that $\ldots x \ldots$ satisfies ϕ at j. Since $x \sim_i y$ implies $x \sim_j y$, it follows by inductive hypothesis that $\ldots y \ldots$ satisfies ϕ at j. Since this holds for every j such that Sij, $\ldots y \ldots$ satisfies $\Delta\phi$ at i.

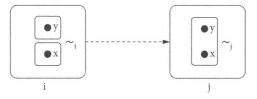

Figure 16.1. A model of borderline identity.

could close up Evans' argument and show that there cannot be any vague identities. However, I think that following this route to Evans' conclusion would be unwise: the principle B is responsible for many of the paradoxes of higher-order vagueness (as I argue in Bacon [4]).

Does Evans' result impose any limitations on the possibility of vague identity? Notice that, in the model shown in Figure 16.1, we have that even though it is borderline at i whether $x = y$, it is also borderline whether it's borderline whether $x = y$—indeed, it can be verified that, relative to i, it is borderline at all orders whether $x = y$. This is no accident: we can prove that any case of a borderline identity will be borderline at all orders. By an analogue of the proof of the necessity of identity one can prove that if $x = y$ then it's determinate that $x = y$.

1. $x = y \rightarrow \Delta x = y$.

For by Leibniz's law we know that $x = y \rightarrow (\Delta x = x \rightarrow \Delta x = y)$, and since it's surely true that $\Delta x = x$, the above conditional follows. Now combining this result with the factivity of determinacy we have:

2. $x = y \leftrightarrow \Delta x = y$.

Thus, $x = y$ is logically equivalent to $\Delta x = y$. Substituting $x = y$ for $\Delta x = y$ twice in succession shows that $x = y$ is also logically equivalent to $\Delta \Delta x = y$ (by the substitution of logical equivalents). We can repeat this as often as we like to see that $x = y$ is logically equivalent to $\Delta^n x = y$ for any n. Thus, from the tautology $\nabla x = y \rightarrow \nabla x = y$ we can infer, by the substitution of logical equivalents, that:

3. $\nabla x = y \rightarrow \nabla \Delta^n x = y$.

By a similar, albeit slightly more complicated argument we can also prove:[9]

4. $\nabla x = y \rightarrow \nabla \nabla^n x = y$.

---

[9] Suppose, for induction, that $\nabla x = y \rightarrow \nabla^n x = y$. On the one hand we have $\nabla^n x = y \rightarrow \neg \Delta \neg \nabla^n x = y$, a simple consequence of the factivity axiom, so $\nabla x = y \rightarrow \neg \Delta \neg \nabla^n x = y$. On the other hand we must appeal to the obvious (but fiddly to prove) principle $\Delta^n x = y \rightarrow \neg \nabla^n x = y$, the determinacy of which permits the conditional $\neg \Delta \neg \Delta^n x = y \rightarrow \neg \Delta \nabla^n x = y$. Then we have the conditionals: $\nabla x = y \rightarrow \nabla \Delta^n x = y$, which we already proved, $\nabla \Delta^n x = y \rightarrow \neg \Delta \neg \Delta^n x = y$ by definition, and the conditional we just appealed to $\neg \Delta \neg \Delta^n x = y \rightarrow \neg \Delta \nabla^n x = y$. Thus by transitivity of $\rightarrow$: $\nabla x = y \rightarrow \neg \Delta \nabla^n x = y$. Putting both of these results together gets us the desired result that $\nabla x = y \rightarrow \nabla \nabla^n x = y$. Here we have freely appealed to the principles of the modal logic KT.

What can we say about the possibility of vague identities? Assuming that vagueness is a barrier to knowledge and, thus, appropriate assertion, it might seem like the answer would be not much: if it's borderline whether x is identical to y, then it's borderline whether it's borderline whether x is y. Thus, if x is indeed borderline identical to y then we cannot be in a position to assert that they are borderline identical: to do so would be to assert something that's itself borderline. Unfortunately, we can't even assert that x and y are borderline borderline identical either: this is because it's also third-order borderline whether $x = y$. More generally, even though x and y are borderline at any given order, we aren't in a position to assert that they are borderline at that order because it's also borderline whether they are borderline at that order (because it is borderline at that order plus one). Similarly, even though it is borderline at all orders whether $x = y$, to assert this would be to assert infinitely many borderline claims (indeed, assuming that borderlineness begets ignorance, for all we know it's determinate at all orders that $x = y$).

In sum, there are no particular examples of borderline identities that we can know about. It is natural to wonder if the situation here is the same as with the existence of boundaries for vague predicates: we can't know of any particular number, N, that it is the last small number, but we can still know that there *is* a last small number, because it's a determinate truth that there's a last small number (as we know by the usual classical reasoning). Even if we can't know any particular objects are borderline identical, could we know the existential claim that there *are* borderline identities? What we need to be able to know is $\exists x \exists y \nabla x = y$, and in order for this to be knowable it had better be determinate, so what we require is a model of:

$$\Delta \exists x \exists y \nabla x = y.$$

It is possible to find a model in which this is true as follows. This time we have three indices, i, j, k, such that i sees j and j sees k and each index sees itself. The domain D consists of three objects x, y, z, such that the following hold. At i, all three objects are identical to themselves only (they bear $\sim_i$ only to themselves). At j, x and y are identical to each other but not identical to z. And at k they are all identical to each other (see Figure 16.2).

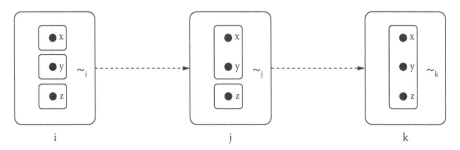

Figure 16.2. A model of the determinate existence of borderline identities. At i, it is determinate that there is a borderline identity, although it's indeterminate whether it is $x = y$ or $y = z$ that is borderline.

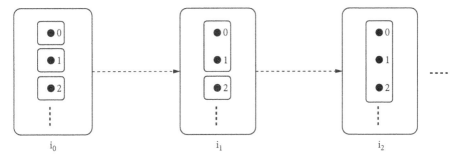

Figure 16.3. A model of the determinate* existence of borderline identities. At i, it is determinate at all orders that there is a borderline identity: for some n it is borderline whether $x_n = x_{n+1}$, although it is vague which n this holds for.

At i, x is borderline identical to y and, at j, x is identical to y and they are borderline identical to z. Since i and j are the only indices accessible to i, it's determinate there are borderline identities (although vague which things figure in those borderline identities). Note, however, that there are no borderline identities at k (everything is determinately identical to everything else, as it were). So the above is not a model of $\triangle\triangle\exists x\exists y\nabla x = y$. Assuming that we adopt the theory of assertion defended in chapter 7, this should not in itself present a barrier to asserting that there are vague identities.

Indeed, to get a model where it's determinately determinate that there are vague identities, we could add another object to our domain, and add another index to the end of our model in the obvious way. The resulting model will validate $\triangle\triangle\exists x\exists y\nabla x = y$ but not $\triangle\triangle\triangle\exists x\exists y\nabla x = y$. This is no accident—one can in fact prove (although we will not do so here):

> If there are at most n things, then it's not determinate^n that there are borderline identities.
>
> $\exists_{\leq n} x\, x = x \rightarrow \neg\triangle^n\exists x\exists y\nabla x = y.$

To get a model where it is determinate at all orders that there are vague identities, one needs infinitely many individuals, x_n, and infinitely many indices, i_n, for each $n \in \mathbb{N}$. Let i_n see i_m iff $m = n + 1$, and let $x_m \sim_{i_n} x_k$ iff $m = k$ or $m, k \leq n$. With a little validation, one can see that x_n and x_{n+1} are borderline identical at the index i_n, so the claim $\exists xy\nabla x = y$ is true at every index in our model, and thus is determinately true at all orders relative to every index in our model.

It is thus consistent that there are vague identities, and indeed consistent to assume that it's determinate at all orders that there are vague identities. Let us suppose for a moment, however, that Evans is correct: that even though it is not logically guaranteed—there are in fact no vague identities. Does it follow, as Evans maintains, that there are no vague objects? That is, could there be vague objects even if there were no vague identities?

The question may be terminological to an extent: I would venture to say that we have little pretheoretic grip on the notion of a vague object, so the answers to such questions will be partly shaped by whether there is a reasonably systematic theory that brings together the various theses people associate with the idea. But I think we can at minimum agree that the existence of vague identities is at least one way for there to be vague objects, if not the only one.

16.3 Vague Parthood

With this in mind, let us examine another criterion that is often cited as sufficient for the existence of vague objects: the existence of objects standing in borderline parthood relations.[10] As a criterion for being a vague object, much needs to be clarified here, but let us begin by getting a handle on the phenomenon.

Following McGee [103], let us suppose that Mt Kilimanjaro is a vague object. It's not a precise matter where the mountain begins and ends, it's not a precise matter exactly how many feet above sea level it is, it's not precise which rocks and stones are parts of it, and so on. It's possible that Mt Kilimanjaro also figures in vague identities: you might think that there is something that's determinately 19,341 feet tall, but that Mt Kilimanjaro, being indeterminately 19,341 feet tall, is indeterminately colocated with that thing. If you also thought that, determinately, colocation of distinct things was impossible, then you might think that Mt Kilimanjaro is indeterminately distinct from that thing. Indeed, one might go further and attempt to explain the indeterminacies about height, boundary, and parthood in terms of the indeterminate identities. Perhaps there are a bunch of entities with precise boundaries and heights and it is indeterminate which of them Mt Kilimanjaro is identical to.[11] But it is unclear how far this strategy will carry us towards the view that Mt Kilimanjaro is uncontroversially a vague object. The picture described seems to be one in which, for all we know, there are no vague objects. For, according to the picture, it's borderline whether our putative examples of vague objects just *are* precise objects. For example, it is borderline whether Mt Kilimanjaro is identical to a precise object (and thus not a vague object after all).

Another strategy would be to attempt to explain the vagueness in Mt Kilimanjaro in mereological terms. Brian Weatherson, however, has argued that, when combined with the assumptions of classical mereology, it's unclear whether this idea is coherent (Weatherson [150]). Let me illustrate some of these difficulties with a simplifying

---

[10] Much of this problem derives from the problem of the many: see Unger [145]. See also Barnes and Williams [8], Lewis [94], McGee [103], Weatherson [150].

[11] Mt Kilimanjaro can't be identical to any of these entities, because they all have precise boundaries, but as we've seen it could still be distinct from them but indeterminately identical to them.

assumption: that there are finitely many objects. Now, given classical mereology we know that the structure of the parthood relation is severely constrained. Indeed, any two scenarios with the same finite number of objects must be mereologically isomorphic: for instance, any two worlds that contain exactly three atoms—a, b, and c—must be the same in the sense that they contain the fusions $a + b$, $b + c$, $a + c$, and $a + b + c$, and that these entities stand in the same configuration of parthood relations. One version of the problem can be stated as follows: assuming that it is not vague what exists and not vague what is identical to what, then it is a determinate matter how many objects there are (since finite numerical claims can be stated using only the quantifiers, identity, and connectives, which we may assume all to be precise). So every admissible precisification must agree on the number of individuals (in keeping with this literature I will temporarily adopt the supervaluationist terminology; all of this can be rephrased in our terminology of accessible indices too). Assume also that classical mereology is determinate, and so true at every admissible precisification: then, by the above argument they must each precisify the parthood relation in an isomorphic way. It follows that the parthood relation imposes a rigid structure on the individuals: a structure that can't vary between precisifications. It seems, then, that vagueness in the parthood relation is severely limited.

However, as Barnes and Williams [8] note, this does not necessarily mean that there can't be any vague parthood at all. Although every precisification has to be mereologically isomorphic in a finite world, there could still be vague parthood because it could be vague which object occupies which mereological role even when the mereological roles form a fixed structure: it could be determinate that there are three objects, a, b and c, that obey classical mereology, but borderline whether a is a part of c because, relative to one precisification, a and b are atoms and c is their fusion, and, relative to another, b and c are atoms and a is their fusion. Matters change also when one moves to the infinite case. One can construct, consistently with standard set theory (i.e. ZFC), two precisifications that agree about how many objects there are, agree that classical mereology is satisfied, and even agree that everything is composed of atoms, but disagree about how many atoms there are, making the two precisifications non-isomorphic.[12] In these cases, even the structure of the parthood relation isn't rigid.

It seems unlikely, however, that these limited kinds of vagueness about parthood could support the sorts of things that those who posit vagueness in parthood typically want to say. It's unclear, for example, how to model the following idea, which is surely consistent: that Mt Kilimanjaro is mostly precise except for a single atom which is such that it is borderline whether it is part of Mt Kilimanjaro. On the Barnes and Williams

---

[12] Indeed, Cohen's original forcing argument was a model in which $2^{\aleph_0} = 2^{\aleph_1} (= \aleph_2)$ (even though $\aleph_0 < \aleph_1$!). In this universe, the powerset of $\aleph_0$ and of $\aleph_1$ are both classical models of mereology with the same number of objects, but are non-isomorphic because they have a different number of atoms. (See also Easton's theorem [37].)

model, such vagueness can only arise at the expense of vagueness elsewhere—if the atom isn't a part of Mt Kilimanjaro relative to some precisification, some other object has to be substituted in its stead. And it is even less clear how the models that play on the strange properties of infinities in consistent extensions of standard set theory help us here.

An extremely natural response to these sorts of worries is to relax classical mereology. Indeed, the puzzle we have been considering above is formally closely analogous to famous temporal and modal puzzles about mereology (see the problem of the statue Goliath and the lump of clay that constitutes it, Gibbard [63], or Geach's puzzle of temporary coincidence outlined in Wiggins [151]). In that context it is not unusual to relax classical mereology to something weaker; however, this is not the place to explore those options in more detail.[13]

It seems, then, that borderline identity and borderline parthood are both at least coherent. However, even granting their coherence it is unclear how either of these things delivers an unambiguous criterion for vague objecthood. The issue is that both notions are relational—vagueness associated with them usually involves at least two objects—and so it is unclear how to identify the source of the vagueness. If it is borderline whether x is identical to y, does it follow that x is a vague object, or that y is a vague object, or both? On a view in which Mt Kilimanjaro is a vague object that is borderline identical to a precise object, X, there can simply be no identity-theoretic answer to this question. Identity is a symmetric relation after all: whatever makes Kilimanjaro the vague object and X the precise one can't be spelled out in terms of identity. A similar question arises for parthood, even though it is not symmetric: if x is a borderline part of y, does that make y the vague object or x, or both?

The general moral follows from a principle we discussed at the end of section 16.1: when it's borderline whether a complex proposition is true, that's due to vagueness in one of the things that compose it. Thus, in particular, for a binary relation R, if it's borderline whether Rxy, then either R is vague, or x is vague, or y is vague. Knowing that a proposition involving parthood or an identity claim is borderline gives us no clue as to what the *source* of the vagueness is. Even assuming we know it's not in the identity or parthood relations, we do not know whether the vagueness is due to the first or second relatum.

Notice, moreover, that there are other plausible sufficient conditions for an object to be vague which involve only monadic properties. For example, if it is borderline whether an object is exactly 19, 341 feet high, then it is also natural to think that that's because the object in question is vague. Or, alternatively, there are relational properties where at least one of the relata is the sort of thing that is (intuitively) not normally

---

[13] Two ways to weaken standard mereology naturally present themselves: weaken the principle of anti-symmetry (explored in Cotnoir [27], and Cotnoir and Bacon [28]) or weaken the supplementation principles (explored in Goodman [66], for example).

vague. It could, for instance, be vague where Mt Kilimanjaro is located.[14] It is not plausible to blame the vagueness here on the location relation, or on the space–time regions that stand in this relation to Mt Kilimanjaro. What we are lacking, however, is some kind of theory that ties all these different kinds of vagueness together: we may have identified some plausible sufficient conditions, but we have not identified a single overarching feature of this sort of vagueness that unifies them as being distinctively to do with vague objects.

16.4 Vague Objects

So far we have made little progress in elucidating the idea of a vague object. Even if it turns out that the notion of a vague object is conceptually basic, it would still be nice to have some sort of independent test for objectual vagueness that might placate those who insist that the notion is too obscure to theorize about.

In the following, I shall rely on the fact that we have a reasonably firm pretheoretic handle on the distinction between vague and precise properties. Unlike the somewhat murky objectual notion, the difference between, say, the property of being bald and the property of having 2,330 hairs or less is manifest to anyone familiar with the phenomenon of vagueness. In that spirit, one might propose the following test for being a vague object:

OBJECTUAL PRECISION: An object x is precise if and only if Fx is precise, whenever F is precise.

More informally, a precise object is one that converts every precise property into a precise proposition, whereas a vague object converts some precise property into a vague proposition.[15] (The test is most intuitive when we focus on properties—functions of type $e \rightarrow t$—but in practice it is sometimes convenient to test for vagueness with functions from e to other types.)

OBJECTUAL PRECISION is a consequence of the account of vagueness at higher types I proposed in section 16.1. To show the left-to-right direction, suppose that o is a precise object, and let F be any precise property. Our definition of property vagueness says that F is precise only if it maps every precise object to a precise proposition. Since o is precise, so is Fo. To show the right-to-left direction we shall show that if o is vague, then there is some precise property that converts o into a vague proposition (a precise property F such that Fo is vague). This direction requires the (by this point uncontroversial) assumption that there is at least one vague proposition

---

[14] Indeed, perhaps one can have vagueness about an object's location that does not derive from vagueness concerning the object's parts—imagine that Mt Kilimanjaro is (determinately) an extended simple, but that it is vague which exact region of space–time it is located at.

[15] As mentioned earlier, a slight modification is needed if it is contingent which properties exist. See footnote 5.

H—the proposition that Harry is bald, say. The argument can be given a model theoretic justification, if we assume that the domain of the type of properties is *full*: for any function f from individuals to propositions there is a corresponding property F—a property such that for any individual a the proposition that a is F is $f(a)$. Just consider the property F, corresponding to any function that maps each precise object o' to some precise truth, such as the contradiction $\bot$, and which maps the vague object o to some vague proposition, H. It follows by our theory of property precision in section 16.1 that F is precise, since $Fo' = \bot$ and is thus precise, whenever o' is precise. It also follows that Fo is vague since $Fo = H$. (Turning this into an argument in the object language argument is a bit delicate, but can be done under reasonable assumptions.[16])

Our principle brings to salience an obvious question. Given that we can identify the precise objects once we know which properties are precise, it's natural to wonder whether we should be taking property precision as basic instead of objectual precision, as we did in section 16.1. Indeed, it's technically possible to reduce everything to property precision, since a proposition p is precise if and only if the constantly p function, $\lambda x.p$, is a precise property. With objectual and propositional precision defined, precision at all other types follows as in section 16.1.[17] Be that as it may, such a reduction is not forced on us. While OBJECTUAL PRECISION may be a convenient way to explain the notion of a precise object to someone who has the notion of a precise property, this is not necessarily the direction of reduction.

We can now check that our test delivers the correct verdict in the case of Mt Kilimanjaro. The property of being exactly 19,341 feet tall is precise, for example, yet applying this property to Mt Kilimanjaro delivers the seemingly borderline and, thus, vague proposition that Mt Kilimanjaro is exactly 19,341 feet tall. It follows by OBJECTUAL PRECISION that Mt Kilimanjaro is not a precise object.

---

[16] The most uncontroversial way to do this is to use Church's $\lambda\delta$-calculus (Church [26]): the system you get from the λ-calculus by adding a conditional operation for making definitions by cases. More precisely, for each type σ and τ, you add a constant δ of type $\sigma \to \sigma \to \tau \to \tau \to \tau$ which is defined so that $\delta abcd$ is equal to c if $a = b$ and is equal to d otherwise. F can then be identified with $\lambda x \delta o x H \bot$.

Another way to do this assumes the L-necessity of distinctness in the sense that $x \neq y \to Lx \neq y$ (this is controversial because it implies the determinacy of distinctness, invalidated by the models of vague identity discussed in section 16.2 above). Define $F := \lambda x(x = o \land H)$. Fo is vague since, by the definition of F, $Fo = (o = o \land H) = H$ (assuming the Boolean identities and the L-necessity of self-identity: $(o = o) = \top$). On the other hand, if o' is precise, then $o \neq o'$ since o is vague, so by the necessity of distinctness, $Lo \neq o'$, which by definition of L means that $(o \neq o') = \top$. Applying negation to both sides of the identity gives us $(o = o') = \bot$. From the definition of F, we have that $Fo' = (o' = o \land H)$, and by the Boolean identities and the fact that $(o = o') = \bot$ this is equal to $\bot$, which is precise. Thus, Fo' is precise whenever o' is precise, and, so, F is precise.

[17] Note, however, that under some combinations of views this definition is inadequate. If there can be vague existence, and one subscribes to a negative free logic in which predicates always output a falsehood when provided with a non-existent object, then the predicate $\lambda x.p$ could have borderline cases even when p is precise: if a exists but it's borderline whether it exists, and p is a precise truth then a witnesses the claim $\exists x \nabla (\lambda x.p)a$.

We established this by appealing to a particular precise property. We shall now see that many of the cases of interest discussed in section 16.2 turn out to be particular applications of our test. Identity is a logical relation and so, presumably, a precise one. It follows that if there were vague identity statements, there would have to be vague objects, vindicating Evans' claim that the existence of vague identities is at least a sufficient condition for the existence of vague objects. To establish this, recall that a binary relation such as identity is represented by an element of the type $e \rightarrow (e \rightarrow t)$, taking an individual x to the monadic property of being equal to x, which we write $x =$. If the object x is precise, it follows, by OBJECTUAL PRECISION, that the property of being equal to x is precise (letting F be the identity relation mapping x to $x =$). Since $x =$ is precise, it follows that, whenever y is precise, $x = y$ is also precise, applying the test again. And so, finally, if every object is precise, the proposition that $x = y$ is precise for *every* x and y. Contraposing, we get that if there is a vague identity statement—an x and y such that $x = y$ is not precise—then there is a vague object (either x, y, or both). By completely analogous reasoning we can also establish that if there is a vague parthood statement, there is at least one vague object. Here, we rely instead on the premise that parthood is a precise relation. Indeed, most of the features connected with vague objects—having a vague location, or a vague boundary, or a vague height, and so on—can be seen to fall out as applications of our test, for locative properties and properties to do with boundaries and heights are plausibly precise.

A little care is needed when evaluating the claim, asserted above, that parthood and identity are precise relations. It is common in this literature to talk as if the considerations we've raised about Mt Kilimanjaro and other vague objects can be seen as establishing that the parthood and identity relations are vague. Note, by contrast, that most would consider it a mistake to conclude that conjunction was vague from the observation that there are vague conjunctions. The proposition that Harry is bald and tall, formalized $p \wedge q$, is vague not because conjunction is but because the arguments p and q are. Conjunction is a binary relation between propositions, but we also make similar distinctions for properties and relations that operate on type e. We similarly don't conclude that the property of being exactly 19,341 feet tall is vague, from the vagueness of the claim that Mt Kilimanjaro is exactly 19,341 feet tall. By parity of reasoning, we should not conclude that parthood is vague from the vagueness of the proposition that a particular rock is part of Mt Kilimanjaro, or that identity is vague from the vagueness of some proposition stating the identity of two objects. The most we can conclude is that the source of the vagueness is *either* in the identity or parthood relation or in the objects flanking the relation. But given that our putative examples of vague identity and parthood statements seem all to involve apparently vague objects, the conclusion that identity and parthood are vague relations is not substantiated. Talk of 'vague identity' and 'vague parthood' in this context are thus a bit of a misnomer—this way of talking might more accurately be replaced with talk of vague identity statements and vague parthood statements.

As I warned at the beginning of the chapter, one question I have not attempted to answer here is whether there are any vague objects. Indeed, I suspect that the answers to these questions will involve engaging in some substantial metaphysics—it would be too much to expect the answers to fall out of a general theory of vagueness of the sort I have been attempting to provide here. What I have offered, however, is a framework in which one can sensibly raise the question of whether vague objects exist; this is something which the linguistic theorist, by contrast, has not succeeded in doing.

17

Beyond Vagueness

There are lots of different ways in which a portion of language can end up being defective or indeterminate. In the case of vagueness, I have argued that the source of the indeterminacy is to be found in the kind of proposition that borderline sentences express, rather than in the sentence itself. There are other phenomena within this broad category of 'defective language' that cannot be treated in a similar way. For example, sentences containing failed demonstratives are defective, but on many theories this is simply because they do not express a proposition at all—not because they express a defective proposition. This kind of defectiveness calls out for a linguistic diagnosis, not a propositional one.

Even though the idea that all defective sentences should receive similar diagnoses is not at all plausible, there are several phenomena that appear to bear a close relation to vagueness that one might think ought to be treated similarly. Indeterminacy that arises due to the liar paradox, in certain counterfactuals, in mathematically indeterminate statements (such as, perhaps, the continuum hypothesis), and in the open future all bear a close family resemblance to vagueness.[1] It is natural to want to extend the non-linguistic analysis of vagueness defended here to these other examples of indeterminacy.

Problematic for this project is the existence of putative examples of *genuine semantic indecision*, examples that seem to call out for a linguistic explanation. The position of this book has been that vagueness is not an example of genuine semantic indecision, but that does not preclude the possibility that the phenomenon exists. In section 17.1, I will firstly consider several putative examples of semantic indecision, and outline what a non-linguistic approach to them might look like. In section 17.2, I ask whether it is possible to treat these cases completely analogously to vagueness. I raise some problems for this project and note the need for an account of semantic indecision in addition to propositional vagueness. In section 17.3, I give an account of semantic indecision in terms of my theory of propositional vagueness—one that situates it within the theory of vagueness I have been developing without compromising it.

---

[1] For the liar paradox, see Field [54], or in the classical setting Bacon [7]; for counterfactuals, see Stalnaker [139]; for the continuum hypothesis, see Field [52]; and for the open future, see Belnap and Green [15].

17.1 Semantic Indecision

The first putative example of semantic indecision is adapted from Brandom [19], although I shall be following Field's presentation in Field [51]. Field begins by inviting us to imagine that at some point before the sixteenth century, a community of English speakers had been separated from the rest of us and that they had then been reintegrated with us at the present day. Upon reintegration, we find out that their mathematicians had also discovered the complex numbers. We also discover that their dialect of English is much like ours, except that in introducing names for the two square roots of −1 they used the symbols '/' and '\', instead of 'i' and '−i'. We may suppose that the two dialects of English become integrated, but the four names for the two square roots of −1 remain. Now it is easy to show that i and −i are the only two square roots of −1, and since / refers to a square root of −1, so we can prove that either '/' refers to i or '/' refers to −i. But which of these two possibilities obtains? The dominant view seems to be that there is no fact of the matter; the names '/' and '\' are referentially indeterminate (symmetrical reasoning suggests a similar conclusion for 'i' and '−i'). This is because there is an automorphism of the complex numbers which maps i to −i, so there is no '/'-free sentence that the isolated community could utter that would distinguish / from i without it also distinguishing it from −i.

To turn this example of referential indeterminacy into an example of a semantically undecided sentence, consider the sentence:

/ is positive

where an imaginary number is taken to be positive if it is a positive real multiple of i, and negative if it is a negative real multiple of i. We assume that 'positive' will not be a notion the separated community will have until they are reintegrated. This sentence presumably either means that i is positive (making it true) or that −i is positive (making it false), but there is nothing that settles which of these two things it might mean.

The second putative example of genuine semantic indecision arises from considering incomplete definitions. Here is an example taken from Fine [56]. Suppose that I stipulate that a natural number is nice* if it is less than 15 and that it's not nice* if it is larger than 15, and that is all that I stipulate about the meaning of the word nice*. Assuming that this stipulation is good, we can go on to say many contentful things about being nice*. For example, we may say that there are nice* primes, or that there are at least 10 nice* numbers. It seems that being nice* is a perfectly legitimate property, which some numbers have and others don't. The difficulty concerns cases like the following:

15 is nice*.

What is the status of this sentence? Of course, one could question whether the stipulation succeeded at all, and that our uses of 'nice*' ever express a property. However, it is common practice to make use of a word without there being explicit necessary and sufficient conditions for its application in all cases, and the stipulation we made seems similar enough to this phenomenon to make it hard to see how it might differ. For example, even in mathematics, we freely make use of exponentiation even though there are many functions, f, which agree with our use of exponentiation but differ over the value of $f(0,0)$. As before, this looks like a case of semantic indecision. The candidate properties 'nice*' might denote appear to be the property of being at most 15 and the property of being at most 14. Both properties are determinate, and the indeterminacy is just a matter of which of the two properties the word 'nice*' picks out; the indeterminacy appears to arise from semantic indecision stemming from our incomplete definition of the word 'nice*'.

The last putative example of semantic indecision arises in relation to questions about what to make of scientific terms after our scientific theories have undergone some kind of radical change. In Field [47], Field considers the example of the word 'mass' before and after Newtonian mechanics was replaced by special relativity. The theory of special relativity reveals that there are in fact two closely related properties, relativistic mass and proper mass, that, at ordinary speeds, play pretty much exactly the same role. Before the discovery of special relativity, however, it seemed as though it was indeterminate which property people were using the word 'mass' to refer to— as Field [47] puts it, 'there are two physical quantities that each satisfy the normal criteria for being the denotation of the term'. Assertions like 'mass is conserved in all interactions' appear to be indeterminate depending on which of the properties we take 'mass' to mean.

In each case, the phenomenon bears a family resemblance with vagueness, yet it is hard to imagine that we could explain these phenomena without appealing to the way that language is used.

17.2 Can We Get By without Semantic Indecision?

Let us begin by asking what would happen if we attempted to assimilate these cases to our theory of vagueness. The relevant move, in each case, would be to insist that the linguistic items in question are not semantically undecided about which of a number of determinate propositions to express. Rather, it is semantically decided that they express a single indeterminate proposition. In each case, then, this requires positing a further fine-grained proposition over and above the relevant precise propositions. This move is not entirely ad hoc, for many of the arguments we have adduced in favour of the non-linguistic theory of vagueness extends to the apparent examples of semantic indecision. Since, for example, it appears as if nobody knows whether 15 is nice* or not, despite knowing that it is at most as big as 15 and knowing that it is not at most as big as 14, there needs to be a more fine-grained proposition to represent our uncertainty and ignorance about this fact. Moreover, just as we argued in chapter 4, we need some explanation for this ignorance.

How might we apply our framework to each of the above examples? The first example looks like it lends itself naturally to an analysis in terms of symmetries. The fact that there is an automorphism of the complex numbers that maps i to $-i$ is effectively a symmetry of the complex numbers—one that preserves all mathematical relations between complex numbers.

If we think of propositions as being determined by sets of indices, and that the indeterminacy described above is a source of fine-grainedness, then the picture will be as follows. All the indices will agree that there are two square roots of -1, although according to some i and / will be identified with the same root (and thus $-i$ and \ will be identified with the other root), and according to others i and \ will be identified with the same root (and $-i$ and / to the other).

Given this picture, an automorphism of propositions is induced by the permutation of indices that maps an index in which i and \ are identified to the index where i and / are identified but otherwise agrees about the other facts. We can think of these kinds of automorphisms either as switching / for \ while keeping i and $-i$ fixed, or as switching i and $-i$ while keeping / and \ fixed. Either way, there is an extremely natural analogy with the automorphism of complex numbers defined earlier and, like that automorphism, all mathematical relations will be preserved by this permutation. Although all mathematical relations are preserved, non-mathematical relations need not be—for example, being positive, being identical to i, and similar properties are not preserved. Nonetheless, it's natural to think that these automorphisms preserve rational credences and the things you can reasonably care about: there is something incoherent about caring about things like whether / and i are the same or not, and it also seems that it would be irrational to be more confident that i is / than that it is \. It is therefore not too far-fetched to think that this permutation determines a rational symmetry of the kind defined in section 13.2. Thus, anything of importance to communication, decision making, and so on is preserved if we uniformly replace the concept of i with $-i$ in everything we think and say.

In our second example we stipulated that a number is nice* if it is less than 15 and not nice* if it is greater. To see how this can modelled using symmetries, let us suppose that these stipulations allowed us to refer to an indeterminate property—the property of being nice*—which is neither identical to the property of being less than 15 nor the property of being greater than 15. If I can do this, then it seems that I could now introduce, via exactly the same method, a name for another vague property, being nice$_*$, as follows:

A number is nice$_*$ iff it is either less than 15, or equal to 15 and not nice*.

Notice that a number is nice$_*$ if it is less than 15, it is not nice$_*$ if it is greater than 15, and 15 is nice$_*$ if and only if 15 is not nice*. There is a symmetry in these definitions which isn't immediately apparent. For if we had started out with the notion of being nice$_*$, instead of being nice*, we could define nice* as follows:

A number is nice* iff it is either less than 15, or equal to 15 and not nice$_*$.

Now we can easily see that the proposition that 15 is nice* is distinct from the proposition that 15 is nice*: the truth of one proposition implies the falsity of the other and at least one of them must be true, so they cannot be identical. Similarly, neither proposition is identical to the proposition that 15 is less than or equal to 15, or the proposition that 15 is less than or equal to 14, since we presumably don't know whether 15 is nice* or nice*, but we surely know that 15 is less than or equal to 15 but not less than or equal to 14. That each proposition implies the negation of the other seems to be the only important difference between them. Anything else that we might think or say to distinguish them would involve the new propositions in some way or another. The proposition that 15 is nice* and the proposition that 15 is nice* can't be distinguished in terms of the role they play in our mental lives. They bear exactly the same evidential relations to other propositions and cannot be treated asymmetrically by a rational agent within her system of beliefs and preferences.

The process by which we expand the space of propositions to include the new proposition that 15 is nice* can be modelled using a well-known mathematical method. Let us suppose that we begin with a complete Boolean algebra of propositions, B, containing no indeterminate or vague propositions (let's suppose they are isomorphic to sets of possible worlds), and we want to see what happens when we 'introduce' a new indeterminate proposition p (that 15 is nice*). The mathematical way to do this is to consider the Boolean algebra, $B[p]$, *freely generated* by adding p to B. This algebra is the result of adding p to B, closing under negations, conjunctions, and disjunctions with p, and subjecting B to no other identities between the new propositions other than those imposed by the axioms of a complete Boolean algebra (see section 3.2). The resulting space of propositions contains two *pure* indeterminate propositions: p (that 15 is nice*) and $\neg p$ (that 15 is nice*). The remaining new propositions are conjunctions and disjunctions of these two pure indeterminate propositions with the old determinate propositions.

Freely generated finite extensions of a complete Boolean algebra, like $B[p]$, have a particularly simple characterization in terms of automorphisms. Let G be the set of automorphisms of $B[p]$ that fix B. Then, the set of elements of $B[p]$ that are fixed by every element of G is just B itself. (More generally, if A is a subalgebra of B, then B is a *Galois* extension of A iff A is the set of elements of B fixed by all automorphisms that fix A.[2])

The set of automorphisms of $B[p]$ that fix B are plausibly symmetries in the sense of chapter 13 (they preserve rational probabilities and values). Indeed, this group is particularly simple—it only contains *two* automorphisms: the identity automorphism, and an automorphism that switches the proposition that 15 is nice* with the proposition that 15 is nice*, but leaves logically independent propositions alone.

Applying our definitions of precision and vagueness, we get that neither the proposition that 15 is nice* nor the proposition that / is positive are precise. Both

---

[2] This notion is central in Galois theory which focuses on the automorphism groups of field extensions.

of these can be mapped, via a symmetry, to a distinct proposition and so are not fixed by all symmetries. Moreover, both propositions are indeterminate. The claim that it's determinate that p is just the conjunction of all propositions p is mapped to under symmetries. Thus, the claim that it's determinate that 15 is nice* is just the conjunctive proposition that 15 is both nice* and nice$_*$, which is always false. Analogously, the claim that it's determinate that $i = /$ ends up entailing both $i = /$ and $i = \backslash$ and so must be false, since $/ \neq \backslash$.

At any rate, insofar as our formal framework goes, the first two examples of semantic indecision lend themselves quite naturally to our model theory that is spelt out in terms of symmetries. Unfortunately, I think that there are philosophical reasons to be sceptical of this account (although I have by no means made up my mind on this point).

The first problem is that it seems that examples analogous to the above examples suggest that there are actually a large number of indeterminate propositions mapped by symmetries to the proposition that 15 is nice* or that $/$ is positive. For instance, suppose that instead of two splintered mathematical communities there were 20 who each introduce their own names for the roots of -1. Are we to think that once the 20 communities reintegrate, there will be 2^{20} propositions expressed by the various combinations of identity statements between the newly introduced names? If so, how is it that these propositions came about?

Note that this situation is unlike the evidential roles that we took to be sufficient for the postulation of a vague proposition in chapter 6—these were determined by the kinds of effects inexact evidence can have on our credences in the precise, and so there are no analogous arguments for multiplying vague propositions in the same way.

Anyway, the crucial question for this picture is to explain how propositions can be multiplied so easily. One could say that the propositions only existed because of the linguistic practices of the 20 different communities, and had there only been two communities there would have only been 2 relevant propositions. On this story, the existence of propositions is highly dependent on the contingent practices of linguistic communities. An account of propositions along these lines, however, would not be friendly to the non-linguistic theory of propositions I have been endorsing here—propositions would be language-dependent in a way that strips the theory of its primary advantages.

A better thing to say, then, is that the propositions would have existed whether or not the communities had introduced names for the roots of -1 in the way they did. But then, presumably, there will be much more than 2^{20} propositions—there will be at least as many propositions as there could be possible mathematical communities that diverge and reconverge in the way discussed. This will presumably be some large infinite cardinal, or, perhaps, something too large to be assigned a set theoretic cardinality. Although messy, I do not take this view to be completely hopeless. There is the charge of arbitrariness to contend with: whatever cardinality we do assign to the number of propositions corresponding under symmetry to the proposition that / is positive, we can always ask why that cardinality and not another. However, the problem of arbitrariness with respect to cardinality questions is, I think, something we must make our peace with in other areas.[3]

While a non-linguistic account of these examples is not hopeless, both the examples discussed above will find a natural analysis in terms of genuine semantic indecision which I shall sketch later. What of examples involving scientific terms? In broad outline a non-linguistic account of this example would posit, in addition to the properties of proper mass and relativistic mass, a third vague property, Newtonian mass. The value of an object's Newtonian mass is always either its proper mass or its relativistic mass, but when these differ it is always indeterminate which.

A couple of other features of Newtonian mass seem reasonable. Firstly, it is natural to think that there is a penumbral connection between the Newtonian mass of any two objects: determinately, if the Newtonian mass of the first object is its proper mass then the Newtonian mass of the second is its proper mass, and similarly for relativistic masses. Secondly, it is also natural to think, given that Newtonian mass turned out not to be a fundamental quantity, that the values of the Newtonian masses of objects supervene on the values of the more fundamental quantities of proper and relativistic mass: either it's necessary that the Newtonian mass of every object is identical to its proper mass or it's necessary that the Newtonian mass of every object is its relativistic mass. Although Newtonian mass is necessarily coextensive with another more fundamental quantity, it is indeterminate which of these quantities it is coextensive with, and, therefore, there is ignorance about which of these quantities Newtonian mass coincides with. Thus, Newtonian mass might have a distinctive conceptual role which neither proper mass nor relativistic mass has: perhaps someone who was fairly, if not fully, confident that Newtonian physics is the correct description of reality, and therefore also pretty confident that Newtonian, proper, and relativistic mass are coextensive, might nonetheless treat the first differently from either of the other two, conditional on the supposition that the theory of relativity is true.

At any rate, on this way of understanding the example, it was determinate that Newton was neither giving a theory of proper mass nor of relativistic mass—he was giving a theory of Newtonian mass. As we later discovered, some of the consequences of his theory are determinately false: they neither hold of proper nor relativistic mass and thus do not hold of Newtonian mass. Other consequences are indeterminate: they would have been true, or approximately true, if he had been talking about proper mass and not relativistic mass, or vice versa. When we eventually did discover the nature of relativity, our use of the word 'mass' changed. It no longer referred to Newtonian mass, and we introduced disambiguations of the word 'mass' to refer to the two properties which we believe do exist.

---

[3] See, for example, Hawthorne and Uzquiano [70], Sider [132].

The picture described seems at least coherent, but is it plausible? One source of implausibility is the idea that physicists frequently find themselves theorizing about vague properties, instead of the perfectly natural and fundamental properties we typically take them to be talking about. Presumably, this issue isn't confined just to previous physical theories we know to be false—it also applies to our present physical theories which, for all we know, are false. Presumably, the most we know about our present theories is that they are approximately true, but this allows for the possibility that our present physical concepts refer to indeterminate concepts much like the notion of 'mass' did amongst physicists before the theory of relativity. Note, however, that if physicists working with false theories are theorizing about properties at all, it is independently natural to think that they are not theorizing about fundamental properties. Note also that we have already cast doubt on the hypothesis that physicists usually find themselves theorizing about precise properties. In section 12.2, for example, I argued that even physical properties like being an electron are vague.

The story here is also subject to the multiplicity worry raised in connection with the first two examples. To prevent the number of vague properties depending on the number of incorrect physical theories proposed throughout history, we would have to posit an abundance of vague properties to account for all of the possible incorrect physical concepts that could be introduced by false physical theories. Perhaps in this case there is more hope of delineating a fixed class of conceptual roles, and arguing that there is a property for each of those conceptual roles, thus answering the arbitrariness worry. However, I do not know whether the mechanism by which we posit such properties seems different enough from the case of vagueness to warrant treating this example in a different way.

17.3 An Account of Semantic Indecision

Each of the examples mentioned looks as though it ought to be amenable to an analysis in terms of semantic indecision. Semantic indecision, unlike my account of vagueness, is something that only makes sense when it is attributed to a linguistic item, such as a sentence or a predicate. But what exactly is semantic indecision? According to McGee and McLaughlin [104], prominent defenders of the semantic indecision account of vagueness, it is semantically decided that an object falls under a predicate when 'the thoughts, experiences, and practices of the speakers of a language determine the conditions of application of its predicates, and a predicate definitely applies to an object just in case the facts about the object determine that these conditions are met'.

Let us consider how we might apply this to the word 'bald'. According to the proposal, we are to suppose that there is no property—or 'condition' in McGee and McLaughlin's lingo—which is determined by the linguistic practices of the speakers to be the meaning of 'bald'. Despite this, there might well be several properties that are *candidate* meanings for the word bald: a property which is not determined to be not meant by the word 'bald' by the linguistic practices of English speakers. With this in place, McGee and McLaughlin go on to develop a broadly supervaluationist account of semantic indecision.

But what exactly is the notion of 'determining' that is being invoked here? Perhaps it can be spelt out in terms of metaphysical necessity by some kind of supervenience thesis. P is determined by present linguistic practices to be the meaning of the word 'bald' if and only if it's necessary that when the linguistic practices of English are as they in fact are, the word 'bald' means P. On the other hand, if no meaning for 'bald' is determined by linguistic practices, P is a candidate meaning iff it's metaphysically possible that linguistic practices be as they in fact are and for 'bald' to mean P.

This account of 'determining' will not do, for McGee and McLaughlin want their theory to be consistent with the supervenience of meaning on use, yet such a supervenience thesis would ensure that the meaning of 'bald' was always determined by the usage facts. Moreover, even if meaning does not supervene purely on use—perhaps the existence of reference magnetism, or something similar, prevents this—it would not be in the spirit of their view to think that the precise meanings of vague predicates *are* determined once you add these extra facts to the supervenience base. Presumably, meaning ought to supervene on the totality of physical facts, for example, but in the relevant sense of 'determine' the extension of the 'bald' is surely not determined by the totality of physical facts for, otherwise, its extension would be determined by completely precise matters.

Perhaps the notion of determining at hand is an epistemic one. Although the linguistic usage facts necessarily imply that the word 'bald' means what it means, and thus 'determines' in this same sense what its cutoff point is at each world, neither the cutoff nor the necessary implication between the usage facts and meaning facts are knowable to us. The sense in which the meaning of 'bald' is undetermined thus might be an epistemic one: there are several properties that, for all we know, are the meanings of the word 'bald'. We don't know which because, although the meaning of 'bald' supervenes on its use, we do not know *how* the meaning of 'bald' supervenes on its use—thus we are just as ignorant about what 'bald' means as we are about how this meaning gets fixed by usage.

This is, of course, just a version of the epistemic theory of vagueness—a theory that McGee and McLaughlin, and other sympathetic writers, take care to distinguish from semantic indecision.[4] If we are to explicate the notion of semantic indecision in terms of how linguistic practices fail to determine meanings of vague words, we need a non-epistemic account of this 'determining' relation. Moreover, the determining relation cannot be spelt out in terms of semantic decidedness itself. For not only would this be circular, the determining relation is something that holds between two types of fact, whereas semantic decidedness is a feature of linguistic items.

---

[4] Indeed, Williamson turns these kinds of observations into an objection to McGee and McLaughlin's theory in Williamson [162].

Luckily, if we accept the theory of vagueness that I have been endorsing so far, then there is a perfectly suited non-linguistic notion of borderlineness, which is neither contingency nor ignorance, that can be used to explicate the sense in which the usage facts can fail to determine the meanings of certain words: when the usage facts leave it indeterminate or borderline what a word means.

On this account, the linguistic facts determine that 'F' means P when it is necessarily determinate that if the linguistic facts are as they in fact are, 'F' means that P.[5] The linguistic facts leave it open whether 'F' means that P iff it's possibly not determinate that 'F' means something else while the linguistic facts are as they in fact are.

Assuming the supervenience of meaning on use, the 'possibly's and 'necessary's can be eliminated from these definitions. Thus we can give simple definitions of semantic definiteness and undecidedness: say that a sentence is semantically definite iff every proposition it doesn't determinately not express is true, and semantically undecided if neither it nor its negation is semantically definite. Given fairly uncontroversial assumptions, a sentence is semantically undecided if there's both a true and a false proposition it doesn't determinately not express.[6]

Two things about this account are worth pointing out at this juncture. Firstly, this account of semantic indecision clearly cannot be accepted by someone who does not already accept the ideology of a non-linguistic determinacy operator. On this account, the borderlineness operator must be taken to be more basic than semantic indecision, and so the latter cannot be used to explicate the former, *pace* McGee and McLaughlin. Secondly, I do not claim to be analysing some pretheoretic notion of 'semantic indecision' that McGee and McLaughlin, or anyone else has in mind. It suffices for my purposes to note that if we have a non-linguistic borderlineness operator, then the above definitions are perfectly sound. Thus anyone accepting a borderlineness operator can make sense of the concepts—I am simply calling the concept 'semantic indecision' due to its resemblance to the notion McGee and McLaughlin are trying to capture.[7]

Of course, it is controversial whether the examples I listed in section 7.1 are semantically undecided according to this account of semantic indecision. Nonetheless, semantic indecision so defined undoubtedly exists—semantic properties and relations, like most non-fundamental properties and relations, are vague. In particular the linguistic meaning relation is vague and has borderline cases. To see this note that

---

[5] The definition in terms of 'necessary determinacy' rather than 'determinate necessity' seems to be the more natural to me; although given their potential distinctness, it is worth noting that variant accounts of 'determining' could be formulated in terms of 'determinate necessity' or both.

[6] The latter way of expressing semantic undecidedness is perhaps preferable because it can be applied to languages that do not contain a negation operator.

[7] Michael Caie, in Caie [23], for example, has recently argued for a similar account of semantic indecision in terms of a non-linguistic notion of indeterminacy. However, unlike me, he takes this to be what vagueness consists in. Although Caie is explicit about his commitment to a non-linguistic notion of indeterminacy, many supervaluationists implicitly seem to commit themselves to this kind of view by talking about vagueness as though there were many candidate meanings which a vague word indeterminately refers to.

words clearly change their meanings over time. For example, the phrase 'a kleenex' went from meaning a particular brand of tissue to being a general term for a tissue: presumably there was a time at which it was borderline which of the two things this phrase meant. Here is an example from the philosophical literature, which we will appeal to later: consider a sorites in which a person's use of 'water' switches from referring to H_2O to another water-like substance, XYZ: perhaps I have moved to twin earth and my uses of the word 'water' are slowly changing their meaning to refer to the local colourless, odourless watery stuff. I start off clearly referring to water, and end up clearly referring to XYZ, but presumably there will be some point in between where it's borderline whether my use of 'water' refers to water or XYZ. It will therefore be borderline which proposition sentences involving the word 'water' express. Sentences involving the word 'water' can be semantically undecided according to my view.

It is important to distinguish semantic indecision from linguistic borderlineness. As I argued in chapter 4, even if one accepts a non-linguistic account of vagueness there is quite clearly a distinction between sentences to be captured, even if it is not taken to be basic: the difference between the sentence 'Harry is bald' and 'Patrick Stewart is bald', for example. The former is linguistically borderline, I argued, because it expresses a borderline proposition, whereas the latter is not linguistically borderline, because it expresses a determinately false proposition.

Note, however, that a sentence can be linguistically borderline without being semantically undecided. Suppose that there is a proposition, P, which the sentence 'Harry is bald' determinately expresses. Presumably P would be the proposition that Harry is bald, and since this is borderline, it follows that the sentence 'Harry is bald' is linguistically borderline. However, it is not semantically undecided: if Harry is bald, then every proposition the sentence 'Harry is bald' doesn't determinately not express is true—there's only one such proposition and that's the proposition that Harry is bald. If Harry is not bald, then by analogous reasoning there's a proposition the sentence 'Harry is bald' determinately expresses and it's false: thus 'Harry is bald' is either semantically definite, or its negation is semantically definite (although it is borderline which).

Conversely, a sentence can be semantically undecided without being linguistically borderline. For example, suppose that it is determinate that XYZ does not contain hydrogen as a chemical component. H_2O, on the other hand, determinately does contain hydrogen as a chemical component. Now, even if it is borderline whether 'water' refers to H_2O or XYZ, the sentence 'water contains hydrogen' is not linguistically borderline. In fact, it's determinate that 'water contains hydrogen' expresses a non-borderline proposition, because it either expresses the determinately true proposition that H_2O contains hydrogen or the determinately false proposition that XYZ contains hydrogen. However, 'water contains hydrogen' is semantically undecided because there's a true and a false proposition it doesn't determinately not express.

The concept I have introduced under the name 'semantic indecision', I think, does a good job of accounting for the examples introduced in this chapter. In section 17.2,

I considered the suggestion that sentences involving the words 'nice*', '/' and 'mass' often expressed vague propositions, and that this could be used to explain the apparent indeterminacy of these examples. In other words, I was suggesting that sentences like '15 is nice*' are linguistically borderline: they express borderline propositions. However, there are some puzzles for this view and I think a much more natural way to model the above examples would be in terms of semantic indecision.

Let us focus on the example involving incomplete definitions. According to this analysis, the word 'nice*' expresses either the property of being at most 14 or the property of being at most 15, although it is borderline which. Thus the sentence '15 is nice*' is semantically undecided: it doesn't determinately fail to express the true proposition that 15 is at most 15, and it doesn't determinately fail to express the false proposition that 15 is at most 14. Note also that neither of the candidate propositions for '15 is nice*' are borderline. Thus, even though this is sentence semantically undecided, it is not linguistically borderline. Presumably on this view there is also a penumbral connection between what 'nice*' refers to and what 'nice$_*$' refers to, namely, the former refers to whichever property the latter doesn't.

It is independently plausible that it is borderline which property the word 'nice*' picks out. If there was a sorites sequence starting with ways of introducing the word 'nice*' so that it refers to the property of being at most 15 and ending with ways of introducing the word 'nice*' so that it ends up referring to the property of being at most 14, you might expect the incomplete definition of 'nice*' given in section 17.1 to be one of the cases in the middle of the sorites sequences—one of the cases where it's borderline which of the two properties it refers to.

Presumably, in this case, I can still talk about the property of being nice*, but when I do I end up referring either to the property of being at most 15 or the property of being at most 14 (and it's propositionally borderline which). That is to say, the expression 'the property of being nice*' inherits semantic indecision from the subexpression 'nice*'. The crucial thing about this view is that, although there is the property of being nice*, it is a precise property for it is either the property of being at most 15 or the property of being at most 14. Both of these properties are precise—the property of being nice* isn't a third vague property distinct from either of these two properties. This is the crucial difference between the semantic indecision account and the account in which '15 is nice*' is linguistically borderline. There is therefore no problem of property multiplication to be levelled against this account in the way that there was according to the latter account.

A similar story can be told about the other two examples. According to the semantic indecision account, the pre-relativistic uses of 'mass' refer either to proper mass or relativistic mass. However, the practices of the people using the word 'mass' leave it borderline whether they meant proper mass or relativistic mass by it, for they never really had to apply the word in the cases where the meanings came apart. Again, one can justify the ascription of borderlineness in meaning by considering a sorites starting off with a community of people who are disposed, after learning about relativity, to apply the word 'mass' to the proper mass of objects, and ending with a community disposed to apply the word 'mass' to relativistic mass upon learning about relativity. One would assume that the way people actually were disposed to apply the word 'mass', once the special theory of relativity was discovered, is one where it's pretty much borderline whether we would want to apply the word 'mass' to proper mass or relativistic mass.[8] On this account, then, we do not need to posit this strange third quantity, 'Newtonian mass'. It was simply borderline whether 'mass' as it was used then referred to proper mass or relativistic mass.

As for the example involving the two mathematical communities, the semantic indecision account would say that the sentence '$i = /$' either expresses a necessarily false proposition saying of one of the square roots of -1 that it is identical to the other, or a necessarily true proposition saying of one of the square roots that it is identical to itself. That is to say, letting x and y be the square roots of -1, the candidate propositions would be the singular propositions that $x = x$, that $y = y$, that $x = y$, and that $y = x$. There are thus between two and four candidate propositions depending on how finely we individuate, and they are all perfectly precise. Again, we can contrast this with the earlier view in which '$i = /$' did not express a singular proposition at all. On that view the contribution of '$/$' and 'i' was roughly the same as a vague name such as 'Everest' to the sentence 'Everest is 11,000ft' (see chapter 4). The proposition that '$i = /$' expressed was a borderline proposition due to names like 'i' and '$/$' introducing vague individual concepts to the proposition.

The account of semantic indecision I have outlined here assumes that it is sometimes borderline which proposition I have expressed with a particular sentence. There is, however, a contrary argument that it's never borderline what proposition a sentence expresses which I therefore need to address. Taking the example of the sentence '15 is nice*' the argument starts with a disquotational premise:

1. It's determinate that '15 is nice*' expresses the proposition that 15 is nice* and nothing else.
2. Therefore it's not borderline what '15 is nice*' expresses.

The argument is indeed valid, assuming existential generalization, and the conclusion excludes semantic indecision for the sentence '15 is nice*'.

A little thought, however, demonstrates that this argument overgenerates quite radically. For if the above argument is fine, then one can argue quite generally that, given that 'P' determinately expresses the proposition that P and nothing else, it is not borderline what 'P' expresses. But this conclusion is absurd for arbitrary sentences, for it shows that the expressing relation has no borderline cases. The relation of semantic expressing is clearly as vague as any other non-fundamental relation, and therefore is just as prone to having borderline cases. It is a matter of routine to describe a sorites

[8] This is not entirely uncontroversial. John Earman, for example, argues that it's not actually borderline how we use the word 'mass'.

in which an expression begins with a particular meaning but over time changes its meaning as people begin to use it in incrementally different ways. As with any sorites, we should conclude that there will be borderline cases in the middle of this sequence.

The solution is to reject the determinacy of the disquotational principle we began with. The idea, in the present case, is that it is borderline whether '15 is nice*' expresses the proposition that 15 is nice* or the proposition that 15 is nice$_*$, but determinate that it expresses one of them. What, then, is to be made of our intuitions in favour of the determinacy of the disquotational principle? What is it that seems to be special about disquotational sentences like '"15 is nice*" expresses the proposition that 15 is nice*'?

In chapter 5, I cast some doubt on the determinacy of the disquotational principles in our discussion of Williamson's metalinguistic safety account of vagueness. However, there is a status that they enjoy which could easily be taken to explain the intuitions we have in favour of them: they are determinately true. This point requires distinguishing sharply between the following two principles:

It's determinate that '15 is nice*' expresses the proposition that 15 is nice*.
The sentence '"15 is nice*" expresses the proposition that 15 is nice*' is determinately true.

The former principle I am denying. The latter, however, is true, which can be reasoned out from the following principles.

1. It's determinate that either '15 is nice*' expresses the proposition that 15 is nice* or the proposition that 15 is nice$_*$.
2. It's determinate that the word 'expresses' expresses the expressing relation.
3. The semantic principle of compositionality is determinate.

The first premise presents us with two cases to consider. Suppose that '15 is nice*' expresses the proposition that 15 is nice*. Then the sentence '"15 is nice*" expresses the proposition that 15 is nice*' expresses the proposition that ['15 is nice*' expresses the proposition that 15 is nice*] by the second premise and compositionality (I am using square brackets here simply to indicate scope). The proposition expressed is true by hypothesis. For the second, non-disquotational, case suppose that '15 is nice*' expresses the proposition that 15 is nice$_*$. This time, the sentence '"15 is nice*" expresses the proposition that 15 is nice*' expresses the proposition that ['15 is nice*' expresses the proposition that 15 is nice$_*$] by the second premise and compositionality. Again, by the description of the case, this proposition is true. So either way the proposition '15 is nice*' expresses is true. Since each of the premises of this argument is determinate, we can determinize our conclusion: '15 is nice*' is determinately true.[9]

---

[9] Superficially, this argument used reasoning by cases, but it is easy to reformulate so that it can be proved just using the logic **KT**.

What is interesting about this case is that the two propositions that the disquotational principle is associated with are both borderline. As I argued earlier, the proposition that ['15 is nice*' expresses the proposition that 15 is nice*] is borderline, as is the proposition that ['15 is nice*' expresses the proposition that 15 is nice$_*$]. Nonetheless, the indices at which the disquotational principle expresses the former proposition happen to be indices at which that proposition is true, and similarly for the indices at which it expresses the latter proposition. There is a penumbral connection between what the sentence expresses and the truth value of the proposition it expresses. This is why it is possible for it to be determinate that the disquotational principle expresses some true proposition, even though everything it doesn't determinately not express is borderline.

This situation might seem pathological, but can be constructed easily in other cases. Let's suppose I stipulatively introduce a name, 'Fred', for the tallest short person.[10] There is plausibly a penumbral connection between the individual 'Fred' refers to and the cutoff point for being short. The sentence 'Fred is short' is thus surely determinately true, since we know that whoever 'Fred' refers to, they're going to be a short person (the tallest one, in fact). But all of these candidate referents of 'Fred' are people who are borderline short. Thus we can also be sure that whoever 'Fred' refers to, the proposition expressed by 'Fred is short' is borderline.

17.4 Concluding Remarks

It is not uncommon for proponents of semantic indecision to gloss the phenomenon as follows: an expression is semantically undecided if and only if there are several candidate precise meanings that it could express, but it is indeterminate which of those meanings it expresses.[11] This characterization, however, invokes an indeterminacy operator that is not available without embracing a theory of propositional indeterminacy. For if it is indeterminate whether a sentence S expresses a proposition P, then the proposition that S expresses P is a vague proposition. Indeed, it is somewhat striking on this view that, even though there are vague propositions as just evidenced, the candidate contents of vague sentences are only ever precise.

Some supervaluationists will no doubt maintain that the use of an indeterminacy operator in this formulation is just a manner of speaking. However, it is far from clear that the linguistic theorist can reformulate this slogan in more appropriate linguistic terminology. Flat-footedly translating the gloss using linguistic idioms delivers the following: an expression E is semantically undecided if and only if there are several meanings that are candidates for satisfying the predicate 'is expressed by E', and the

---

[10] I am treating 'Fred' here as a descriptive name, much like the well-known example of 'Julius' for the inventor of the zip.

[11] This sort of view is implicit, for example, in Keefe's supervaluationist account of speech reports in Keefe [79]. According to that view, when someone utters the sentence 'Harry is bald', it is indeterminate which of several candidate precise propositions has been said.

application of 'is expressed by E' to any of them is semantically undecided. (Due to issues relating to 'quantifying in' we have formulated this in terms of a linguistic predicative undecidedness instead of sentential borderlineness; see section 4.4.) On this picture, the semantic undecidedness of an expression E is explained in terms of the semantic undecidedness of another metalinguistic expression, 'is expressed by E'. Regress looms.

Ironically, the non-linguistic theory of vagueness adopted here allows us to take this account of semantic indecision at face value. On this picture, even semantic indecision cannot purely be explained by appealing solely to linguistic practices of those using the language: one must also invoke a borderlineness operator that is not explained in terms of language.

18

Appendices

18.1 Appendix A

Here, I prove that CS and PMP commit you to sharp cutoff points. I shall work in a propositional language consisting of $\wedge$, $\vee$, $\neg$, and $\rightarrow$. For readability, I shall be lax about parentheses. Let S_n denote the proposition that the nth person is rich. According to our definition in chapter 1, we can say what it means for n to be a cutoff point by the sentence $\neg(S_n \rightarrow S_{n+1})$. In addition to CS we assume the following background logic:

MP $\quad A, A \rightarrow B \vdash B$.

MT $\quad \neg B, A \rightarrow B \vdash \neg A$.

CC $\quad A \rightarrow B, A \rightarrow C \vdash A \rightarrow (B \wedge C)$.

CE $\quad (A_1 \wedge \ldots \wedge A_n) \rightarrow A_i, 1 \leq i \leq n$.

DM $\quad \neg(A_1 \wedge \ldots \wedge A_n) \vdash (\neg A_1 \vee \ldots \vee \neg A_n)$.

As we shall see, the last axiom, DM, is not that important for the main upshot of the argument.

The argument then proceeds as follows:

1. $((S_0 \rightarrow S_1) \wedge \ldots \wedge (S_{999999} \rightarrow S_{1000000})) \rightarrow ((S_0 \rightarrow S_1) \wedge (S_1 \rightarrow S_2))$ by CE.
2. $((S_0 \rightarrow S_1) \wedge (S_1 \rightarrow S_2)) \rightarrow (S_0 \rightarrow S_2)$ an instance of CS.
3. $((S_0 \rightarrow S_1) \wedge \ldots \wedge (S_{999999} \rightarrow S_{1000000})) \rightarrow (S_0 \rightarrow S_2)$ by the transitivity of $\rightarrow$. (A consequence of CS, MP and CE.)
4. $((S_0 \rightarrow S_1) \wedge \ldots \wedge (S_{999999} \rightarrow S_{1000000})) \rightarrow (S_2 \rightarrow S_3)$ by CE.
5. $((S_0 \rightarrow S_1) \wedge \ldots \wedge (S_{999999} \rightarrow S_{1000000})) \rightarrow ((S_0 \rightarrow S_2) \wedge (S_2 \rightarrow S_3))$ by CC from 3 and 4.
6. $((S_0 \rightarrow S_2) \wedge (S_2 \rightarrow S_3)) \rightarrow (S_0 \rightarrow S_3)$ CS.
7. $((S_0 \rightarrow S_1) \wedge \ldots \wedge (S_{999999} \rightarrow S_{1000000})) \rightarrow (S_0 \rightarrow S_3)$ by transitivity of $\rightarrow$.

. . . and so on.

8. $((S_0 \rightarrow S_1) \wedge \ldots \wedge (S_{999999} \rightarrow S_{1000000})) \rightarrow (S_0 \rightarrow S_{1000000})$.
9. $\neg(S_0 \rightarrow S_{1000000})$ (premise).
10. $\neg((S_0 \rightarrow S_1) \wedge \ldots \wedge (S_{999999} \rightarrow S_{1000000}))$ by MT.
11. $\neg(S_0 \rightarrow S_1) \vee \ldots \vee \neg(S_{999999} \rightarrow S_{1000000}))$ by DM.

The conclusion is just the claim that either 0 is a cutoff point, or 1 is, or 2 is or ... or 100000. Thus, we have proved that there is a cutoff point. Notice that the deMorgan law, DM, is applied only at the last step. Thus, even without DM one must accept the penultimate step of this argument, which many might think is just as bad as accepting a sharp cutoff.

The argument against PMP proceeds similarly, although this time we must assume the transitivity of the conditional instead of modus ponens:

TR $A \rightarrow B, B \rightarrow C \vdash A \rightarrow C$.

(Indeed, MP could have been substituted for TR in the last argument as well. Both of these arguments therefore work for the logic **LP** which does not have modus ponens.)

Here is the argument:

1. $S_0 \wedge \bigwedge(S_n \rightarrow S_{n+1}) \rightarrow (S_0 \wedge (S_0 \rightarrow S_1))$ by CE.
2. $(S_0 \wedge (S_0 \rightarrow S_1)) \rightarrow S_1$ instance of PMP.
3. $S_0 \wedge \bigwedge(S_n \rightarrow S_{n+1}) \rightarrow S_1$ from 1 and 2 by TR.
4. $S_0 \wedge \bigwedge(S_n \rightarrow S_{n+1}) \rightarrow (S_1 \rightarrow S_2)$ by CE.
5. $S_0 \wedge \bigwedge(S_n \rightarrow S_{n+1}) \rightarrow (S_1 \wedge (S_1 \rightarrow S_2))$ by 3 and 4 and CC.
6. ...
7. $S_0 \wedge \bigwedge(S_n \rightarrow S_{n+1}) \rightarrow S_{1000000}$.
8. $\neg S_{1000000}$.
9. $\neg(S_0 \wedge \bigwedge(S_n \rightarrow S_{n+1}))$ by modus tollens.
10. $\neg S_0 \vee \neg(S_0 \rightarrow S_1) \vee \ldots \vee \neg(S_{999999} \rightarrow S_{1000000})$ by DM.

Here $\neg S_0$ is also a disjunct in our conclusion, but since this theorist accepts S_0, this is of little solace.

18.2 Appendix B

The **Principle of Plenitude** says that, for any function t from the set of maximally specific precise propositions to $[0, 1]$, there is a proposition p_t such that for every conceptually coherent prior Pr, and maximally specific precise proposition, w, $Pr(p_t \mid w) = t(w)$ provided $Pr(w) > 0$.

As with any existence postulate, such as Lewis' plenitude principle for possible worlds or the naïve comprehension principle in set theory, there is a question about its consistency. Here, I show that it is in fact consistent.

Theorem 18.2.1. *Suppose that $\mathbb{B}$ is a complete atomic Boolean algebra with countably many atoms. Then there is an extension of $\mathbb{B}$, $\mathbb{B}' \geq \mathbb{B}$, such that for every function t : $Atom(\mathbb{B}) \rightarrow [0, 1]$ there is a proposition in $\mathbb{B}'$, p_t, with the following property:*

Any countably additive probability function Cr over $\mathbb{B}$ can be extended to a probability function Cr' over $\mathbb{B}'$ in a way such that:

1. *$Cr'(p) = Cr(p)$ for each $p \in \mathbb{B}$.*
2. *$Cr'(p_t \mid w) = t(w)$ for each $w \in Atom(\mathbb{B})$, provided $Cr(w) > 0$.*

Proof. Let $W := Atom(\mathbb{B})$. Without loss of generality we may identify $\mathbb{B}$ with $\mathcal{P}(W)$. Given $p \subseteq W \times [0,1]$, let $p_w := \{x \mid \langle w, x \rangle \in p\}$. We define the expansion as follows:

- $\mathbb{B}' := \mathcal{P}(W \times [0,1])$.
- $Cr'(p) = \sum_{w \in W} \lambda(p_w).Cr(w)$ where λ is the Lebesgue measure.
- $p_t := \bigcup_w (\{w\} \times [0, t(w)])$.

$\mathbb{B} \leq \mathbb{B}'$ via the map $h : p \to p \times [0,1]$. Note that $Cr'(p \times [0,1]) = \sum_{w \in W} \chi_p(w)$. $Cr(w) = \sum_{w \in p} Cr(w) = Cr(p)$, where χ_p is the characteristic function of p. This shows that (1) is satisfied.

Also $Cr'(p_t \mid w) = \frac{Cr'((\bigcup_{w \in W} \{w\} \times [0, t(w)]) \cap (\{w\} \times [0,1]))}{Cr'(\{w\} \times [0,1])} = \frac{Cr'(\{w\} \times [0, t(w)])}{Cr'(\{w\} \times [0,1])} = \frac{t(w).Cr(w)}{Cr(w)} = t(w)$. So, (2) is satisfied. $\square$

The consistency of **Plenitude** is a simple corollary of this result. We simply identify the set of conceptually coherent priors with the set of probability measures on $\mathbb{B}'$ which extend regular probability measures over $\mathbb{B}$ and satisfy $Pr(p_t | w) = t(w)$ for each atom w of $\mathbb{B}$.

Next, I consider a worked example to illustrate how **Plenitude** can deliver propositions that play the right sort of evidential role. If our evidence in the scenario where we have seen a tree from a distance is a vague proposition, as I have argued it is, there ought to be a proposition such that the result of updating on it results in the kind of credences you'd expect to have; something like the credence distribution in Figure 6.4. At the end of section 6.3, I argued that the **Principle of Plenitude** provides us with such a proposition.

Suppose, as in section 6.3, that my prior credence over the possible heights of the tree less than 1000cm is uniform, and is given by $Cr(w_i) = \alpha$, where w_i is the proposition that the tree is $i$cm, and my posterior credence function after seeing the tree is given by $Cr'(w_i)$. By the **Principle of Plenitude**, there is a proposition p_t such that $Cr(p_t \mid w_i) = Cr'(w_i)$ for each i, which can be gotten from the **Principle of Plenitude** by choosing a function that maps each maximally strong consistent precise proposition consistent with w_i to $Cr'(w_i)$. I'll show that p_t is the required update to get our credences from a uniform distribution to a distribution looking like Figure 6.4. I.e. $Cr(\cdot \mid p_t) = Cr'(\cdot)$.

1. $Cr(w_i \mid p_t) = Cr(p_t \mid w_i).\frac{Cr(w_i)}{Cr(p_t)}$ by Bayes' theorem.
2. $Cr(w_i \mid p_t) = t(w_i).\frac{Cr(w_i)}{\sum_j Cr(p_t | w_j).Cr(w_j)}$
3. $Cr(w_i \mid p_t) = t(w_i).\frac{\alpha}{\sum_j t(w_j).\alpha}$
4. $Cr(w_i \mid p_t) = t(w_i).\frac{1}{\sum_j t(w_j)}$
5. $Cr(w_i \mid p_t) = t(w_i)$ since $\sum_j t(w_j) = \sum_j Cr'(w_j) = 1$.
6. Thus $Cr(\cdot \mid p_t) = Cr'(\cdot)$ since they coincide on the worlds.

Bibliography

[1] Ernest Wilcox Adams. *The Logic of Conditionals: An Application of Probability to Deductive Logic*. D. Reidel Publishing Co., 1975.

[2] A. J. Ayer. *Language, Truth, and Logic*. Gollancz, 1936.

[3] Andrew Bacon. The broadest necessity. *Journal of Philosophical Logic*, (2017). https://doi.org/10.1007/s10992-017-9447-9

[4] Andrew Bacon. Vagueness at every order. Unpublished, 2009.

[5] Andrew Bacon. A new conditional for naive truth theory. *Notre Dame Journal of Formal Logic*, 54(1): 87–104, 2013.

[6] Andrew Bacon. Giving your knowledge half a chance. *Philosophical Studies*, 171(2): 373–97, November 2014.

[7] Andrew Bacon. Can the classical logician avoid the revenge paradoxes? *Philosophical Review*, 124(3): 299–352, 2015.

[8] Elizabeth Barnes and J. R. G. Williams. Vague parts and vague identity. *Pacific Philosophical Quarterly*, 90(2): 176–87, June 2009.

[9] Elizabeth Barnes and J. R. G. Williams. A theory of metaphysical indeterminacy. In Karen Bennett and Dean W. Zimmerman, editors, *Oxford Studies in Metaphysics volume 6*, pages 103–48. Oxford University Press, 2011.

[10] David Barnett. Vagueness and rationality. Unpublished.

[11] David Barnett. Is vagueness sui generis? *Australasian Journal of Philosophy*, 87(1): 5–34, 2009.

[12] David Barnett. Does vagueness exclude knowledge? *Philosophy and Phenomenological Research*, 82(1): 22–45, January 2011.

[13] J. C. Beall. Vague intensions: A modest marriage proposal. In Richard Dietz and Sebastiano Moruzzi, editors, *Cuts & Clouds: Vagueness, Its Nature, and Its Logic*, page 187. Oxford University Press, 2010.

[14] J. C. Beall and Greg Restall. *Logical Pluralism*. Oxford University Press, 2006.

[15] Nuel Belnap and Mitchell Green. Indeterminism and the thin red line. *Philosophical Perspectives*, 8: 365–88, 1994.

[16] Ethan D. Bolker. A simultaneous axiomatization of utility and subjective probability. *Philosophy of Science*, 34(4): 333–40, December 1967.

[17] Richard Bradley. Radical probabilism and bayesian conditioning. *Philosophy of Science*, 72(2): 342–64, April 2005.

[18] Ross Brady. *Universal Logic*. CSLI Publications, 2006.

[19] Robert Brandom. The significance of complex numbers for Frege's philosophy of mathematics. In *Proceedings of the Aristotelian Society*, volume 96, pages 293–315, 1996.

[20] David Braun and Theodore Sider. Vague, so untrue. *Noûs*, 41(2): 133–56, 2007.

[21] T. Burge. Individualism and the mental. *Midwest Studies in Philosophy*, 4(1): 73–121, 1979.

[22] Michael Caie. Vagueness and semantic indiscriminability. *Philosophical Studies*, 160(3): 365–77, 2012.

[23] Michael Caie. Metasemantics and metaphysical indeterminacy. In Alexis Burgess and Brett Sherman, editors, *Metasemantics: New Essays on the Foundations of Meaning*, pages 55–96. Oxford University Press, 2014.

[24] Rudolf Carnap. *The Logical Syntax of Language*. London: Kegan Paul, Trench, Trubner, & Co, 1937.

[25] Rudolf Carnap. *Logical Foundations of Probability*. University of Chicago Press, 1950.

[26] Alonzo Church. *The Calculi of Lambda-Conversion*. Princeton University Press, 1941.

[27] Aaron J. Cotnoir. Anti-symmetry and non-extensional mereology. *Philosophical Quarterly*, 60(239): 396–405, 2010.

[28] Aaron J. Cotnoir and Andrew Bacon. Non-wellfounded mereology. *Review of Symbolic Logic*, 5(2): 187–204, 2012.

[29] M. J. Cresswell. Propositional identity. *Logique et Analyse*, 10(39/40): 283–92, 1967.

[30] Mark Crimmins and John Perry. The prince and the phone booth: Reporting puzzling beliefs. *The Journal of Philosophy*, 86(12): 685–711, 1989.

[31] David Deutsch. Quantum theory of probability and decisions. *Proceedings: Mathematical, Physical and Engineering Sciences*, 455(1988): 3129–37, 1999.

[32] Richard Dietz. Betting on borderline cases. *Philosophical Perspectives*, 22(1): 47–88, 12, 2008.

[33] Cian Dorr. The expressivist and the relativist. Unpublished.

[34] Cian Dorr. Vagueness without ignorance. *Philosophical Perspectives*, 17(1): 83–113, 2003.

[35] Cian Dorr. Iterating definiteness. In Richard Dietz and Sebastiano Moruzzi, editors, *Cuts & Clouds: Vagueness, Its Nature, and Its Logic*, pages 550–86. Oxford University Press, 2010.

[36] James Dreier. Meta-ethics and the problem of creeping minimalism. *Philosophical Perspectives*, 18(1): 23–44, 2004.

[37] William B. Easton. Powers of regular cardinals. *Annals of Mathematical Logic*, 1(2): 139–78, 1970.

[38] Dorothy Edgington. On conditionals. *Mind*, 104(414): 235–329, 1995.

[39] Dorothy Edgington. Vagueness by degrees. In Rosanna Keefe and Peter Smith, editors, *Vagueness: A Reader*, pages 294–316. The MIT Press, 1997.

[40] Matti Eklund. What vagueness consists in. *Philosophical Studies*, 125(1): 27–60, 2005.

[41] Adam Elga. Subjective probabilities should be sharp. *Philosophers' Imprint*, 10(5): 1–11, 2010.

[42] Gareth Evans. Can there be vague objects? *Analysis*, 38(4): 208, 1978.

[43] Hugh Everett. The theory of the universal wave function. In Bryce S. DeWitt and Neill Graham, editors, *The Many-Worlds Interpretation of Quantum Mechanics*, pages 3–140. Princeton Unversity Press, 1973.

[44] Jeremy Fantl and Matthew McGrath. *Knowledge in an Uncertain World*. Oxford University Press, 2009.

[45] Delia Graff Fara. Shifting sands: An interest relative theory of vagueness. *Philosophical Topics*, 28(1): 45–81, 2000.

[46] Delia Graff Fara. Gap principles, penumbral consequence, and infinitely higher-order vagueness. *Liars and Heaps: New Essays on Paradox*, pages 195–222, 2003.

[47] Hartry Field. Theory change and the indeterminacy of reference. *The Journal of Philosophy*, 70(14): 462–81, 1973.

[48] Hartry Field. Logic, meaning, and conceptual role. *The Journal of Philosophy*, 74(7): 379–409, 1977.

[49] Hartry Field. A note on Jeffrey conditionalization. *Philosophy of Science*, 45(3): 361–7, 1978.

[50] Hartry Field. Disquotational truth and factually defective discourse. *The Philosophical Review*, 103(3): 405–52, 1994.

[51] Hartry Field. Some thoughts on radical indeterminacy. *The Monist*, 81(2): 253–73, 1998.

[52] Hartry Field. Which undecidable mathematical sentences have determinate truth values. In H. G. Dales and Gianluigi Oliveri, editors, *Truth in Mathematics*, pages 291–310. Oxford University Press, 1998.

[53] Hartry Field. Indeterminacy, degree of belief, and excluded middle. *Noûs*, 34(1): 1–30, 2000.

[54] Hartry Field. *Saving truth from paradox*. Oxford University Press, 2008.

[55] Hartry Field. This magic moment: Horwich on the boundary of vague terms. In Richard Dietz and Sebastiano Moruzzi, editors, *Cuts and Clouds: Vaguenesss, Its Nature, and Its Logic*, pages 200–8. Oxford University Press, 2010.

[56] Kit Fine. Vagueness, truth and logic. *Synthese*, 30(3): 265–300, 1975.

[57] Kit Fine. The question of realism. *Philosophers' Imprint*, 1(2): 1–30, 2001.

[58] Kit Fine. The impossibility of vagueness. *Philosophical Perspectives*, 22(1): 111–36, 2008.

[59] Dov M. Gabbay and Valentin B. Shehtman. Products of modal logics, part 1. *Logic Journal of IGPL*, 6(1): 73–146, 1998.

[60] Daniel Gallin. *Intensional and Higher-Order Modal Logic: With Applications to Montague Semantics*. North-Holland, Amsterdam, 1975.

[61] André Gallois. *Occasions of Identity: A Study in the Metaphysics of Persistence, Change, and Sameness*. Oxford University Press, 1998.

[62] Daniel Garber. Field and Jeffrey conditionalization. *Philosophy of Science*, 47(1): 142–5, 1980.

[63] Allan Gibbard. Contingent identity. *Journal of Philosophical Logic*, 4(2): 187–222, 1975.

[64] Allan Gibbard. *Thinking How to Live*. Harvard University Press, 2003.

[65] Jeremy Goodman. Knowledge and vagueness. Unpublished.

[66] Jeremy Goodman. Matter and mereology. Unpublished.

[67] Hillary Greaves. Understanding Deutsch's probability in a deterministic multiverse. *Studies in History and Philosophy of Science Part B: Studies in History and Philosophy of Modern Physics*, 35(3): 423–56, 2004.

[68] John Hawthorne. Epistemicism and semantic plasticity. *Metaphysical Essays*, 1(9): 185–211, 2006.

[69] John Hawthorne. *Metaphysical Essays*, chapter Three-dimensionalism, pages 85–110. Oxford University Press, 2006.

[70] John Hawthorne and Gabriel Uzquiano. How many angels can dance on the point of a needle? Transcendental theology meets modal metaphysics. *Mind*, 120(477): 53–81, 2011.

[71] Paul Horwich. The nature of vagueness. *Philosophy and Phenomenological Research*, 57(4): 929–35, 1997.

[72] Paul Horwich. The sharpness of vague terms. *Philosophical Topics*, 28(1): 83–92, 2000.

[73] Richard C. Jeffrey. *The Logic of Decision*. University of Chicago Press, 1990.

[74] Richard C. Jeffrey. *Probability and the Art of Judgment*. Cambridge University Press, 1992.

[75] James M. Joyce. A nonpragmatic vindication of probabilism. *Philosophy of Science*, 65(4): 575–603, 1998.

[76] Hans Kamp. Two theories about adjectives. In E. Keenan, editor, *Formal Semantics of Natural Language*, pages 123–55. Cambridge University Press, 1975.

[77] Stephen Kearns and Ofra Magidor. Epistemicism about vagueness and meta-linguistic safety. *Philosophical Perspectives*, 22(1): 277–304, 2008.

[78] Rosanna Keefe. *Theories of Vagueness*. Cambridge University Press, 2000.

[79] Rosanna Keefe. Supervaluationism, indirect speech reports, and demonstratives. In Richard Dietz and Sebastiano Moruzzi, editors, *Cuts and Clouds: Vaguenesss, Its Nature, and Its Logic*. Oxford University Press, 2010.

[80] Rosanna Keefe and Peter Smith. *Vagueness: A Reader*. The MIT Press, 1997.

[81] William Kneale. Modality de dicto and de re. In E. Nagel, P. Suppes, and A. Tarski, editors, *Logic, Methodology and Philosophy of Science. Proceedings of the 1960 International Congress*, pages 622–33. Stanford University Press, 1962.

[82] Daniel Z. Korman. The argument from vagueness. *Philosophy Compass*, 5(10): 891–901, 2010.

[83] Charles H. Kraft, John W. Pratt, and Abraham Seidenberg. Intuitive probability on finite sets. *The Annals of Mathematical Statistics*, 30(2): 408–19, 1959.

[84] Saul A. Kripke. A completeness theorem in modal logic. *The Journal of Symbolic Logic*, 24(1): 1–14, 1959.

[85] Saul A. Kripke. Outline of a theory of truth. *The Journal of Philosophy*, 72(19): 690–716, 1975.

[86] Ágnes Kurucz. Combining modal logics. *Studies in Logic and Practical Reasoning*, 3: 869–924, 2007.

[87] Ágnes Kurucz and Michael Zakharyaschev. A note on relativised products of modal logics. *Advances in Modal Logic*, 4: 221–42, 2002.

[88] David Lewis. Desire as belief. *Mind*, 97(418): 323–32, 1988.

[89] David K. Lewis. General semantics. *Synthese*, 22(1): 18–67, 1970.

[90] David K. Lewis. *Counterfactuals*. Blackwell, 1973.

[91] David K. Lewis. A subjectivist's guide to objective chance. *Studies in Inductive Logic and Probability*, 2: 263–93, 1980.

[92] David K. Lewis. Logic for equivocators. *Noûs*, 16(3): 431–41, 1982.

[93] David K. Lewis. *On the Plurality of Worlds*. Blackwell Publishers, 1986.

[94] David K. Lewis. Many, but almost one. In Keith Cambell, John Bacon, and Lloyd Reinhardt, editors, *Ontology, Causality, and Mind: Essays on the Philosophy of D. M. Armstrong*, pages 23–38. Cambridge University Press, 1993.

[95] Steffen Lewitzka. Denotational semantics for modal systems S3–S5 extended by axioms for propositional quantifiers and identity. *Studia Logica*, 103(3): 507–44, 2015.

[96] Luc Van Lier. A simple sufficient condition for the unique representability of a finite qualitative probability by a probability measure. *Journal of Mathematical Psychology*, 33(1): 91–8, 1989.

[97] John MacFarlane. Fuzzy epistemicism. In Richard Dietz and Sebastiano Moruzzi, editors, *Cuts and Clouds: Vaguenesss, Its Nature, and Its Logic*, pages 438–63. Oxford University Press, 2010.

[98] D. M. Mackinnon, F. Waismann, and W. C. Kneale. Symposium: Verifiability. *Aristotelian Society Supplementary Volume*, 19(1): 101–64, 1945.

[99] Anna Mahtani. The instability of vague terms. *The Philosophical Quarterly*, 54(217): 570–6, 2004.

[100] Norman Malcolm. Are necessary propositions really verbal? *Mind*, 49(194): 189–203, 1940.

[101] Maarten Marx and Yde Venema. *Multi-Dimensional Modal Logic*. Kluwer Academic Publishers, 1997.

[102] Vann McGee. *Truth, Vagueness, and Paradox: An Essay on the Logic of Truth*. Hackett Publishing Company Inc., 1990.

[103] Vann McGee. Kilimanjaro. *Canadian Journal of Philosophy*, 27(sup1): 141–63, 1997.

[104] Vann McGee and Brian McLaughlin. Distinctions without a difference. *The Southern Journal of Philosophy*, 33(S1): 203–51, 1995.

[105] Christopher J. G. Meacham and Jonathan Weisberg. Representation theorems and the foundations of decision theory. *Australasian Journal of Philosophy*, 89(4): 641–63, 2011.

[106] Henry Mehlberg. *The Reach of Science*. University of Toronto Press, 1958.

[107] Richard Montague. Syntactical treatments of modality, with corollaries on reflexion principles and finite axiomatizability. *Acta Philosophica Fennica*, 16: 153–67, 1963.

[108] Richard Montague. The proper treatment of quantification in ordinary English. In Patrick Suppes, Julius Moravcsik, and Jaakko Hintikka, editors, *Approaches to Natural Language*, pages 221–42. Dordrecht, 1973.

[109] Sarah Moss. *Probabilistic Knowledge*. Oxford University Press, forthcoming.

[110] Francis Jeffry Pelletier. The not-so-strange modal logic of indeterminacy. *Logique et Analyse*, 27(8): 415–22, 1984.

[111] Graham Priest. The logic of paradox. *Journal of Philosophical Logic*, 8(1): 219–41, 1979.

[112] Arthur N. Prior. *Objects of Thought*. Oxford University Press, 1971.

[113] Arthur N. Prior. Changes in events and changes in things. In Robin Le Poidevin and Murray MacBeath, editors, *The Philosophy of Time*, pages 35–46. Oxford University Press, 1993.

[114] Willard Van Orman Quine. Three grades of modal involvement. *Proceedings of the XIth International Congress of Philosophy*, 14: 65–81, 1953.

[115] Diana Raffman. Vagueness without paradox. *The Philosophical Review*, 103(1): 41–74, 1994.

[116] Frank Plumpton Ramsey. *On Truth: Original Manuscript Materials (1927-1929) from the Ramsey Collection at the University of Pittsburgh*, volume 16. Springer, 1991.

[117] Agustín Rayo. A metasemantic account of vagueness. In Richard Dietz and Sebastiano Moruzzi, editors, *Cuts and Clouds: Vagueness, Its Nature, and Its Logic*, pages 23–45. Oxford University Press, 2010.

[118] Stephen Read and Crispin Wright. Hairier than Putnam thought. *Analysis*, 45(1): 56–8, 1985.

[119] Giuliana Regoli. Comparative probability orderings. Preprint on the Imprecise Probabilities Project website: http://ippserv.rug.ac.be, 1998.

[120] Mark Richard. *Propositional Attitudes: An Essay on Thoughts and How we Ascribe them*. Cambridge University Press, 1990.

[121] Bertil Rolf. A theory of vagueness. *Journal of Philosophical Logic*, 9(3): 315–25, 1980.

[122] Gideon Rosen. Nominalism, naturalism, epistemic relativism. *Philosophical Perspectives*, 35(15): 69–91, 2001.

[123] Jeffrey Sanford Russell. Possible worlds and the objective world. *Philosophy and Phenomenological Research*, 90(2): 389–422, 2015.

[124] Nathan Salmon. *Frege's Puzzle*. The MIT Press, 1986.

[125] Nathan Salmon. Vagaries about vagueness. In Richard Dietz and Sebastiano Moruzzi, editors, *Cuts and Clouds: Vaguenesss, Its Nature, and Its Logic*. Oxford University Press, 2010.

[126] Stephen Schiffer. Vagueness and partial belief. *Philosophical Issues*, 10(1): 220–57, 2000.

[127] Stephen Schiffer. Vague properties. In Richard Dietz and Sebastiano Moruzzi, editors, *Cuts and Clouds: Vagueness, Its Nature, and Its Logic*, pages 109–30. Oxford University Press, 2010.

[128] Mark Schroeder. *Expressing Our Attitudes: Explanation and Expression in Ethics*. Oxford University Press, 2015.

[129] Moritz Schulz. Wondering what might be. *Philosophical Studies*, 149(3): 367–86, 2010.

[130] Adam Sennet. Semantic plasticity and epistemicism. *Philosophical Studies*, 161(2): 273–85, 2012.

[131] Stewart Shapiro. *Vagueness in Context*. Oxford University Press, 2006.

[132] Theodore Sider. Williamson's many necessary existents. *Analysis*, 69(2): 250–8, 2009.

[133] Theodore Sider. *Writing the Book of the World*. Oxford University Press, 2011.

[134] Nicholas J. J. Smith. Degree of belief is expected truth value. In Richard Dietz and Sebastiano Moruzzi, editors, *Cuts and Clouds: Vagueness, Its Nature, and Its Logic*, pages 491–507, Oxford University Press, 2010.

[135] Nicholas J. J. Smith. Vagueness, uncertainty and degrees of belief: Two kinds of indeterminacy—one kind of credence. *Erkenntnis*, 79(5): 1027–44, 2014.

[136] Scott Soames. *Beyond Rigidity: The Unfinished Semantic Agenda of Naming and Necessity*. Oxford University Press, 2002.

[137] Roy Sorensen. Vagueness. In Edward N. Zalta, editor, *The Stanford Encyclopedia of Philosophy*. Metaphysics Research Lab, Stanford University, winter 2016 edition, 2016.

[138] Roy A. Sorensen. *Blindspots*. Oxford University Press, 1988.

[139] Robert Stalnaker. A defense of conditional excluded middle. In William Harper, Robert Stalnaker, and Glenn Pearce, editors, *Ifs*, pages 87–104, D. Reidel Publishing Co., 1980.

[140] Robert Stalnaker. *Inquiry*. The MIT Press, 1984.

[141] Robert Stalnaker. Responses to Stanley and Schlenker. *Philosophical Studies*, 151(1): 143–57, 2010.

[142] Jason Stanley. 'Assertion' and intentionality. *Philosophical Studies*, 151(1): 87–113, 2010.

[143] Roman Suszko. Identity connective and modality. *Studia Logica*, 27(1): 7–41, 1971.

[144] Peter Unger. There are no ordinary things. *Synthese*, 41(2): 117–54, 1979.

[145] Peter Unger. The problem of the many. *Midwest Studies in Philosophy*, 5(1): 411–68, 1980.

[146] Bas C. van Fraassen. Singular terms, truth-value gaps, and free logic. *The Journal of Philosophy*, 63(17): 481–95, 1966.

[147] Bas C. van Fraassen. Probabilities of conditionals. In William L. Harper and Clifford Allan Hooker, editors, *Foundations of Probability Theory, Statistical Inference, and Statistical Theories of Science*, volume 1, pages 261–308. D. Reidel Publishing Co., 1973.

[148] Bas C. van Fraassen. Representational of conditional probabilities. *Journal of Philosophical Logic*, 5(3): 417–30, 1976.

[149] David Wallace. Everettian rationality: Defending Deutsch's approach to probability in the Everett interpretation. *Studies in the History and Philosophy of Modern Physics*, 34(3): 415–39, 2003.

[150] Brian Weatherson. Many many problems. *The Philosophical Quarterly*, 53(213): 481–501, 2003.

[151] David Wiggins. On being in the same place at the same time. *Philosophical Review*, 77(1): 90–5, 1968.

[152] J. Robert G. Williams. Degree supervaluational logic. *Review of Symbolic Logic*, 4(1): 130–49, 2011.

[153] J. Robert G. Williams. Indeterminacy and normative silence. *Analysis*, 72(2): 217–25, 2012.

[154] J. Robert G. Williams. Decision-making under indeterminacy. *Philosophers' Imprint*, 14(4): 1–34, 2014.

[155] Timothy Williamson. Inexact knowledge. *Mind*, 101(402): 217–42, 1992.

[156] Timothy Williamson. *Vagueness*. Routledge, 1994.

[157] Timothy Williamson. Putnam on the sorites paradox. *Philosophical Papers*, 25(1): 47–56, 1996.

[158] Timothy Williamson. Imagination, stipulation and vagueness. *Philosophical Issues*, 8: 215–28, 1997.

[159] Timothy Williamson. On the structure of higher-order vagueness. *Mind*, 108(429): 127–43, 1999.

[160] Timothy Williamson. *Knowledge and its Limits*. Oxford University Press, 2000.

[161] Timothy Williamson. Everything. *Philosophical Perspectives*, 17(1): 415–65, 2003.

[162] Timothy Williamson. Reply to McGee and McLaughlin. *Linguistics and Philosophy*, 27(1): 113–22, 2004.

[163] Timothy Williamson. Why epistemology cannot be operationalized. In Quentin Smith, editor, *Epistemology: New Essays*, pages 277–300. Oxford University Press, 2008.

[164] Timothy Williamson. Probability and danger. *The Amherst Lecture in Philosophy*, 4: 1–35, 2009.

[165] Timothy Williamson. *Modal Logic as Metaphysics*. Oxford University Press, 2013.

[166] Jessica M. Wilson. A determinable-based account of metaphysical indeterminacy. *Inquiry*, 56(4): 359–85, 2013.

[167] Crispin Wright. On being in a quandary: Relativism, vagueness, logical revisionism. *Mind*, 110(437): 45–98, 2001.

[168] Juhani Yli-Vakkuri. Epistemicism with worldly vagueness. Unpublished, 2011.

[169] Elia Zardini. Squeezing and stretching: How vagueness can outrun borderlineness. *Proceedings of the Aristotelian Society*, 106: 421–8, 2006.

Index

Printed and bound by CPI Group (UK) Ltd, Croydon, CR0 4YY